CAMBRIDGE TRACTS IN
MATHEMATICS

General Editors

B. BOLLOBAS, P. SARNAK, C.T.C. WALL

112   Schur algebras and
       representation theory

STUART MARTIN

Magdalene College, Cambridge

Schur algebras and representation theory

CAMBRIDGE
UNIVERSITY PRESS

CAMBRIDGE UNIVERSITY PRESS
Cambridge, New York, Melbourne, Madrid, Cape Town, Singapore, São Paulo, Delhi

Cambridge University Press
The Edinburgh Building, Cambridge CB2 8RU, UK

Published in the United States of America by Cambridge University Press, New York

www.cambridge.org
Information on this title: www.cambridge.org/9780521415910

First published 1993
This digitally printed version 2008

A catalogue record for this publication is available from the British Library

ISBN 978-0-521-41591-0 hardback
ISBN 978-0-521-10046-5 paperback

To the clan Martin

In my books I am solemn, sweet, refined; in real life I am rather vehement, sharp and contemptuous, a busy mocker. But I am also something of a fatalist. However, I am going to try to leave the Long free for writing and to have a subject ready to begin upon....I think I ought to be able to write rather a good story— if I weren't really so *lazy*: that is my main trouble, my hurried exhuberance.

A. C. Benson (1862–1925), *Diary, December 21, 1913*

Contents

Introduction

Between the years 1896 and 1900, G. Frobenius [1896], [1900] invented and mapped out the theory of complex representations of finite groups, paying special attention to representations of the symmetric group Σ_r. Thus, given a finite group we want to determine all group homomorphisms from our group into $GL_m(\mathbf{C})$ for arbitrary positive integers m. Around this time Frobenius suggested to his pupil Issai Schur that he might examine the representation theory of the infinite group $\Gamma_{\mathbf{C}} = GL_n(\mathbf{C})$. The subsequent investigations appeared in Schur's [1901] beautiful doctoral thesis of 1901. Schur studied $\Gamma_{\mathbf{C}}$ by means of the $\mathbf{C}$-space $A_{\mathbf{C}}(n, r)$ of r-homogeneous polynomial functions in the n^2 coordinate functions on $\Gamma_{\mathbf{C}}$. In particular he showed that the isomorphism types of irreducible representations of $\mathbf{C}\Gamma_{\mathbf{C}}$ with a given degree of homogeneity r are in one-to-one correspondence with the partitions of r into at most n parts; he also showed that the character of an irreducible module indexed by such a partition λ is a certain symmetric function, now called a 'Schur function', s_λ. In order to exploit Frobenius' work on Σ_r he set up (in modern parlance) an equivalence between the category of polynomial representations and the module category of Σ_r.

In another paper, Schur [1927] took a rather different approach. He analysed the actions of both Σ_r and $\Gamma_{\mathbf{C}}$ on the rth tensor power of the natural module for $\Gamma_{\mathbf{C}}$ (the space of n-columns over $\mathbf{C}$). Tensor space is then a permutation module and, using this instead of the function space, he thereby obtained shorter and very elegant proofs of results in his thesis. The procedure received publicity in Weyl's great treatise [1973], first published in 1939.

It is natural to present a characteristic-free treatment of general linear and symmetric groups. The subject began with the work of Thrall [1942] and led to the fundamental paper of Carter and Lusztig [1974].

They took up the representation theory of $\Gamma_K = \mathrm{GL}_n(K)$ over an infinite field, K, of arbitrary characteristic. This time the main tool is the hyperalgebra $\mathcal{U}_K$ constructed out of the Kostant $\mathbf{Z}$-form of the universal enveloping algebra of the general linear Lie algebra over $\mathbf{Q}$. They produce explicit polynomial generators for the centre of $\mathcal{U}_{\mathbf{Q}}$, with nice multiplicative properties, and these are used to produce the required generalisations of Schur's classical results. In particular they constructed the 'Weyl modules' as certain subspaces of tensor space, showed they were defined over $\mathbf{Z}$ and hence could be regarded as reductions modulo p of the modules studied by Schur in characteristic zero. Another approach was pioneered by Clausen [1979, 1980] whose main tool was the letter place algebra.

Green's monograph appeared in 1980 (Green [1980]). He placed these developments in the context of certain finite-dimensional K-algebras, which he christened Schur algebras, and clarified the connections between representations of Γ_K and the symmetric group. In this scenario the Schur algebra is the dual of the 'Schur coalgebra' $A_K(n,r)$ of r-homogeneous polynomial functions on Γ_K, and may be identified with the centralising algebra of the Σ_r-action on r-tensor space of the natural Γ_K-module. The influence of these hundred or so pages is hard to over-emphasise: throughout the eighties most work on Schur algebras was inspired by this presentation. His basic approach was combinatorial, with no overt appeal to algebraic group theory, and as such is in the same spirit as Schur's original exposition.

My intention in this book is to expand on Green's treatment of Schur algebras, and also to write a fairly full account of the exciting developments which have occurred in the time since Green's work was first published. I would like to mention a selection of these developments straight away. The work of Cline, Parshall and Scott, together with results of Ringel [1991], on highest weight categories (the module categories for so-called quasi-hereditary algebras) has been very influential in recent years. These ideas were arrived at by abstraction from the classical representation theory of semisimple algebraic groups and Lie algebras. In some sense the abstract versions are 'characteristic-free,' the unifying theme being an axiom which postulates the existence of certain filtrations which are the natural analogue of composition series in characteristic zero. Parshall [1989] proved in that the Schur algebra, $S_K(n,r)$, is quasi-hereditary, hence its module category is a highest weight category. In a series of papers, Green [1990b], [1991a], [1991b] and [1992] gave elegant constructions of the costandard modules (called

Schur modules in this text) and the standard modules (called here the Weyl modules) for this category. Meanwhile Donkin [1981] had also proved that the module category for $S_K(n,r)$ was a highest weight category, and in a remarkable set of papers (Donkin [1986], [1987], [1992], [1993]) he uses the machinery of algebraic group theory to prove deep theorems about the modular representation theory and associated block theory of the Schur algebras. The interplay between the block theory of symmetric groups and the Schur algebras has received quite a lot of attention recently. A study of Schur algebras of finite representation-type was initiated by Erdmann [1993a], and has now been developed by various people in several directions, see Erdmann, Martin and Scopes [1993], Erdmann [1993b] and Donkin and Reiten [1993].

Donkin was one of several authors to consider the cohomology theory of the Schur algebra; in particular he give a direct proof that the Schur algebras have finite global dimension (before the birth of quasi-hereditariness), which at the time was a great surprise. Such characteristic-free themes have also been employed by Akin *et al.*, (see, for example, Akin and Buchsbaum [1985], [1988]) in constructing explicit projective resolutions of Weyl modules.

Finally, a long collection of papers by Dipper and James [1986], [1987], [1989], [1991] is devoted to the study of Hecke algebras of type A and the action of them on tensor space; in the process, they introduce the q-Schur algebras, the centralising algebra of this action. The module category of this algebra is equivalent to the category of polynomial representations of a certain quantum group; putting $q = 1$ we re-obtain the classical situation. But more is true: if q is a prime power and coprime to p we obtain strong information on the p-modular representations of the finite reductive group $\mathrm{GL}_n(q)$. Indeed the calculation of decomposition numbers for the various structures mentioned above might be considered the central theme of the whole text.

My presentation is designed as a shop window for these new ideas: I hope that the fairly comprehensive bibliography at the end will aid the reader wanting to go into a topic more deeply. Let me now advertise the forthcoming attractions chapter by chapter.

Chapter 1 opens with a précis of Schur's thesis topic. This conveniently allows us to define frequently used notation and explain the underlying combinatorics necessary for passage to finite characteristic. In Theorem 1.6.1 we show that the isomorphism types of the simple r-homogeneous polynomial modules are in bijective correspondence with the set $\Lambda^+(n,r)$ of 'dominant weights', i.e. the set of partitions of r

into at most n parts. The proof requires little beyond familiarity with symmetric functions and characters.

These preliminaries complete, in Chapter 2 we introduce the Schur algebra, $S_K(n,r)$. We characterise this finite-dimensional K-algebra as the centralising algebra of the right action of the symmetric group Σ_r on r-tensor space $E^{\otimes r}$, where E is the natural module of n-columns for Γ_K. In Theorem 2.2.7 we give the important result that the category $\mathcal{P}_K(n,r)$ of r-homogeneous polynomial representations of Γ_K is equivalent to the module category for the Schur algebra. The second part of this chapter is devoted to proving that certain elements of the coalgebra $A_K(n,r)$, called standard bideterminants, form a basis of $A_K(n,r)$. The basic combinatorial input is the famous Straightening Formula, an old result in multilinear algebra with a long history extending back to ideas of Young, and indeed beyond into the mists of time to the founders of invariant theory. Since $A_K(n,r)$ is the K-dual of the Schur algebra, so we find that in the process a basis of 'standard codeterminants' is obtained for $S_K(n,r)$.

In Chapter 3 the search begins in earnest for the simple objects of $\mathcal{P}_K(n,r)$. We show that given any $\lambda \in \Lambda^+(n,r)$ there exist $S_K(n,r)$-modules $V(\lambda)$ and $M(\lambda)$, which I shall call Weyl modules and Schur modules respectively, both equipped with bases of bideterminants or co-determinants. In section 3.3 we discuss in some detail the fundamental notion of quasi-hereditary algebras and prove that Schur algebras have this property by exhibiting an explicit defining chain of ideals. It is seen that the Schur modules and Weyl modules mentioned above are precisely the costandard and standard modules in the highest weight category $\mathcal{P}_K(n,r)$. An easy consequence is the conclusion that Schur algebras have finite global dimension. The cohomology of Schur algebras deserves a book to itself, and all we do here is summarise some of the recent endeavours towards the construction of explicit resolutions and other aspects of cohomology theory.

Chapter 4 brings symmetric groups into the fray. If one assumes that $r \leq n$ then existence of a weight $\omega = (1^r)$ is guaranteed. In turn, the idempotent $e = \xi_\omega \in S_K(n,r)$ defined by this weight induces a functor f_e between the category $\mathcal{P}_K(n,r)$ and the module category for $K\Sigma_r$ (map a module in the first category to its ω-weight space). It is precisely this functor that gave Schur a category equivalence over $\mathbf{C}$, but there are complications in finite characteristic; for example, complete reducibility does not carry over. By analysing the images of Schur modules and the Weyl modules under f_e, one can construct the Specht modules for

$\mathbf{Q}\Sigma_r$ and the p-modular irreducibles. The Schur functor is used to outline some interesting conjectures about modular irreducibles for both
symmetric groups and general linear groups, and also it features in the
treatment of James' theorem in section 4.4: for $r \leq n$, this exhibits the
decomposition matrix for Σ_r as a submatrix of that for Γ_K. The final
two sections define and study Young modules. These are certain trivial
source modules for Σ_r, and a knowledge of how $E^{\otimes r}$ decomposes into
Young module components is seen as equivalent to a knowledge of Γ_K
decomposition numbers. Use is made of the existence of certain filtrations whose sections are all Schur modules—such '∇-good filtrations' are
a by-product of the representation theory of the quasi-hereditary algebra
$S_K(n, r)$.

   Throughout, a common theme is to proceed by analogy with Brauer's
theory of modular representations of finite groups and, as noted above,
we see in Chapter 4 that one has a concept of modular reduction for
Γ_K and hence p-modular decomposition numbers. Another important
aspect of Brauer's theory hinges on the notion of a p-block. Donkin
[1980] investigated block decompositions for the reductive group Γ_K and
the semisimple simply connected algebraic groups like SL_n, where the
affine Weyl group determines the blocks. One might expect that the
Schur algebras have similar properties, and indeed this is the case. A
rule akin to the famous 'Nakayama Conjecture' for Σ_r has been proved
by Donkin [1992] to give the blocks of $S_K(n, r)$. In general this amounts
to stripping off hooks of p-power length, instead of the more familiar
case of Σ_r, where one strips off p-hooks. As already suggested, this
proof involves some algebraic group theory, the basics of which appear
in summary in the Appendix. This done, we prove a finiteness theorem
by showing that in fact there are only finitely many Morita equivalence
classes of blocks of $S_K(n, r)$ for given n, r: this is an analogue of a result
for G, due to Scopes [1991]. Some examples of blocks of $S_K(n, r)$ of
finite type are given.

   The last two chapters are designed as an introduction to the theory of
q-Schur algebra. I have settled on an informal survey of this very active
area of research. We begin by defining a q-deformation of the coordinate
ring of $n \times n$ matrices to obtain a graded bialgebra; on taking its rth
homogeneous component there is produced a coalgebra $A_q(n, r)$, whose
dual is defined to be the q-Schur algebra. Specialising to $q = 1$ we
obtain the usual Schur algebra $S_K(n, r) = A_K(n, r)^*$. At present there
are several versions of the q-Schur algebra in the literature (Dipper and
James [1989], Dipper and Donkin [1991], Parshall and Wang [1991]) and

we spend most of Chapter 6 showing that the definitions are equivalent. We will identify $S_q(n,r)$ and $\mathrm{End}_{\mathcal{H}_q}(\oplus M_q^\lambda)$, where $\mathcal{H}_q$ is a certain Hecke algebra reducing to $K\Sigma_r$ if $q = 1$, and M_q^λ ($\lambda \in \Lambda^+(n,r)$) is a q-version of the well-known permutation module M^λ on cosets of Young subgroup Σ_λ. The direct sum of the M_q^λ over all compositions of r is a version of tensor space, and $\mathcal{H}_q$ acts on this space as described in section 6.6; factoring out the kernel gives the q-Schur algebra as originally defined; see Theorem 6.6.7.

We produce a q-version of the coordinate ring $K[\Gamma]$ of Γ_K; there emerges an algebra, $K[\Gamma_q]$, which it is convenient to think of as the coordinate ring of a 'quantum group' Γ_q over K. It is shown that the category of polynomial $K[\Gamma_q]$-modules is equivalent to the module category for $S_q(n,r)$, and since $S_q(n,r)$ is quasi-hereditary, this module category is a highest weight category. Thus there are costandard modules (the q-Schur modules) and standard modules (the q-Weyl modules). The study of these modules we adopt is combinatorial, with little use of the now extensive theory of quantum groups developed by Manin [1988], Parshall and Wang [1991] and Andersen *et al.* [1991]. The q-Schur algebras $S_q(n,r)$ originally were studied by Dipper and James [1989] in order to link up the p-modular representations of Σ_n, representations of $\mathrm{GL}_n(K)$ and representations of $\mathrm{GL}_n(q)$ over fields whose characteristic is coprime to q. Once again there are q-Young modules, and a knowledge of their multiplicities as summands of M_q^λ is equivalent to a knowledge of the decomposition numbers of $S_q(n,r)$, i.e. of the multiplicities of the simple modules in the q-Weyl modules: a positive solution to this problem would in turn give the decomposition numbers for all the structures just mentioned.

Notation and Conventions. The following notation is fixed throughout.

- We will use $\mathbf{N}$, $\mathbf{N}_0$, $\mathbf{Z}$, $\mathbf{Q}$, $\mathbf{R}$ and $\mathbf{C}$ for the sets of natural numbers, non-negative integers, integers, rationals, reals and complex numbers, respectively.
- $X \subseteq Y$ means that X is a subset of Y, $X = \emptyset$ or Y being included.
- $V \oplus W$ stands for the direct sum of the vector spaces V and W; there are obvious extensions to finite and arbitrary direct sums.
- $\langle e_1, \ldots, e_n \rangle$ stands for the vector space spanned by $e_1, \ldots, e_n$.
- $\square$ will signify the end of a proof.

Lists of all other commonly used notation appear in indexes at the end of the book. The numbering inside the chapters is as follows: equations referred to in the text are labelled by bracketed numbers, thus 'as seen in (4.6)' refers to equation (4.6) contained within Chapter 4. Sections, lemmas and theorems etc. are identified by unbracketed numbers, thus 'as proved in 4.6 and see also 4.6.1' refers to section 4.6 in Chapter 4, and to a result labelled 4.6.1 contained therein. The exception to this schema is that Section 4 in the Appendix will be referred to as A.4.

Acknowledgements. This work was carried out while the author held an Official Fellowship at Magdalene College, Cambridge and thanks are due to the Master and Fellows for employing me. On a personal level, I am grateful to all the following people. Firstly, my warmest thanks go to Sandy Green who has carefully read several versions of the typescript and has taught me many things about Schur algebras (and much else besides), and suggested endless improvements; to Karin Erdmann who, as my research supervisor kindled my interest in group representations and symmetric groups; to Steve Donkin, Steve Doty and Gordon James for making available many preprints of their work and for many rewarding discussions on algebraic groups and Schur algebras; to John Fitzgerald for proofreading in the Joycean tradition. Any gaffes that remain are of course my own responsibility. It is conventional to thank David Tranah of Cambridge University Press for all his help and endurance, and I am more than pleased to continue the tradition. I should also mention the fact that this document has been prepared by the author using the CUPBOOK implementation of LaTeX provided by the Press.

The fire of the genesis of this work has been nurtured and stoked by many friends and colleagues, of whom Dave Benson, Colin Duff, Tim Harper, Sandeep Kapur, Lelia Roeckell and Wilson Sutherland must be singled out.

1

Polynomial functions and combinatorics

1.1 Introductory remarks

This chapter opens with a brief summary of Schur's work on the complex polynomial representations of the general linear group $GL_n(\mathbf{C})$ of $n \times n$ nonsingular matrices with entries in $\mathbf{C}$; we then proceed to analyse ways of extending his ideas to representations over an infinite field K of arbitrary characteristic. Thus, Schur considered a polynomial matrix representation of $M \in GL_n(\mathbf{C})$, i.e. a group homomorphism $T : GL_n(\mathbf{C}) \to GL_m(\mathbf{C})$ such that each entry of $T(M)$ could be expressed as a polynomial with complex coefficients in the entries of M; however we shall study a $KGL_n(K)$-module V whose coefficient space lies in the bialgebra $A_K(n)$ generated by the n^2 coordinate functions. In this way V affords a matrix representation whose entries are polynomials over K in the original entries. We shall then take the rth homogeneous component, $A_K(n,r)$, of $A_K(n)$ and consider the category of all representations having coefficient space lying in this component. Basic properties of the coalgebra $A_K(n,r)$ are catalogued in 1.3.

Throughout the text the methods employed use a blend of the standard combinatorial machinery. Nowadays I can safely assume that the combinatorics of partitions and their associated Young diagrams is well known, so no more than a cursory treatment is given in 1.4. This leads to one definition of a weight (as a certain n-tuple of integers) and we connect this idea with the algebraic group concept of weight (arising from the action of the torus) in 1.5. At the same time we provide a short discussion on formal and natural characters of representations of $GL_n(K)$ and compute these for a few accessible but important examples. Technical matters relating to the representation theory of algebraic groups are safely relegated to the Appendix. Notwithstanding some simple lem-

1

mas on weight spaces in 1.6, we already have sufficient material to show
that the isomorphism classes of the irreducible objects in the category
$A_K(n,r)$ **mod** of finite-dimensional left $A_K(n,r)$-modules are in one-to-
one correspondence with partitions of r into at most n parts. Concrete
constructions of these modules are made in the next two chapters.

1.2 Schur's thesis

We choose $2n^2$ independent variables $a_{\alpha\beta}$, $b_{\alpha\beta}$, where α and β run
independently through the set $\mathbf{n} = \{1, 2, \ldots, n\}$; we may regard these as
entries in the two $n \times n$ matrices $A = [a_{\alpha\beta}]$, $B = [b_{\alpha\beta}]$. Let $T = T(A) =$
$[t_{\gamma\delta}]$ be some $m \times m$ matrix ($m \in \mathbf{N}$) with entries $t_{\gamma\delta}$ belonging to the
C-algebra $\mathbf{C}[a]$, generated by the $a_{\alpha\beta}$ (α, $\beta \in \mathbf{n}$). We observe therefore
that each $t_{\gamma\delta}$ is a polynomial with complex coefficients in the n^2 entries
$a_{\alpha\beta}$ of the matrix A.

Suppose that $C = AB$, that is $C = [c_{\alpha\beta}]$ with

$$c_{\alpha\beta} = \sum_{\mu \in \mathbf{n}} a_{\alpha\mu} b_{\mu\beta}. \tag{1.1}$$

Thus the entries of C lie in the polynomial algebra $\mathbf{C}[a, b]$ generated
by the $a_{\alpha\beta}$ and $b_{\alpha\beta}$. Now Schur was interested in the case where T is
invariant, that is, T has the following property:

$$T(A)\, T(B) = T(C).$$

Simple examples are $T(A) = A$ with $m = n$, or if $m = 1$ we might take
$T(A) = [\det A]$.

If $g, h \in \Gamma_{\mathbf{C}} = \mathrm{GL}_n(\mathbf{C})$, then by (1.1) we have $T(g)T(h) = T(gh)$,
and so $T : \Gamma_{\mathbf{C}} \to \mathrm{GL}_m(\mathbf{C})$ is a matrix representation of $\Gamma_{\mathbf{C}}$ of dimension
m, provided we assume additionally that $T(I_n) = I_m$. We shall call T a
complex polynomial representation of $\Gamma_{\mathbf{C}}$ of degree m.

Remark. Actually, Schur started by considering *rational representations*, i.e. those $T(A)$ whose entries were rational functions of the entries
of the matrix A. In this case we may write $T(A) = p(A)q(A)^{-1}T'(A)$,
where $A \mapsto p(A)$, $A \mapsto q(A)$ are maps into $\mathbf{C}[a]$, with p and q polynomial
in the entries of A, and $T'(A)$ is polynomial as above. Hence it is no loss
to consider those representations T, as mentioned above, whose entries
lie in the ring $A_{\mathbf{C}}(n)$ of all polynomial functions $\mathrm{GL}_n(\mathbf{C}) \to \mathbf{C}$.

Definition 1.2.1 The invariant matrices T and U of degree m are *equiv-*

alent if there exists $P \in \mathrm{GL}_m(\mathbf{C})$ such that $P^{-1}TP = U$. We write $T \simeq U$ in this case. We say T is *reducible* (over $\mathbf{C}$) if

$$T \simeq \begin{bmatrix} T_1 & V \\ 0 & T_2 \end{bmatrix},$$

where T_i is an invariant matrix of degree m_i and V is some $m_1 \times m_2$ matrix with entries in $A_{\mathbf{C}}(n)$. If T cannot be written in this way it is termed *irreducible*. Finally, if T is equivalent to a matrix in block diagonal form in which each diagonal block matrix is irreducible, then T is said to be *completely reducible*. These definitions also apply to matrix representations of groups over arbitrary fields.

Suppose that $T(A)$ is invariant, and x is some variable independent of the $a_{\alpha\beta}$. Then $T(xA) = T_0(A) + xT_1(A) + \cdots + x^r T_r(A)$ for some $r \in \mathbf{N}$. Each $T_i(A)$ is itself invariant. Further, T is equivalent to the block diagonal matrix $\mathrm{diag}(T_0, \ldots, T_r)$. To see this, let $E_i = T_i(I_n)$. Then $T(xI_n) = E_0 + xE_1 + \ldots + x^r E_r$ for all $x \in \mathbf{C}$. The invariance condition $T(xI_n)T(A) = T(xA) = T(A)T(xI_n)$ forces

$$(E_0 + xE_1 + \cdots + x^r E_r)(T_0(A) + T_1(A) + \cdots + T_r(A)) =$$
$$T_0(A) + xT_1(A) + \cdots + x^r T_r(A)$$

for all $x \in \mathbf{C}$. Equating coefficients implies $E_i T_i(A) = T_i(A)E_i$. Now from linear algebra we may find $P \in \mathrm{GL}_m(\mathbf{C})$ such that, for each i, $P^{-1}E_iP = E_i'$, where E_i' has m_i consecutive 1s down the diagonal and 0s everywhere else, such that the sum of the E_i' equals I_n. Here m_i is the order of T_i (so the sum of the m_i is the order of T). Writing $P^{-1}T_i(A)P = T_i'(A)$ we obtain the relations $E_i'T_i'(A) = T_i'(A)E_i'$ for all i. Hence $T_i'(A)$ has zero entries everywhere except in those rows and columns in which E_i' has 1s. We conclude that $P^{-1}T(A)P$ is the diagonal sum of the matrix blocks $T_i'(A)$ as required.

Now the entries in each T_p are homogeneous polynomials of the same degree p in the entries of A, that is, $T_p(xA) = x^p T_p(A)$. Thus we lose nothing in assuming henceforth that T is already homogeneous of some degree, say r. Equivalently, we may assume that each entry $t_{\gamma\delta}$ of T lies in $A_{\mathbf{C}}(n, r)$, the subspace of $A_{\mathbf{C}}(n)$ generated by functions that are homogeneous of total degree r in the n^2 variables $a_{\alpha\beta}$.

Notation

Before we can describe a basis for $A_C(n, r)$, we list some notation, which, in fact, will hold throughout the text.

- If K is an arbitrary infinite field and S is a K-algebra, we write $_S\mathbf{mod}$ for the category of all finite-dimensional left S-modules; $\mathbf{mod}_S$ will be the category of finite-dimensional right S-modules. If $V, W \in {}_K\mathbf{mod}$ we usually write $V \otimes W$ for $V \otimes_K W$.

- If H is a finite group, we denote its group algebra by KH.

- $I = I_n = I(n, r) = \{\text{maps from } \mathbf{r} \text{ to } \mathbf{n}\} = \{i = (i_1, \ldots, i_r) : i_\rho \in \mathbf{n},\ \forall \rho \in \mathbf{r}\}$ is the set of *multi-indices* (which some authors write in the power notation $\mathbf{n^r}$).

- $G = \Sigma_r$, the symmetric group on $\mathbf{r}$; more generally, if $X \subseteq \mathbf{r}$, we write Σ_X for the subgroup of G fixing every element outside X. G acts naturally on the right on I by place permutations:

$$i\pi = (i_{\pi(1)}, \ldots, i_{\pi(r)}) \qquad (i \in I,\ \pi \in G).$$

G also acts on $I \times I$ as follows:

$$(i, j)\pi = (i\pi, j\pi) \quad (i \in I,\ j \in I,\ \pi \in G).$$

- We write $i \sim j$ when $i,\ j \in I$ are in the same G-orbit. Similarly, we write $(i, j) \sim (p, q)$ when $(i, j), (p, q) \in I \times I$ are in the same G-orbit, i.e. if $p = i\pi$ and $q = j\pi$ for some $\pi \in G$.

- If $i \in I$ we write $G_i = \{\pi \in G : i\pi = i\}$ for the stabiliser of i.

Evidently, $A_C(n, r)$ has a **C**-basis consisting of all monomials of degree r:

$$a_{i,j} = a_{i_1 j_1} \cdots a_{i_r j_r} \quad (i_\rho,\ j_\rho \in \mathbf{n}\ \forall \rho \in \mathbf{r}). \tag{1.2}$$

Notice our convention in writing $a_{i,j}$ (with a comma) when the subscripts are multi-indices, but a_{pq} (without a comma) if the subscripts are integers. A given monomial may have two different expressions of the form (1.2). Indeed for $i, j, i', j' \in I$,

$$a_{i,j} = a_{i',j'} \Leftrightarrow (i, j) \sim (i', j').$$

We now have a basis for $A_C(n, r)$:

Theorem 1.2.2 *Choose arbitrarily a set Ω of representatives of the G-orbits of $I \times I$. Then*

$$\{a_{i,j} : (i,j) \in \Omega\}$$

is an irredundantly described basis of $A_{\mathbf{C}}(n,r)$. *Hence*

$$\dim A_{\mathbf{C}}(n,r) = \binom{n^2 + r - 1}{r}.$$

Proof The number of distinct monomials of degree r in the n^2 variables $a_{\alpha\beta}$ is easily seen to be given by the quoted binomial coefficient. $\square$

With the above notation established, we return to the general problem of finding conditions on the $m \times m$ matrix $T(A)$ that make it invariant. The entries of $T(A)$ lie in $A_{\mathbf{C}}(n,r)$, so using the basis provided by 1.2.2 we may write

$$T(A) = \sum_{(i,j)\in\Omega} a_{i,j} M_{i,j}$$

where each $M_{i,j}$ is some $m \times m$ matrix with complex entries. Thus

$$T(A)T(B) = \sum_{(i,j)\in\Omega} \sum_{(k,l)\in\Omega} a_{i,j} b_{k,l} M_{i,j} M_{k,l}.$$

By the definition of invariance this must be equal to

$$T(C) = \sum_{(p,q)\in\Omega} c_{p,q} M_{p,q},$$

where, from (1.1), the coefficients satisfy

$$c_{p,q} = c_{p_1 q_1} \cdots c_{p_r q_r} = \prod_{\rho} \sum_{h_\rho \in \mathbf{n}} a_{p_\rho h_\rho} b_{h_\rho q_\rho} = \sum_{h \in I} a_{p,h} b_{h,q}.$$

Certainly $(p,q) \in \Omega$, but there is no reason why either of the pairs (p,h) and (h,q) have to lie in Ω. We may write the last formula as

$$c_{p,q} = \sum_{(i,j)\in\Omega} \sum_{(k,l)\in\Omega} \gamma(i,j,k,l,p,q) a_{i,j} b_{k,l},$$

where, for any (p,q), (i,j), $(k,l) \in \Omega$,

$$\gamma(i,j,k,l,p,q) = |\{h \in I : (p,h) \sim (i,j), \ (h,q) \sim (k,l)\}|.$$

By comparing coefficients, we conclude that $T(A)T(B) = T(C)$ if and only if

$$M_{i,j} M_{k,l} = \sum_{(p,q)\in\Omega} \gamma(i,j,k,l,p,q) M_{p,q}$$

for all (i,j), $(k,l) \in \Omega$. Experts will recognise this somewhat unwieldy expression as Schur's famous 'Product Rule'.

It is hoped that the reader will now find it quite natural to define a linear $\mathbf{C}$-algebra $S_\mathbf{C}(n,r)$ (cf. Schur [1901, p. 29]) having as basis the symbols $\xi_{i,j}$, $(i,j) \in \Omega$, and multiplication defined by

$$\xi_{i,j}\xi_{k,l} = \sum_{(p,q)\in\Omega} \gamma(i,j,k,l,p,q) \xi_{p,q} \qquad (1.3)$$

for all (i,j), $(k,l) \in \Omega$.

Definition 1.2.3 $S = S_\mathbf{C}(n,r)$ is the *complex Schur algebra* for r and n.

By 1.2.2,

$$\dim_\mathbf{C} S = \binom{n^2 + r - 1}{r}.$$

Visibly, there is a one-to-one correspondence $R \longleftrightarrow R_0$ between

$$\left\{ \begin{array}{c} \text{invariant matrices} \\ T(A) = \sum a_{i,j} M_{i,j} \end{array} \right\} \longleftrightarrow \left\{ \begin{array}{c} \text{matrix representations of } S \\ \xi_{i,j} \mapsto M_{i,j} \end{array} \right\}$$

respecting the representation-theoretic properties of equivalence and irreducibility. Using this principle, Schur showed that if $n \geq r$, there are further correspondences $R_0 \leftrightarrow R_1$ between the representations R_0 of S and representations R_1 of Σ_r. The complex representations of Σ_r, determined by Frobenius [1896], [1900], together with the correspondences $R_0 \leftrightarrow R \leftrightarrow R_1$, were used by Schur to show that every invariant matrix was completely reducible. In modern terminology, the main problem he posed and solved was the following.

Given n, find a complete set of irreducible polynomial representations of degree r for $\mathrm{GL}_n(\mathbf{C})$.

He showed that the set of such matrix representations was in one-to-one correspondence with the set $\Lambda^+(n,r)$ of partitions of r into at most n parts. Further, given some such matrix, say $T = T_\lambda(A)$ ($\lambda \in \Lambda^+(n,r)$), its trace, that is the character of the corresponding representation (which Schur termed the *characteristic*) was shown by an 'irrelevance argument' to be a certain symmetric function $\zeta_1, \ldots, \zeta_n$ of the eigenvalues of A. Using the above correspondences, he ordered the partitions and determined the characters by induction on the ordering. The character corresponding to $\lambda = (\lambda_1, \ldots, \lambda_n)$ is known as the *Schur function of type* λ, as in Macdonald [1979, I§3]. It is defined by the following quotient of a Vandermonde determinant

$$s_\lambda(\zeta_1, \ldots, \zeta_n) = \frac{|\zeta_i^{\lambda_i + n - j}|}{|\zeta_i^{n-j}|}.$$

Remarks.

(1) For more on the rich history of the development of the subject, the survey article of Dieudonné [1981] is recommended. For comments on the links with invariant theory, see the notes in Weyl [1973] and the exposition of Crawley-Boevey [1990].

(2) Schur's irrelevance argument effectively amounts to restricting the representation T of Γ_C to the diagonal subgroup of Γ_C: see 1.5.5 below.

Example. To illustrate some of these ideas, let us consider the following simple situation. We take $n = r = 2$ and let $A = \begin{bmatrix} a_{11} & a_{12} \\ a_{21} & a_{22} \end{bmatrix}$. There are two types of irreducible invariant matrices.

(a) If $\lambda = (2, 0)$, we have a 3-dimensional irreducible representation

$$T_{(2,0)}(A) = \begin{bmatrix} a_{11}^2 & a_{11}a_{12} & a_{12}^2 \\ 2a_{11}a_{21} & a_{11}a_{22} + a_{12}a_{21} & 2a_{12}a_{22} \\ a_{21}^2 & a_{21}a_{22} & a_{22}^2 \end{bmatrix}.$$

Putting $A = \text{diag}(\zeta_1, \zeta_2)$ in $T_{(2,0)}(A)$, we obtain its character

$$s_{(2,0)}(\zeta_1, \zeta_2) = \zeta_1^2 + \zeta_1\zeta_2 + \zeta_2^2.$$

(b) The other case is to take $\lambda = (1, 1)$, which gives

$$T_{(1,1)}(A) = [a_{11}a_{22} - a_{12}a_{21}]$$

(that is, the determinant of A). This is a 1-dimensional representation with character

$$s_{(1,1)}(\zeta_1, \zeta_2) = \zeta_1\zeta_2.$$

Remarks.

(1) These representations are afforded by special cases of the Schur modules constructed in Chapter 3.

(2) Most of the definitions and concepts outlined above carry over in an

obvious way if $\mathbf{C}$ is replaced by K, an infinite field of arbitrary characteristic. In fact, as we shall see in Chapter 2, the correspondence
$R \leftrightarrow R_0$ between polynomial representations of $\mathrm{GL}_n(K)$ of degree r and
representations of the Schur algebra $S_K(n, r)$ is still valid; however, the
correspondence $R_0 \leftrightarrow R_1$ with Σ_r is not so precise, for example, complete reducibility fails. Nevertheless, one can progress quite far. The
irreducible polynomial representations may be constructed and are still
labelled by $\Lambda^+(n, r)$, but the characters are as yet unknown for general
n, r. We return to this point several times in the sequel.

Integral forms. One of the cornerstones of R. Brauer's modular representation theory of finite groups is the concept of p-modular reduction.
In the study of invariant matrices T_λ we say that T_λ is *integral* if all its
entries are polynomials in the $a_{\alpha\beta}$ in which the coefficients of the polynomials are integers. Hence we can interpret them as representations of
$\mathrm{GL}_n(K)$ over any field K, or indeed over any commutative domain R.
As examples both $T_{(2,0)}$ and $T_{(1,1)}$ of the last example are integral. Actually, thinking of these as representations of the Schur algebra S via the
correspondence $R \leftrightarrow R_0$, it is a splendid fact that one can always find
matrix representations for S which are integral. These matters were first
investigated by Carter and Lusztig [1974, 3.5] who gave explicit integral
forms for the Weyl modules (the duals of the Schur modules referred
to above). See also Clausen [1979] and Akin and Buchsbaum [1985] for
other aspects of the problem. As usual there are many possible choices
for a $\mathbf{Z}$-form:

Example. Returning to the above example, the representation

$$T'_{(2,0)}(A) = \begin{bmatrix} a_{11}^2 & 2a_{11}a_{12} & a_{12}^2 \\ a_{11}a_{21} & a_{11}a_{22} + a_{12}a_{21} & a_{12}a_{22} \\ a_{21}^2 & 2a_{21}a_{22} & a_{22}^2 \end{bmatrix}$$

is integral and is equivalent over $\mathbf{C}$ to $T_{(2,0)}$, but as representations of
$\mathrm{GL}_2(K)$ $T_{(2,0)}$ and $T'_{(2,0)}$ are not equivalent over fields of characteristic
2—they are not even irreducible!

This is not a serious drawback: we are usually interested only in the
composition factors of these representations. These turn out to be the
same, whichever $\mathbf{Z}$-form is chosen, by a theorem of Brauer and Nesbitt
(see Curtis and Reiner [1981, (16.16)]).

1.3 The polynomial algebra

As a first step in generalising Schur's procedure, discussed in the last section, we replace $\mathbf{C}$ by an arbitrary infinite field K. Consider the set K^Γ of all maps from $\Gamma = \Gamma_K = \mathrm{GL}_n(K)$ to K. It is clear that K^Γ is a commutative K-algebra having addition and multiplication defined pointwise, and with identity $1 : x \mapsto 1_K$. In fact, we have

Lemma 1.3.1 K^Γ *is a* (Γ, Γ)-*bimodule under the actions of left and right translation of functions; that is, for* $f \in K^\Gamma$ *and* $g \in \Gamma$, *define* $g \circ f$ *and* $f \circ g$ *in* K^Γ *by*

$$(g \circ f)(x) = f(xg)$$
$$(f \circ g)(x) = f(gx)$$

for all $x \in \Gamma$. *These actions are linear, multiplicative and the left and right actions commute.*

Definition 1.3.2 For each pair $\alpha, \beta \in \mathbf{n}$, define the *coefficient functions* or *coordinate functions* $c_{\alpha\beta} \in K^\Gamma$ by

$$c_{\alpha\beta}(g) = g_{\alpha\beta}$$

for all $g = [g_{\alpha\beta}] \in \Gamma_K$. The *algebra of polynomial functions* on Γ_K, denoted $A_K(n)$, is the subalgebra of K^Γ generated by the n^2 coefficient functions $c_{\alpha\beta}$ $(\alpha, \beta \in \mathbf{n})$.

The set $\{c_{\alpha\beta} : \alpha\,\beta \in \mathbf{n}\}$ is algebraically independent over K since K is infinite, and so $A_K(n)$ can be regarded as the algebra of all polynomials in the n^2 indeterminates $c_{\alpha\beta}$. The elements of $A_K(n)$ are called *polynomial functions* on Γ_K. We often drop the subscript K when the field is understood.

Definition 1.3.3 For any $r \geq 0$, denote by $A_r = A_K(n, r)$ the K-subspace of $A_K(n)$ generated by those elements in $A_K(n)$ that, considered as polynomials in the $c_{\alpha\beta}$ are homogeneous of total degree r.

Clearly we have a grading

$$A_K(n) = \bigoplus_{r \geq 0} A_K(n, r)$$

by homogeneous degree, where we define $A_K(n, 0) = K1$. The argument justifying 1.2.2 can be generalised in an obvious way to give

Theorem 1.3.4 A_r *has* K-*basis*

$$\{c_{i,j} = c_{i_1 j_1} \cdots c_{i_r j_r} : (i,j) \in \Omega\},$$

where Ω *is a transversal of the set of all* G-*orbits on* $I \times I$; *hence*

$$\dim A_r = \binom{n^2 + r - 1}{r}.$$

The next observation relates A_r to K^Γ.

Lemma 1.3.5 A_r *is a* (Γ, Γ)-*sub-bimodule of* K^Γ *for all* $r > 0$. *In fact, for any* $i, j \in I$ *and* $g, h \in \Gamma_K$,

$$
\begin{aligned}
g \circ c_{i,j} &= \sum_{t \in I} g_{t,j} c_{i,t} \\
c_{i,j} \circ h &= \sum_{t \in I} h_{i,s} c_{s,j}
\end{aligned}
$$

where we write $g_{t,j} = g_{t_1 j_1} \cdots g_{t_r j_r}$ *(with a similar definition for* $h_{i,s}$*). Further,* A_0 *is a* (Γ, Γ)-*bimodule since* $g \circ 1 = 1 = 1 \circ h$.

Proof For any $x, g \in \Gamma$ and $\alpha, \beta \in \mathbf{n}$ we have, using 1.3.1,

$$(g \circ c_{\alpha\beta})(x) = c_{\alpha\beta}(xg) = [xg]_{\alpha\beta} = \sum_{\gamma \in \mathbf{n}} x_{\alpha\gamma} g_{\gamma\beta}$$

and so $g \circ c_{\alpha\beta} = \sum_\gamma c_{\alpha\gamma} g_{\gamma\beta}$. Thus

$$
\begin{aligned}
g \circ c_{i,j} &= g \circ (c_{i_1 j_1} \cdots c_{i_r j_r}) \\
&= \prod_{\rho \in \mathbf{r}} (g \circ c_{i_\rho j_\rho}) \\
&= \prod_{\rho \in \mathbf{r}} \sum_{t_\rho \in \mathbf{n}} g_{t_\rho j_\rho} c_{i_\rho t_\rho} \\
&= \sum_{t \in I} g_{t,j} c_{i,t}.
\end{aligned}
$$

Similarly for the right action. $\qquad\square$

We now extend the actions of Γ to actions of the group algebra $K\Gamma$. (Recall that if H is an infinite group, the group algebra KH is the collection of all $\sum \alpha_g g$, where $\alpha_g = 0$ for all but a finite number of the g.) Now, taking $\kappa = \sum_\Gamma \alpha_g g \in K\Gamma$ and $f \in K^\Gamma$, we put

$$\kappa \circ f = \sum_{g \in \Gamma} \alpha_g (g \circ f) \qquad f \circ \kappa = \sum_{g \in \Gamma} \alpha_g (f \circ g),$$

so that in this way

$$(\kappa \circ f)(\tau) = f(\tau \kappa), \qquad (f \circ \kappa)(\tau) = f(\kappa \tau)$$

for $f \in K^\Gamma$ and $\tau, \kappa \in K\Gamma$. It is worth noting that in general the $K\Gamma$-actions are not multiplicative (unless $\kappa = g \in \Gamma$).

A_r *is a coalgebra*

For the basic theory of coalgebras etc. see Sweedler [1969] or Abe [1980, §2]. There is a K-monomorphism $K^\Gamma \otimes K^\Gamma \to K^{\Gamma \times \Gamma}$ given by sending $f \otimes f'$ to the map $f''(h, h') = f(h)f'(h')$ $(f, f' \in K^\Gamma,\ h, h' \in \Gamma)$; thus we identify $K^\Gamma \otimes K^\Gamma$ with a K-subspace of $K^{\Gamma \times \Gamma}$. Define two K-algebra maps Δ and ε by

$$
\begin{array}{rcl}
\Delta : K^\Gamma & \longrightarrow & K^{\Gamma \times \Gamma} \\
f & \longmapsto & (\Delta f(u, v) = f(uv));
\end{array}
\qquad (1.4)
$$

$$
\begin{array}{rcl}
\varepsilon : K^\Gamma & \longrightarrow & K \\
f & \longmapsto & f(1_\Gamma)
\end{array}
\qquad (1.5)
$$

for all $f \in K^\Gamma$, $u, v \in \Gamma$.

Lemma 1.3.6 *$A(n)$ is a bialgebra, and A_r is a sub-coalgebra of $A(n)$.*

Proof Restricting the maps of (1.4) and (1.5) on K^Γ to $A(n)$ defines a comultiplication

$$\Delta(c_{i,j}) = \sum_{h \in I} c_{i,h} \otimes c_{h,j} \qquad (i, j \in I,\ r \geq 0),$$

and an augmentation

$$\varepsilon(c_{i,j}) = \delta_{i,j} = \delta_{i_1 j_1} \cdots \delta_{i_r j_r}$$

on $A(n)$. Thus endowed, $(A(n), \Delta, \varepsilon)$ becomes a coalgebra, and one makes the additional checks necessary to ensure that we have the required conditions for a bialgebra. Also, A_r is a sub-coalgebra of $A(n)$. $\quad\square$

The category $\mathcal{P}_K(n, r)$

We now define two categories, $\mathcal{P}_K(n)$ and $\mathcal{P}_K(n, r)$. The objects are all those finite-dimensional left-$K\Gamma$-modules V whose coefficient space lies in $A(n)$, respectively A_r. Let $\{v_b : b \in B\}$ be a basis of V, and suppose V is defined by the matrix representation $\rho : g \mapsto [\alpha_{ab}(g)]$ where

$$gv_b = \sum_{a \in B} \alpha_{ab}(g)v_a \quad (g \in \Gamma, \ b \in B). \tag{1.6}$$

The elements $\alpha_{ab} \in K^\Gamma$ span a K-space which we call the *coefficient space*, $\mathrm{cf}(V)$, of V. This space is independent of the basis $\{v_b\}$. The matrix $[\alpha_{ab}(g)]$ is called the *invariant matrix*. Therefore, for membership of $\mathcal{P}_K(n)$, we are requiring that $\mathrm{cf}(V) \subseteq A(n)$, while to be an object of $\mathcal{P}_K(n, r)$ we need $\mathrm{cf}(V) \subseteq A_r$.

The morphisms $\phi : V \to V'$ in the respective categories are simply the $K\Gamma$-homomorphisms. By an obvious generalisation of the argument following 1.2.1 or by results of Green [1976, 1.6], we have the following decomposition:

Theorem 1.3.7 *Let $V \in \mathcal{P}_K(n)$. Then $V = \bigoplus_{r \geq 0} V_r$, where for each r V_r is a submodule of V with $V \in \mathcal{P}_K(n, r)$. Hence, each indecomposable $V \in \mathcal{P}_K(n)$ is homogeneous of some degree.*

Henceforth we may concentrate on modules lying in $\mathcal{P}_K = \mathcal{P}_K(n, r)$ for fixed r. Clearly any sub- or factor module of an object in $\mathcal{P}_K$ also lies in $\mathcal{P}_K$; so too does any finite direct sum of modules in $\mathcal{P}_K$.

Definition 1.3.8 The object V of $\mathcal{P}_K(n, r)$ is a *polynomial module*, and the associated representations are called *polynomial representations* of $GL_n(K)$.

The categories of right modules $\mathcal{P}'_K(n)$ and $\mathcal{P}'_K(n, r)$ are defined analogously.

Remark. At times one can profitably shift one's viewpoint to think of $A_K(n)$ as the coordinate ring of the affine algebraic monoid $M = \mathrm{Mat}_n(K)$ consisting of all $n \times n$ matrices over K. In this scenario, polynomial representations of Γ are the same as rational representations of M: this approach is illustrated by our treatment of Doty's conjecture in 4.2.

1.4 Combinatorics

Underlying all that we do is the idea of a partition and its Young diagram. We now assemble all the relevant notation associated with these concepts.

Partitions and weights

Definition 1.4.1 A *partition* λ of r is a sequence $(\lambda_1, \lambda_2, \ldots)$ of non-negative weakly decreasing integers $\lambda_1 \geq \lambda_2 \geq \cdots$, such that $\sum \lambda_i = r$. The set of all partitions of r is denoted by $\Lambda^+(r)$. The λ_i are the *parts* of the partition. In the case $\lambda_{n+1} = \lambda_{n+2} = \cdots = 0$, we say λ *has length at most* n, and denote the set of partitions of r of length at most n by $\Lambda^+(n, r)$.

Dropping the condition that the λ_i are decreasing, we say that λ is a *composition* of r, and write $\Lambda(r)$ for the set of all compositions of r; we write $\Lambda(n, r)$ for the set of all compositions of r into at most n parts. Thus $\Lambda(n, r)$ is the set of all sequences of non-negative integers $(\lambda_1, \lambda_2, \ldots)$ such that $\sum \lambda_i = r$ and $\lambda_{n+1} = \lambda_{n+2} = \cdots = 0$.

If $\lambda \in \Lambda(n, r)$, the *conjugate partition* $\lambda' = (\lambda'_1, \ldots, \lambda'_n)$ has parts λ'_i satisfying $\lambda'_i = |\{j : \lambda_j \geq i\}|$. We note that $\lambda' \in \Lambda^+(m, r)$ for some $m > 0$.

Two further terms are important later. Let $p \in \mathbf{N}$. Call $\lambda \in \Lambda^+(r)$ *p-regular* if there does not exist an i such that $\lambda_{i+1} = \cdots = \lambda_{i+p} > 0$; denote the set of such partitions by $\Lambda_{(p)}(r)$. Otherwise, λ is *p-singular*. Finally λ is *p-restricted* if $0 \leq \lambda_i - \lambda_{i+1} \leq p - 1$ $(1 \leq i \leq r - 1)$ and $0 \leq \lambda_r \leq p - 1$; the set of such partitions will be denoted $\Lambda^{(p)}(r)$.

Remarks.
(1) A curiosity: p does not need to be prime for the above definitions to make sense! However, in practice p will be the characteristic of K. In this context, we often drop the p if it is understood.
(2) One tends to omit zero parts, and to write repeated parts as powers in a partition; e.g. we write $(5, 4, 2, 2, 2, 1, 1, 0, 0, \ldots) \in \Lambda^+(17)$ as $(5, 4, 2^3, 1^2)$.

Orderings of weights and partitions are crucial in future deliberations. The natural partial order on $\Lambda^+(r)$ is known as the *dominance order* (James and Kerber [1981, p. 23]). Formally,

Definition 1.4.2 If $\lambda, \mu \in \Lambda^+(r)$, we say that λ *dominates* μ, and write $\lambda \trianglerighteq \mu$, provided that for all j

$$\sum_{i=1}^{j} \lambda_i \geq \sum_{i=1}^{j} \mu_i.$$

It is also convenient to have a total order on the set of partitions of r.

Definition 1.4.3 If λ, $\mu \in \Lambda^+(r)$, write $\lambda \geq \mu$ if and only if $\lambda = \mu$ or the least j for which $\lambda_j \neq \mu_j$ is such that $\lambda_j > \mu_j$. This is called the *dictionary* or *lexicographic order* on partitions.

It is trivial to observe that the dictionary order contains the dominance order, in the sense that $\lambda \trianglelefteq \mu \Rightarrow \lambda \leq \mu$. We say that $\leq$ *refines* $\trianglelefteq$. Also, the largest partition with respect to $\geq$ is (r), and the smallest is (1^r).

We also have an important link with the multi-index set $I = I(n, r)$ introduced in 1.2.

Definition 1.4.4 We say that the composition $\lambda = (\lambda_1, \ldots, \lambda_n)$ is the *weight* of $i \in I(n, r)$, written $i \in \lambda$, if

$$\lambda_\nu = |\{\rho \in \mathbf{r} : i_\rho = \nu\}|$$

for all $\nu \in \mathbf{n}$.

Example. With $n = 3$ and $r = 5$, $i = (1, 3, 1, 1, 1)$ has weight $\lambda = (4, 0, 1) \in \Lambda(3, 5)$.

If $i \in \lambda$ it is easy to see that $G_i \cong \Sigma_\lambda = \Sigma_{\lambda_1} \times \Sigma_{\lambda_2} \times \ldots$, the (standard) *Young subgroup* associated to λ. Note also that $i, i' \in I$ have the same weight if and only if $i \sim i'$ (see the notation in 1.2). Hence the G-orbits on I are in one-to-one correspondence with the elements of $\Lambda(n, r)$. Therefore we may freely identify G-orbits and elements of $\Lambda(n, r)$, and call elements of both sets weights.

One can order I by stipulating that, for all $\rho \in \mathbf{r}$,

$$i \leq j \Leftrightarrow i_\rho \leq j_\rho.$$

It is immediate that if $i \leq j$ and $i \in \lambda$, $j \in \mu$, then $\lambda \trianglerighteq \mu$.

Now $W = \Sigma_n$ acts on I on the left via $wi = (w(i_1), \ldots, w(i_r))$ for all $w \in W$ and $i \in I$. The action commutes with the right action of G; hence W acts on $\Lambda(n, r)$ via $w^{-1}\lambda = (\lambda_{w(1)}, \ldots, \lambda_{w(n)})$ for $\lambda \in \Lambda(n, r)$ and $w \in W$. It is clear that each W-orbit contains a unique weight with decreasing parts, that is, an element of $\Lambda^+(n,r)$. We often call $\Lambda^+(n,r)$ the poset of *dominant weights of degree r*.

Diagrams and tableaux

Definition 1.4.5 Let $\lambda \in \Lambda^+(n,r)$ be given. The *Young diagram* for λ is the subset

$$[\lambda] = \{(i,j) : i,j \in \mathbf{N},\ i \geq 1 \text{ and } 1 \leq j \leq \lambda_i\}$$

of $\mathbf{Z}^2$; any map from $[\lambda]$ to a set is called a λ-*tableau.*

Informally, this means that we replace each element of $[\lambda]$ by any element of the set. Notice that we do not stipulate that the map is bijective. If we take our set as $\mathbf{r}$ and decree that the map $T^\lambda : [\lambda] \to \mathbf{r}$ be bijective, then we call the λ-tableau *basic*. Choosing one such bijection arbitrarily, say $T^\lambda(p,q) = a(p,q)$, $((p,q) \in [\lambda])$, we may write T^λ as

$$T^\lambda : \begin{matrix} a(1,1) & a(1,2) & \cdots\cdots\cdots & a(1,\lambda_1) \\ a(2,1) & a(2,2) & \cdots\cdots & a(2,\lambda_2) \\ & \cdots & \\ a(n,1) & a(n,2) & \cdots & a(n,\lambda_n) \end{matrix}$$

and call $T = T^\lambda$ the *basic λ-tableau*. The element $a(p,q)$ is said to be in row p and column q of T. For $p \in \mathbf{n}$ let $R_p = R_p(T^\lambda) = \{a(p,k) : k \in \mathbf{N}\}$ (the pth row of T^λ), and similarly for the qth column $C_q = C_q(T^\lambda)$. These are (possibly empty) subsets of $\mathbf{r}$, whose union is $\mathbf{r}$.

Define two subgroups of G as follows: the *row stabiliser* $R(T) = \Sigma_{R_1}\Sigma_{R_2}\ldots$ and the *column stabiliser* $C(T) = \Sigma_{C_1}\Sigma_{C_2}\ldots$, preserve the rows, or respectively columns, of T setwise. These groups depend on the choice of basic λ-tableau; however, the stabilisers corresponding to different basic λ-tableaux are G-conjugate (they are, after all, Young subgroups for λ).

Example. A basic $(3,2,1)$-tableau is

$$T = T^{(3,2,1)} = \begin{matrix} 1 & 2 & 4 \\ 3 & 5 & \\ 6 & & \end{matrix} . \tag{1.7}$$

In this example $R_1 = \{1,2,4\}, R_2 = \{3,5\}, R_3 = \{6\}$, and $C_1 = \{1,3,6\}$, $C_2 = \{2,5\}, C_3 = \{4\}$. Also $R(T) = \Sigma_{\{1,2,4\}} \times \Sigma_{\{3,5\}}$, of order $3!\,2!$, and $C(T) = \Sigma_{\{1,3,6\}} \times \Sigma_{\{2,5\}}$. Visibly $R_j = C_j = \emptyset$ if $j \geq 4$.

The set $I(n,r)$ of multi-indices acts on the λ-tableaux as follows: if $i \in I$ then $i \in \mathbf{n}^r$ and so we may consider the composite λ-tableau $iT^\lambda : [\lambda] \to \mathbf{n}$. Denote this by T_i^λ. Of course, even if T^λ is basic, T_i^λ need not necessarily be basic. We further comment that λ need not be dominant for the definitions of T^λ and T_i^λ to make sense.

We write the composite as

$$
T_i^\lambda : \quad
\begin{matrix}
i_{a(1,1)} & i_{a(1,2)} & \cdots\cdots\cdots & & i_{a(1,\lambda_1)} \\
i_{a(2,1)} & i_{a(2,2)} & \cdots\cdots & & i_{a(2,\lambda_2)} \\
 & & \cdots & & \\
i_{a(n,1)} & i_{a(n,2)} & \cdots & ai_{(n,\lambda_n)} &
\end{matrix}
$$

and call $i_{a(p,q)}$ the *entry in the $a(p,q)$th place.*

Example. In (1.7), taking $i \in I(n,6)$

$$
T_i^{(3,2,1)} = \quad
\begin{matrix}
i_1 & i_2 & i_4 \\
i_3 & i_5 & \\
i_6 & &
\end{matrix}
\quad .
$$

Remarks.

(1) Loosely speaking, T_i^λ is a way of 'writing i according to the shape of λ'.

(2) The notation is consistent: the multi-index $i = (i_1, \ldots, i_r) \in I(n,r)$ may be identified with T_i^λ where $T^\lambda = 1 \cdots r$ is a basic λ-tableau for $\lambda = (r) = (r, 0, 0, \ldots)$.

The next definition goes back to Young's efforts on finding an indexing set for a basis of the Specht module; see Robinson [1977, p. 258].

Definition 1.4.6 For $\lambda \in \Lambda^+(r)$, $i \in I$ and T^λ basic, we call T_i^λ *row semistandard* if the entries in each row (weakly) increase from left to right; T_i^λ is *column standard* if the entries in each column strictly increase from top to bottom. T_i^λ is *standard* if it is both row semistandard and column standard. We write

$$
I^\lambda = \{i \in I(n,r) : \ T_i^\lambda \text{ is standard}\}.
$$

The fundamental connection between dominance and standardness is the the subject of the next result.

Proposition 1.4.7 *If $\lambda \in \Lambda^+(r)$ and $i \in I^\lambda$ has weight μ, then $\lambda \trianglerighteq \mu$.*

Proof If T_i^λ is standard, then all 1s must be in $R_1(T_i^\lambda)$, forcing $\mu_1 \leq \lambda_1$. Again, by standardness, all 1s and 2s must be found in $R_1(T_i^\lambda)$ or $R_2(T_i^\lambda)$, forcing $\mu_1 + \mu_2 \leq \lambda_1 + \lambda_2$. The proof continues in this manner. □

Definition 1.4.8 Each basic λ-tableau T^λ determines an order on $\mathbf{r}$: one simply proceeds down each column starting at the left. This is the T^λ-*order on* r.

Example. If $\lambda = (3,2,1)$, and we choose T^λ as in (1.7), then the $T^{(3,2,1)}$-order on $\mathbf{6}$ is $1 < 3 < 6 < 2 < 5 < 4$.

This induces an order on the T_i^λ $(i \in I)$ as follows.

Definition 1.4.9 For i, $j \in I$ we say that $T_i^\lambda < T_j^\lambda$ if at the first $\rho \in \mathbf{r}$ (in the T^λ-order) where $i_\rho \neq j_\rho$, $i_\rho < j_\rho$ holds.

1.5 Character theory and weight spaces

Recall the combinatorial definition of weight from 1.4.4. Within the realms of our category $\mathcal{P}_K(n,r)$ there is another meaning ascribed to this term, whose provenance is from algebraic group theory. Rather than go through much unnecessary theory at this stage, we refer you to A.3 and to Humphreys [1987] and Jantzen [1987] for the complete technical background.

Let T be the diagonal subgroup of Γ_K, consisting of all matrices of the form

$$d(t) = \operatorname{diag}(t_1, \ldots, t_n) \qquad (1.8)$$

where $t_i \in K^* = K \setminus \{0\}$.

Definition 1.5.1 Given $\lambda = (\lambda_1, \ldots, \lambda_n) \in \Lambda(n,r)$, we define the *multiplicative character* of T to be the map χ_λ defined by

$$\begin{aligned} \chi_\lambda : T &\rightarrow K^* \\ d(t) &\mapsto t_1^{\lambda_1} \ldots t_n^{\lambda_n}, \end{aligned}$$

for $t_i \in K^*$. The set of such characters is called the set of *polynomial weights of degree r* of T. If $\lambda \in \Lambda^+(n,r)$ the corresponding character is termed *dominant polynomial of degree r*.

The full weight lattice $X(T)$ has canonical basis consisting of the functionals ε_i (the restrictions of the coordinate functions c_{ii} to T). We identify $\Lambda(n,r)$ and the set of polynomial weights of degree r by means of the correspondence $\lambda \longleftrightarrow \sum \lambda_i \varepsilon_i$. Similarly we identify $\Lambda^+(n,r)$ with the set of dominant polynomial weights of degree r.

Given $\lambda \in \Lambda(n,r)$ and some $V \in \mathcal{P}_K$, define the (left) λ-*weight space* V^λ of V to be the set

$$V^\lambda = \{v \in V : d(t) \circ v = \chi_\lambda(d(t))v \ \forall \, d(t) \in T\},$$

where $d(t)$ is the toral element of (1.8). Call the elements $v \in V^\lambda$ *weight elements of weight* λ. V^λ is a T-stable subspace of V (maybe zero). In fact only finitely many weight spaces are non-zero: if $\lambda \in \Lambda(n,r)$ is such that $V^\lambda \neq 0$ then λ is a *weight* of V. The integer $\dim V^\lambda$ is called the *multiplicity* of the weight. One says that $\lambda \in \Lambda^+(n,r)$ is the *highest weight* of V if V^λ is 1-dimensional, and all other weights $\mu \in \Lambda(n,r)$ occurring in V satisfy $\lambda \rhd \mu$. Of course one may also define the (right) μ-weight space $^\mu W$ of a $K\Gamma$-module $W \in \mathcal{P}'_K$. I should issue the warning that many authors would denote the right weight space as W^λ and the left weight space as $^\lambda V$.

Example. Let $i, j \in I$ have weights λ, μ respectively. Lemma 1.3.5 implies that $c_{i,j}$ is a right weight element of A_r of right weight λ; consanguineously $c_{i,j}$ is a left weight element of left weight μ. Those who know about representations of G from James [1978a] will recognise $c_{i,j}$ as a λ-tabloid of type μ.

Properties of weight spaces

At this point it is convenient to state a few elementary facts about weight spaces. One can find proofs in most relevant texts, e.g. Green [1980, 3.3]. We shall prove the weight space decomposition of 1.5.2(i) by a Schur algebra method in 2.2.10.

Proposition 1.5.2 *Given* $V \in \mathcal{P}_K(n,r)$ *and* $\lambda \in \Lambda(n,r)$,

(i) $V = \bigoplus_{\mu \in \Lambda(n,r)} V^\mu$ *is a direct sum of its non-zero weight spaces.*

(ii) *If* $\pi \in \Sigma_n$ *then* $V^\lambda \cong V^{\pi(\lambda)}$ *as* K-*spaces.*

(iii) *If* $W \in \mathcal{P}_K(n,s)$ $(s \in \mathbf{N})$ *and* $\mu \in \Lambda(n, r+s)$ *then* $(V \otimes W)^\mu = \bigoplus_{\alpha,\beta} V^\alpha \otimes W^\beta$ *where* $\alpha \in \Lambda(n,r)$, $\beta \in \Lambda(n,s)$ *are such that* $\alpha + \beta = \mu$ *(added componentwise).*

(iv) *Suppose we have a short exact sequence* $0 \to V_1 \to V_2 \to V_3 \to 0$
in $\mathcal{P}_K$. *Then the induced sequence* $0 \to V_1^\lambda \to V_2^\lambda \to V_3^\lambda \to 0$ *is*
exact.

(v) *Let* $L : K$ *be a field extension, and identify* $\mathcal{P}_K$ *as a full sub-*
category of $\mathcal{P}_L$. *Let* $V \in \mathcal{P}_K$, *and suppose* $V_L = V \otimes_K L$; *then*
$V_L \in \mathcal{P}_L$, *and one thinks of* V *as the subset* $V \otimes 1_L$ *of* V_L. *Then*
for any $\lambda \in \Lambda(n, r)$

$$\dim_L V_L^\lambda = \dim_K V^\lambda.$$

Formal characters

The theory of characters is familiar from courses on representations of
finite groups over $\mathbf{C}$. If $V \in \mathcal{P}_K(n, r)$, what might we understand by the
character of V? For polynomial $K\Gamma$-modules there are two meanings in
common currency.

Definition 1.5.3 Given $V \in \mathcal{P}_K(n, r)$ and $\lambda \in \Lambda(n, r)$, we may form
the monomial $\mathbf{X}^\lambda = X_1^{\lambda_1} \cdots X_n^{\lambda_n} \in \mathbf{Z}[X_1, \ldots, X_n]$ of degree r in the n
variables $X_1, \ldots, X_n$ over $\mathbf{Q}$. We define the *formal character* $\mathrm{ch}(V)(\mathbf{X})$
of V to be the r-homogeneous polynomial

$$\mathrm{ch}(V)(\mathbf{X}) = \sum_{\lambda \in \Lambda(n,r)} (\dim_K V^\lambda) \mathbf{X}^\lambda.$$

This is an element of $\mathbf{Z}[X_1, \ldots, X_n]$.

From 1.5.2(ii) we note that $\mathrm{ch}(V)$ is symmetric, indeed one can write

$$\mathrm{ch}(V)(\mathbf{X}) = \sum_{\lambda \in \Lambda^+(n,r)} (\dim_K V^\lambda) m_\lambda(\mathbf{X}),$$

where m_λ is the λth monomial symmetric function (Macdonald [1979,
p. 11]). Proposition 1.5.2 also shows that $\mathrm{ch}(V)$ is additive with respect
to short exact sequences, and is multiplicative with respect to tensoring.
Isomorphic modules have the same character. Since the characters of
all objects of $\mathcal{P}_K$ lie in the ring $\mathcal{S}_{n,r}$ of all symmetric functions of total
degree r in the X_i, by the 'fundamental theorem of symmetric functions'
(Macdonald [1979, (2.4)]), the additive subgroup of $\mathbf{Z}[\mathbf{X}]$ that they generate is precisely $\mathcal{S}_{n,r}$ and hence is independent of K.

Another notion of character may be familiar:

Definition 1.5.4 Suppose that $V \in \mathcal{P}_K(n, r)$ affords the representation $\rho : \Gamma \to \mathrm{GL}(V)$, and let tr be the usual trace function. The *natural character* χ_V of V is given by the formula

$$\chi_V(g) = \mathrm{tr}\,\rho(g)$$

for all $g \in \Gamma$.

It is clear that $\chi_V \in A_r$, being the trace of the invariant matrix afforded by a basis of V. The next result goes back to Schur (Green [1980, p. 42]) and the proof requires us to quote a few things on symmetric functions from Macdonald [1979, Chapter I]. In particular, we recall that for $d \geq 1$ the dth elementary symmetric function, $e_d(X_1, \ldots, X_n)$, is the sum of all possible distinct products, taken d at a time, of $X_1, \ldots, X_n$ (and $e_0 = 1$).

Proposition 1.5.5 *Let $g \in \Gamma$ have eigenvalues $\zeta_1, \ldots, \zeta_n$ (with multiplicities) lying in some algebraic closure of K. Then*

$$\chi_V(g) = \mathrm{ch}(V)(\zeta_1, \ldots, \zeta_n).$$

Proof It is no loss to assume that K is algebraically closed since, by 1.5.2(v), extending the field to L, say, replaces χ_V by a function on Γ_L whose restriction to Γ_K is χ_V.

Let $C = [c_{ab}]$ and t an indeterminate over K. The characteristic polynomial of C may be written

$$\det(tI_n - C) = t^n - a_1 t^{n-1} + \cdots + (-1)^n a_n$$

for some $a_i \in A_r$. Of course $a_d(g) = e_d(\zeta_1, \ldots, \zeta_n)$ for $1 \leq d \leq n$, and we write $\mathrm{ch}(V) = \sum_{\lambda \in \Lambda^+(r)} d^\lambda e_1^{\lambda_1} \cdots e_r^{\lambda_r}$ where $d^\lambda \in \mathbf{Z}$. So, given $g \in \Gamma$,

$$\mathrm{ch}(V)(\zeta_1, \ldots, \zeta_n) = \left(\sum_\lambda (d^\lambda 1_K) a_1^{\lambda_1} \cdots a_r^{\lambda_r} \right)(g) = \psi(g)$$

where $\psi \in A_r$. We make the further assumption that g is diagonalisable, that is, there exists $z \in \Gamma$ such that $zgz^{-1} = \mathrm{diag}(\zeta_1, \ldots, \zeta_n) = d(\zeta)$.

Now, by 1.5.2(i), find a weight space decomposition $V = \bigoplus V^\mu$, and order the basis of V to be compatible with this decomposition. Then $d(\zeta)$ is represented by a diagonal matrix with $\dim V^\mu$ terms equal to $\zeta_1^{\mu_1} \cdots \zeta_n^{\mu_n}$ for each $\mu \in \Lambda(n, r)$. Applying the trace function,

$$\chi_V(g) = \chi_V(zgz^{-1}) = \mathrm{ch}(V)(\zeta_1, \ldots, \zeta_n) = \psi(g).$$

We have now constructed two polynomials in the c_{ab}, namely ψ and χ_V which agree on all semisimple elements of Γ. But every matrix with distinct eigenvalues is semisimple, and so ψ and χ_V agree on all elements of $g \in \Gamma$ such that $\delta(g) \neq 0$ where $\delta \in A_{r(r-1)}$ is the discriminant of the characteristic polynomial. Hence $\psi = \chi_V$ as required. $\qquad\square$

Corollary 1.5.6 *The set of (formal) characters corresponding to a finite set of mutually non-isomorphic absolutely irreducible objects in $\mathcal{P}_K$ constitutes a linearly independent subset of $S_{n,r}$.*

Proof Let $\chi_1, \ldots, \chi_m$ be the natural characters of such a set of simple modules. By the Frobenius-Schur theorem (Curtis and Reiner [1981, (3.41)]) the χs are linearly independent in A_r. Suppose that there is a non-trivial linear relation $\sum_{i \in \mathbf{m}} \alpha_i f_i = 0$ ($\alpha_i \in \mathbf{Z}$) where f_i is the formal character that corresponds to χ_i. We have to suppose that if $\operatorname{char} K = p$ then some α_i are coprime to p. By 1.5.5 the non-trivial relation,

$$(\alpha_1 1_K)\chi_1(g) + \cdots + (\alpha_m 1_K)\chi_m(g) = 0,$$

holds for all $g \in \Gamma$, contradicting the second sentence. $\qquad\square$

Examples

We take this opportunity to introduce some important objects of $\mathcal{P}_K$ and to calculate their characters.

(1) *The natural module.* Let E be an n-dimensional vector space over K. Choose and fix a basis $\{e_1, \ldots, e_n\}$ for E. Then Γ acts (on the left) on E by linear substitution:

$$g \circ e_b = \sum_{a \in \mathbf{n}} c_{ab}(g)e_a$$

for all $g \in \Gamma_K$ and $b \in \mathbf{n}$. People often call E the *natural module* for Γ_K. Note that by definition, $E \in \mathcal{P}_K(n, 1)$. It is clear that $\operatorname{ch}(E)(\mathbf{X}) = e_1(\mathbf{X}) = \sum X_i$. For, if we let $\lambda_a = (0, \ldots, 0, 1, 0, \ldots, 0) \in \Lambda(n, 1)$, with 1 in the ath position, then $E^{\lambda_a} = Ke_a$ for all $a \in \mathbf{n}$.

(2) *Exterior powers.* For each $0 \leq r \leq n$ the rth *exterior power* $\Lambda^r(E)$ is in $\mathcal{P}_K(n, r)$. Let us write P_r for the collection of subsets of $\mathbf{n}$ of cardinality r. If $[i] = \{i_1 < \cdots < i_r\} \in P_r$, then $\Lambda^r(E)$ has basis $\{e_{[i]} = e_{i_1} \wedge \ldots \wedge e_{i_r}\}$, hence is of dimension $\binom{n}{r}$. Viewing $[i]$ as a multi-index of weight $\alpha \in \Lambda(n, r)$, say, $d(t)e_{[i]} = t_1^{\alpha_1} \cdots t_n^{\alpha_n} e_{[i]}$. Moreover

distinct elements of P_r give rise to distinct weights. We conclude that the weight spaces of V all have dimension 1 or 0, namely

$$V^\alpha = \begin{cases} Ke_{[i]} & \text{if } [i] \in P_r \text{ of weight } \alpha \\ 0 & \text{for all other } \alpha \in \Lambda(n,r) \end{cases}.$$

Thus $\text{ch}(\Lambda^r(E)) = e_r$. Finally, $\Lambda^r(E)$ is a simple module of highest weight $\omega_r = \varepsilon_1 + \cdots + \varepsilon_r$, the rth *fundamental representation* of Γ.

(3) *Symmetric powers.* The rth *symmetric power* $S^r(E)$ of E is the space of all homogeneous polynomials of degree r in $K[e_1, \ldots, e_n]$. Here we regard the e_a as commuting indeterminates. Thus it has a basis consisting of all monomials $e^\beta = \prod_{a=1}^n e_a^{\beta_a}$ where $\beta \in \Lambda(n,r)$. Any such β gives the weight space $V^\beta = Ke^\beta$, and so $\text{ch}(S^r(E))(\mathbf{X}) = \sum_{\beta \in \Lambda(n,r)} X_1^{\beta_1} \ldots X_n^{\beta_n}$. This function, denoted $h_r(n)$, is known as the *complete symmetric function.* More generally, given $\lambda \in \Lambda(r)$ one defines $S^\lambda(E) = S^{\lambda_1}(E) \otimes S^{\lambda_2}(E) \otimes \cdots$ and similarly $\Lambda^\lambda(E) = \Lambda^{\lambda_1}(E) \otimes \Lambda^{\lambda_2}(E) \otimes \cdots$.

1.6 Irreducible objects in $\mathcal{P}_K(n,r)$

We close the first chapter with a 'formal' description of the irreducible objects of $\mathcal{P}_K(n,r)$. This is an area to which we return frequently in the sequel, and indeed we shall actually present a far more concrete description of the irreducible modules $L(\lambda)$ in 3.4.2. The result itself, which says that simple objects in $\mathcal{P}_K(n,r)$ are in one-to-one correspondence with elements of $\Lambda^+(n,r)$, the correspondence being given by the highest weight, is one of the highlights of Schur's dissertation for the case $K = \mathbf{C}$.

In what follows, when we refer to the leading term of an element of $\mathbf{Z}[\mathbf{X}]$, we intend the order to be taken relative to the lexicographic ordering of weights $\lambda \in \Lambda(n,r)$ or of monomials $\mathbf{X}^\lambda$.

Theorem 1.6.1 *Given $n \geq 1$, $r \geq 0$, the following hold.*

(i) *For each $\lambda \in \Lambda^+(n,r)$ there exists an absolutely irreducible object $L(\lambda) \in \mathcal{P}_K(n,r)$ affording a character $\text{ch}_{\lambda,K}$ whose leading term is $\mathbf{X}^\lambda$.*

(ii) $\{\text{ch}_{\lambda,K} : \lambda \in \Lambda^+(n,r)\}$ *is a $\mathbf{Z}$-basis for the ring $\mathcal{S}_{n,r}$.*

(iii) *Every irreducible object in $\mathcal{P}_K(n,r)$ is isomorphic to $L(\lambda)$ for precisely one $\lambda \in \Lambda^+(n,r)$.*

Proof For (i), we write $\lambda' = \mu = (\mu_1, \ldots, \mu_r)$ and recall the natural module from the last section. Clearly, the product of exterior powers $V = \Lambda^{\mu_1}(E) \otimes \ldots \otimes \Lambda^{\mu_r}(E)$ has character $\mathrm{ch}(V) = e_1^{\mu_1} \cdots e_r^{\mu_r} = e_\mu$. This has leading term $\mathbf{X}^\lambda$. By 1.5.2(iv), there must exist a composition factor U of V whose character has $\mathbf{X}^\lambda$ as its leading term. Define $U = L(\lambda)$.

By Curtis and Reiner [1981, (3.43)], to show absolute irreducibility we want to show that $\mathrm{End}_K(U)$ is precisely the set of homotheties. By the first paragraph of our proof, $\dim U^\lambda = 1$. Now if $\theta \in \mathrm{End}_K U$ then $\theta : U^\lambda \to U^\lambda$ so that there is some $k \in K$ with $\theta(u) = k.u$ for all $u \in U^\lambda$. But $U' = \{u \in U : \theta(u) = ku\}$ is a submodule of U, thus $U' = U$, hence $\theta = k.1_U$ as required.

For (ii), we express the $\mathrm{ch}_{\lambda,K}$ in terms of monomial symmetric functions and then each m_λ ($\lambda \in \Lambda^+(n,r)$) in terms of e_λs as in the proof in Macdonald [1979, (2.3)]. We thus obtain $\mathrm{ch}_\lambda = e_\lambda + \sum_{\mu < \lambda} a_{\lambda\mu} e_\mu$ for some $a_{\lambda\mu} \in \mathbf{N}_0$, and so the various ch_λ as λ ranges through $\Lambda^+(n,r)$ form a $\mathbf{Z}$-basis of $\mathcal{S}_{n,r}$.

For (iii), let $F : K$ be an extension of fields with F algebraically closed. $L(\lambda)$ is absolutely irreducible, by (i), so $L(\lambda)_F = L(\lambda) \otimes F$ is too. By 1.5.2(v), $\mathrm{ch}_{\lambda,K} = \mathrm{ch}_{\lambda,F}$. Let S be an arbitrary irreducible object in $\mathcal{P}_F$. If $S \not\cong L(\lambda)_F$ for any λ, then the set $\{\mathrm{ch}_S, \{\mathrm{ch}_{\lambda,K}\}_\lambda\}$ is linearly independent by 1.5.6, contrary to (ii). Finally, we assume that T is any irreducible in $\mathcal{P}_K$, and choose a minimal submodule M of T_F. By the above, one can find $\lambda \in \Lambda^+(n,r)$ with $L(\lambda)_F \cong M$. Then $\mathrm{Hom}_{\Gamma_F}(L(\lambda)_F, T_F) \neq 0$ and so $\mathrm{Hom}_{\Gamma_K}(L(\lambda), T) \neq 0$. Hence $L(\lambda) \cong T$, completing the proof. $\square$

Remarks.
(1) Note carefully what this theorem is saying: it asserts that the module $L(\lambda)$ is defined up to isomorphism by the following two properties.

- $L(\lambda)$ is an irreducible object of $\mathcal{P}_K(n,r)$;
- $\mathrm{ch}L(\lambda)$ is the character $\mathrm{ch}_{\lambda,K}$ having leading term $\mathbf{X}^\lambda$.

(2) The proof also makes it clear that $\mathrm{ch}_{\lambda,K} = \mathrm{ch}_{\lambda,F}$ for all field extensions $F : K$, and so $\mathrm{ch}_{\lambda,L}$ is the same for all infinite fields L of the same characteristic. We therefore write $\mathrm{ch}_{\lambda,p}$ for $\mathrm{ch}_{\lambda,K}$ if $p = \mathrm{char}K$. Thus we have a fifth $\mathbf{Z}$-basis $\{\mathrm{ch}_{\lambda,p} : \lambda \in \Lambda^+(n,r)\}$ of $\mathcal{S}_{n,r}$ for each prime p.

We have avoided computing the $\mathrm{ch}_{\lambda,p}$ explicitly. In fact this is an open problem, the subject of the famous Lusztig Conjecture (Lusztig [1980]). On the other hand, the situation in characteristic zero is understood completely. Recall that the Schur function of type λ is $s_\lambda = a_{\lambda+\delta}/a_\delta$ where, for $\alpha \in \Lambda(n,r)$,

$$a_\alpha(\mathbf{X}) = \sum_{\pi \in \Sigma_n} (-1)^\pi \mathbf{X}^{\pi(\alpha)} = \det[X_i^{\alpha_j}]$$

and $\delta = (n-1, n-2, \ldots, 0) \in \Lambda(n, \frac{1}{2}n(n-1))$. If $K = \mathbf{Q}$ the proof of the assertion $\mathrm{ch}_{\lambda,0} = s_\lambda$ was executed by Schur himself, and we refer the interested reader to Green's treatment, [1980, pp. 47–9]. Notice that there are now five bases of the ring $\mathcal{S}_n = \bigoplus_{r \geq 0} \mathcal{S}_{n,r}$ of all symmetric functions in the X_i, namely the monomial, complete and elementary symmetric functions, the Schur functions and the formal characters of the simple polynomial Γ_K-modules.

As an appetizer for future material in Chapter 4, we point the reader to the work of Doty and Walker [1992b] on so-called modular symmetric functions: they attack the problem of computing explicitly the $\mathrm{ch}_{\lambda,p}$ by decomposing tensors of truncated symmetric powers.

(3) Thus far we have not mentioned the finite general linear group $\mathrm{GL}_n(q)$, where q is a prime power. It is due to appear more frequently towards the end of the text, but here is a good place to mention the early work of Steinberg [1951], who produced one ordinary irreducible character (the *unipotent character*) for each partition of n; one should also be aware of Green's calculation (see Green [1955], and Macdonald [1979, Chapter IV]) of the natural characters of $\mathrm{GL}_n(q)$. In contrast with the relatively straightforward computation for the infinite group, the finite case is involved. Indeed taking $p = q$ and working over our infinite field K of characteristic p (the *describing characteristic*), we do not even have satisfactory models for all the irreducible $\mathrm{GL}_n(p)$ modules except for small n: Doty and Walker [1992a] consider the case $n = 3$ (the cases for $n = 1, 2$ being straightforward). A study of representations of $\mathrm{GL}_n(q)$ over fields whose characteristic does not divide q (the *non-describing characteristic*), proceeding by analogy with symmetric groups, appears in James [1984]. This work is the starting point for a whole seam of theory leading to a definition of the q-Schur algebra.

(4) The deeper importance of the module V appearing in the proof of 1.6.1(i) has only recently been fully recognised: direct summands of Vs are the partial tilting objects in $\mathcal{P}_K(n,r)$. We say a little more about this in 4.5 and refer the curious to Donkin [1993] and Ringel [1991].

2

The Schur algebra

This chapter introduces the Schur algebra $S_r = S_r(\Gamma)$. The definition is motivated by consideration of the matrix representation afforded by the $K\Gamma$-module $E^{\otimes r}$, the tensor power of the natural module. Theorem 2.1.3 unearths a characterisation of the Schur algebra as a Hecke algebra, namely as the centralising algebra of the action of $G = \Sigma_r$ on the r-tensor space. This shadows the method of Schur [1927]. The rest of the chapter is devoted to an examination of the basic properties of S_r. We exhibit a basis, and a rule for multiplying two basis elements together that generalises the rule of (1.3). At this stage though, the most important fact is the statement that the module category for S_r is equivalent to the category of homogeneous polynomial representations, $\mathcal{P}_K(n, r)$. There is a simple rule (2.6) allowing one to switch easily from one category to another.

The Schur algebra was defined rather differently in Green's monograph [1980]. There it was defined as the vector space dual of the coalgebra A_r. In 2.3 we shall show how to identify the two algebras. This identification is very important, and will be generalised to q-Schur algebras later.

In part of the next chapter we shall be concerned with the problem of constructing a complete set of irreducible modules for S_r. There are now several approaches available in the literature, most of which involve the coalgebra A_r to some degree. We first obtain a K-basis for A_r, the elements of which consist of certain combinatorial objects now called bideterminants. These objects occur in such diverse topics as invariant theory and Grassmanian geometry. The assertion that any bideterminant is expressible as some integral linear combination of so-called standard bideterminants is a result that has come to be known as the 'Straightening Formula'. This is a fundamental result going back to Mead [1972] and the Rota school (Doubilet *et al.* [1974]) and indeed having its origins in the work of Capelli, Deruyts and Young. We give a version due to Green in 2.5.

Establishing linear independence of the set of standard bideterminants is the harder part of the proof of the basis theorem 2.4.7, whichever approach one might adopt. Recently Green [1992] has introduced elements of S_r called codeterminants. These are certain products of pairs of elements in the canonical basis of S_r, i.e. the one dual to the basis of A_r appearing in 1.3.4; moreover codeterminants are dual to bideterminants in a precise sense. The two sets of standard bi/co-determinants determine a unimodular matrix D, the Gram matrix of the two bases with respect to a certain bilinear form. The nonsingularity of D renders the problem of independence for the set of standard bideterminants and the set of standard codeterminants (almost) trivial. Hence the standard codeterminants form a basis of S_r. These facts will also make our treatment of Schur and Weyl modules for S_r in Chapter 3 much more transparent.

Sources for the Schur algebra are the standard classics of Weyl [1974] and Green [1980]. At times I have also drawn on other sets of lecture notes by Green [1981], [1991a] and [1990b].

2.1 Definition

We recall that $E = E_K$ denotes the natural module for Γ_K, with K-basis $e_1, \ldots, e_n$. On the r-fold tensor product, $E^{\otimes r} = E \otimes \ldots \otimes E$, there is a diagonal action of Γ. Thus, relative to the K-basis $\{e_i = e_{i_1} \otimes \ldots \otimes e_{i_r} : i \in I(n,r)\}$ of $E^{\otimes r}$, the Γ-action may be written as follows: if $j \in I$,

$$
\begin{aligned}
ge_j &= ge_{j_1} \otimes \cdots \otimes ge_{j_r} \\
&= \sum_{i \in I} g_{i_1 j_1} \cdots g_{i_r j_r} e_i.
\end{aligned} \tag{2.1}
$$

We may express this as

$$
ge_j = \sum_{i \in I} c_{i,j}(g) e_i. \tag{2.2}
$$

So, by (2.2) $cf(E^{\otimes r}) = \sum_{i,j \in I} K c_{i,j} = A_K(n,r)$, which demonstrates that $E^{\otimes r}$ is an object of $\mathcal{P}_K(n,r)$.

The symmetric group G, and hence also its group algebra KG, act on the right on $E^{\otimes r}$ by place permutations of the subscripts:

$$e_i \cdot \sigma = e_{i_{\sigma(1)}} \otimes \cdots \otimes e_{i_{\sigma(r)}}$$

for all $i = (i_1, \ldots, i_r) \in I$, $\sigma \in G$, and this commutes with the action of Γ.

Let $\mathrm{Mat}_I(K)$, or respectively $\mathrm{GL}_I(K)$, be the set of all, or respectively non-singular, matrices with entries in K, having rows and columns indexed by multi-indices. By (2.1), each element $g \in \Gamma$ induces a linear endomorphism $T_{n,r}$ on $E^{\otimes r}$, namely $T_{n,r}(g) = [g_{i,j}]_{i,j \in I}$. $T_{n,r} : \Gamma \to \mathrm{GL}_I(K)$ will also denote the corresponding matrix representation of Γ defined by sending g to $[g_{i,j}]_{i,j \in I}$. One can extend $T_{n,r}$ by linearity to a matrix representation of $K\Gamma$:

$$
\begin{aligned}
T_{n,r} : K\Gamma &\to \mathrm{Mat}_I(K) \\
\kappa = \sum_\Gamma \alpha_g g &\mapsto \left[\sum \alpha_g g_{i,j}\right]_{i,j \in I} = [c_{i,j}(\kappa)]_{i,j \in I}
\end{aligned}
\tag{2.3}
$$

(see after 1.3.5 for remarks on linear extension). At last we may introduce the star of the show.

Definition 2.1.1 Given n, r, K as usual, let

$$S_r(\Gamma) = \mathrm{Im}(T_{n,r}) = T_{n,r}(K\Gamma).$$

Call $S_r(\Gamma)$ the *Schur algebra*.

We shall often denote this by S_r, if n and K are clear from the context. Notice that the definition of S_r is as a certain collection of matrices whose rows and columns are indexed by multi-indices. Now, S_r is a K-subalgebra of $\mathrm{Mat}_I(K) \cong \mathrm{Mat}_{n^r}(K)$ so we have

Proposition 2.1.2 S_r *is an associative K-algebra, with identity element* $1_{S_r} = [\delta_{i,j}]$.

The literature is peppered with alternative definitions, so one task will be to show that they are all equivalent. One of these definitions exhibits S_r in the rôle of a centralising algebra for the action of G on tensor space. Define $\mathcal{M}$ to be the linear closure of the group $\{T_{n,r}(g) : g \in \Gamma\}$ (i.e. another way of looking at S_r). Each permutation $\sigma \in G$ induces a linear endomorphism of $E^{\otimes r}$, say $\Pi(\sigma) : x \mapsto x\sigma$ for $x \in E^{\otimes r}$. Now Weyl [1974, pp. 98, 130] proves that if $\theta \in M = \mathrm{End}_K(E^{\otimes r})$ commutes with all $\Pi(\sigma)$, $\sigma \in G$, then $\theta \in \mathcal{M} = S_r$. This demonstrates that S_r

is precisely the set $M^G = \operatorname{End}_{KG}(E^{\otimes r})$ of G-fixed points on the KG-module M with G-action given by $\theta^\sigma = \Pi(\sigma)\theta\Pi(\sigma)^{-1}$.

Theorem 2.1.3 (Schur's Commutation Theorem) *The Schur algebra is precisely the centralising algebra of the right action of G on $E^{\otimes r}$, i.e.*

$$S_r(\Gamma) = \operatorname{End}_{KG}(E^{\otimes r}).$$

The algebra on the right was first studied by Thrall [1942] in the case $r = n$ and K a field of order at least n.

2.2 First properties

Given $f \in K^\Gamma$, there is a unique linear extension $f \in K^{K\Gamma}$: take $\kappa = \sum \kappa_g g \in K\Gamma$ ($\kappa_g \in K$), and define $f(\kappa) = \sum \kappa_g f(g)$ (the 'evaluation of f at κ'). Let V be some $K\Gamma$-module in $\mathcal{P}_K(n, r)$ with finite K-basis $\{v_b : b \in B\}$. If Γ acts as $gv_b = \sum_B \alpha_{ab}(g)v_a$, as per (1.6), then $K\Gamma$ acts as

$$\kappa v_b = \sum \alpha_{ab}(\kappa)v_a, \qquad (2.4)$$

for all $\kappa \in K\Gamma$, $b \in B$. Let $\rho : K\Gamma \to \operatorname{End}_K(V)$ be the representation afforded by V, and let Y be its kernel. The first lemma shows that Y and $\operatorname{cf}(V)$ are 'orthogonal'.

Lemma 2.2.1 *For any $f \in K^\Gamma$ and $\kappa \in K\Gamma$,*

   (i) $\kappa \in Y \Leftrightarrow f(\kappa) = 0 \ \ \forall f \in \operatorname{cf}(V).$
   (ii) $f \in \operatorname{cf}(V) \Leftrightarrow f(\kappa) = 0 \ \ \forall \kappa \in Y.$

Proof For (i),

$$\kappa \in Y \Leftrightarrow \alpha_{ab}(\kappa) = 0 \ \ \forall a, b \in B \Leftrightarrow f(\kappa) = 0 \ \ \forall f \in \operatorname{cf}(V).$$

For (ii), we consider the finite-dimensional K-spaces $Z = \{f \in K^\Gamma : f(Y) = 0\}$ and $N = \rho(K\Gamma)$. Define the non-singular pairing

$$\langle \, , \, \rangle \ : \ Z \times N \longrightarrow K$$
$$\langle f, \nu \rangle \ = \ f(\kappa) \quad (\forall f \in Z, \ \nu = \rho(\kappa) \in N).$$

It is easy to check directly that this is well defined. By (i), $\operatorname{cf}(V) \subseteq Z$. If $0 \neq \nu = \rho(\kappa) \in N$ is such that $\langle f, \nu \rangle = 0$ for all $f \in \operatorname{cf}(V)$ then $\kappa \in Y$, hence $\nu = 0$. Thus $\operatorname{cf}(V) = Z$. $\qquad\square$

Corollary 2.2.2 *If U and V are any objects of $\mathcal{P}_K$ affording the $K\Gamma$-representations μ and ρ then*

$$\mathrm{cf}(U) \supseteq \mathrm{cf}(V) \Leftrightarrow \mathrm{Ker}\mu \subseteq \mathrm{Ker}\rho.$$

To apply this, we take $V \in \mathcal{P}_K$ affording ρ and $E^{\otimes r}$ affording T as above. The following theorem amounts to saying that any polynomial representation of Γ, homogeneous of degree r, factors through T, a fact known to Schur over $\mathbf{C}$.

Theorem 2.2.3 $V \in \mathcal{P}_K(n,r) \Leftrightarrow \mathrm{Ker}\, T_{n,r} \subseteq \mathrm{Ker}\, \rho$.

Proof Since

$$V \in \mathcal{P}_K(n,r) \Leftrightarrow A_K(n,r) = \mathrm{cf}(E^{\otimes r}) \supseteq \mathrm{cf}(V),$$

by 2.2.2, the theorem is proved. □

The proof of 2.2.1 produced a non-singular pairing between $\mathrm{cf}(V)$ and $N = \rho(K\Gamma)$, so that in particular $\dim(\mathrm{cf}(V)) = \dim N$. We may specialise this construction as follows:

Theorem 2.2.4 *There is a non-singular bilinear form $\langle \ , \ \rangle : A_K(n,r) \times S_r(\Gamma) \to K$, given by sending $\langle c, \xi \rangle \mapsto c(\kappa)$, where $\xi = T_{n,r}(\kappa) \in S_r$ and $c \in A_r$. It follows that*

$$\dim S_r = \dim A_r = \left(\begin{array}{c} n^2 + r - 1 \\ r \end{array} \right).$$

Proof Put $V = E^{\otimes r}$ above. □

This proves that S_r and A_r^* are isomorphic as K-spaces. Actually, 2.3.5 below will prove that they are isomorphic as (S_r, S_r)-bimodules. For the moment let us state one further simple result.

Lemma 2.2.5 *Let $\xi, \eta \in S_r$ and $i, j \in I$. Then*

(i) $\langle c_{i,j}, \xi \rangle =$ *the (i,j)th entry of ξ.*

(ii) $\langle c_{i,j}, \xi\eta \rangle = \displaystyle\sum_{h \in I} \langle c_{i,h}, \xi \rangle \langle c_{h,j}, \eta \rangle.$

Proof We observe that (i) is true when $\xi = T_{n,r}(g)$ for some $g \in \Gamma$, because in that case $\langle c_{i,j}, \xi \rangle = c_{i,j}(g) = g_{i,j}$. Hence (i) is true for all $\xi \in \mathrm{Im}T = S$ by linearity. For part (ii) we do a calculation:

$$\langle c_{i,j}, \xi\eta \rangle \ = \ (i,j)\text{th entry of } \xi\eta$$
$$= \ \sum_{h \in I} \xi_{i,h}\eta_{h,j}$$
$$= \ \sum_{h \in I} \langle c_{i,h}, \xi \rangle \cdot \langle c_{h,j}, \eta \rangle.$$

$\square$

A basis for S_r

The obvious basis to be investigated for S_r is that which is dual to the basis $\{c_{i,j} : (i,j) \in \Omega\}$ of A_r with respect to the pairing $\langle \, , \, \rangle$ of 2.2.4. We define $\xi_{i,j} \in S_r$, for given $i, j \in I$, by

$$\langle c_{k,l}, \ \xi_{i,j} \rangle = \begin{cases} 1 & \text{if } (k,l) \sim (i,j) \\ 0 & \text{otherwise} \end{cases} \tag{2.5}$$

for all $k, l \in I$. It is now clear that

$$\xi_{i,j} = \xi_{p,q} \Leftrightarrow (i,j) \sim (p,q),$$

and hence

Theorem 2.2.6 $\{\xi_{i,j} : (i,j) \in \Omega\}$ *is a K-basis of S_r.*

Example. (Action on $E^{\otimes r}$). Let us work out the action of S_r on tensor space. From the definition in 2.1.1, $S_r = \operatorname{Im} T_{n,r}$ acts on tensor space as

$$\xi e_j = \kappa e_j \text{ for } \xi \in S_r, \ \xi = T_{n,r}(\kappa), \ \kappa \in K\Gamma, \ j \in I.$$

This ensures that $E^{\otimes r}$ is an (S_r, KG)-bimodule. Now, from (2.2) we know that $\kappa e_j = \sum_{i \in I} c_{i,j}(\kappa)e_i$; we then have a reformulation of 2.2.5(i):

$$\xi e_j = \sum_{i \in I} \langle c_{i,j}, \xi \rangle e_j.$$

In particular, for $(u, v) \in \Omega$,

$$\xi_{u,v}(e_j) = \sum_{i \in I} \langle c_{i,j}, \xi_{u,v} \rangle e_j.$$

An equivalence of categories

Given $V \in \mathcal{P}_K(n, r)$ we may obtain an $S_r(\Gamma)$-module as follows. Let $\xi \in S_r(\Gamma)$, and find $\kappa \in K\Gamma$ such that $T_{n,r}(\kappa) = \xi$. Then define

$$\xi \circ v = \kappa \circ v \qquad (\forall v \in V). \qquad (2.6)$$

If $\kappa' \in K\Gamma$ is another element whose image under T is ξ, then $\kappa - \kappa' \in$ $\mathrm{Ker}(T_{n,r})$, and hence $\kappa - \kappa' \in \mathrm{Ker}\rho$ by 2.2.3. Conclude that $\kappa \circ v = \kappa' \circ v$, so there is no ambiguity. Thus, if $V \in \mathcal{P}_K$, then it can be made into a left $S_r(\Gamma)$-module. Conversely, if $V \in {}_{S_r(\Gamma)}\mathbf{mod}$, we obtain a polynomial $K\Gamma$-module using (2.6) to define the action of $\kappa \in K\Gamma$, since the left-hand side is known. The representation ρ given by this action clearly satisfies $\mathrm{Ker}T \subseteq \mathrm{Ker}\rho$, and so by 2.2.3, $V \in \mathcal{P}_K$.

We now have a basic result of Schur, who knew it in characteristic zero:

Theorem 2.2.7 *The categories $\mathcal{P}_K(n, r)$ and ${}_{S_r}\mathbf{mod}$ are equivalent; given any object of $\mathcal{P}_K$, we obtain an object of ${}_{S_r}\mathbf{mod}$ using (2.6), and vice versa. Moreover, if W is a K-subspace of V then W is a $K\Gamma$-submodule of V if and only if it is a S_r-submodule of V. Given $V, V' \in \mathcal{P}_K$ then $\alpha \in \mathrm{Hom}_K(V, V')$ is in $\mathrm{Hom}_{K\Gamma}(V, V')$ if and only if it is in $\mathrm{Hom}_{S_r}(V, V')$. Further,*

(i). *$V \in \mathcal{P}_K$ is irreducible if and only if it is irreducible as an S_r-module.*

(ii) *$V, V' \in \mathcal{P}_K$ are isomorphic if and only if they are isomorphic as S_r-modules.*

In particular the irreducible, and hence the projective indecomposable S_r-modules can be indexed by the set $\Lambda^+(n, r)$; cf. 1.6.1.

In view of 2.2.7 one tends to blur the distinction between (left) S_r-modules and polynomial (left) Γ-modules. Of course there are statements, *mutatis mutandis*, for the categories $\mathcal{P}'_K$ of right polynomial $K\Gamma$-modules and $\mathbf{mod}_{S_r}$ of right S_r-modules.

Complete reducibility

Theorem 2.2.8 *Suppose that K has characteristic zero, or has characteristic exceeding r. Then $S_r(\Gamma)$ is semisimple as an algebra. Every* *object of* $_{S_r(\Gamma)}$**mod,** *and hence every object of* $\mathcal{P}_K$, *is completely reducible.*

Proof All parts follow if we can show that S_r is semisimple.

We define the bilinear form b using the usual trace map:

$$b(\xi, \eta) = \text{tr}(\xi\eta)$$

for all $\xi, \eta \in S$. Using 2.2.5,

$$b(\xi, \eta) = \sum_{h \in I}(\xi\eta)_{h,h} = \sum_{h \in I}\langle c_{h,h}, \xi\eta \rangle = \sum_{h \in I}\sum_{p \in I}\langle c_{h,p}, \xi \rangle\langle c_{p,h}, \eta \rangle.$$

Choosing $\xi = \xi_{i,j}$ and $\eta = \xi_{k,l}$ gives

$$b(\xi_{i,j}, \xi_{k,l}) = \sum_{h,p \in I}\langle c_{h,p}, \xi_{i,j} \rangle\langle c_{p,h}, \xi_{k,l} \rangle.$$

It is clear that this is zero unless $(i, j) \sim (l, k)$; in this case the number of non-zero terms in the sum is $|\{(h, p) \in I^2 : (h, p) \sim (i, j)\}|$. But this is just the length of the G-orbit of (i, j). Thus there are $[G : G_{i,j}]$ terms, where $G_{i,j} = G_i \cap G_j$. Hence

$$b(\xi_{i,j}, \xi_{k,l}) = [G : G_{i,j}]1_K \text{ or } 0 \tag{2.7}$$

for $\xi_{i,j} = \xi_{k,l}$ or not respectively.

The main point to observe is that under the stated conditions of the field, the form is non-singular, by (2.7). We choose an element ξ in the radical Rad S of S. Then for any $\eta \in S$, $\xi\eta$ is nilpotent which means $b(\xi, \eta) = 0$. Since b is non-singular, $\xi = 0$ and thus Rad S comprises of just the zero matrix, i.e. S_r is semisimple.   □

Idempotents and weight spaces

Proposition 2.1.2 tells us that S_r is an associative algebra, so we can ask for simple expressions involving the product of two elements. Recall from (1.3) Schur's multiplication rule: we may derive a similar expression for $S_r(\Gamma)$ using our inner product. From 2.2.5(ii) with $\xi = \xi_{i,j}$ and $\eta = \xi_{h,l}$, and (2.5)

$$\xi_{i,j}\xi_{h,l} = \sum_{(p,q) \in \Omega}(\gamma(i,j,h,l,p,q)1_K)\xi_{p,q} \tag{2.8}$$

with $\gamma(i,j,h,l,p,q) = |\{s \in I : (i,j) \sim (p,s) \text{ and } (h,l) \sim (s,q)\}|$. The following lemma is immediate from 2.2.5 and (2.5), or directly via (2.8):

Lemma 2.2.9 *For any* $i,j,k,l \in I$

(i) $\xi_{i,j}\xi_{h,l} = 0$ *unless* $j \sim h$.

(ii) $\xi_{i,i}\xi_{i,j} = \xi_{i,j} = \xi_{i,j}\xi_{j,j}$.

(iii) $\xi_{i,i}^2 = \xi_{i,i}$, $\xi_{i,i}\xi_{j,j} = 0$ *if* $i \nsim j$ *so the* $\xi_{i,i}$ *are mutually orthogonal idempotents.*

Now for $i,j \in I$, and $i \in \lambda$,

$$\xi_{i,i} = \xi_{j,j} \Leftrightarrow (i,i) \sim (j,j) \Leftrightarrow i \sim j$$

and so j has weight λ also. Hence if $i \in \lambda$, define

$$\xi_\lambda = \xi_{i,i}.$$

By the definition of T, $1_S = T_{n,r}(1) = \sum_{\lambda \in \Lambda(n,r)} \xi_\lambda$. Since $\xi_\lambda^2 = \xi_\lambda$ and $\xi_\lambda \xi_\mu = 0$ if $\lambda \neq \mu$ we have that

$$1_S = \sum_{\lambda \in \Lambda(n,r)} \xi_\lambda$$

is an orthogonal idempotent decomposition of the identity. We make no claims about the primitivity of the ξ_λ. We can, however, now prove one of Schur's main results [1901, p. 6], namely that any homogeneous polynomial module is a direct sum of its weight spaces (cf. 1.5.2(i)).

Theorem 2.2.10 *Let* $V \in \mathcal{P}_K(n,r)$. *Then if* $\lambda \in \Lambda(n,r)$, $\xi_\lambda V = V^\lambda$ *is the* λ-*weight space of* V, *and* $V = \bigoplus_{\lambda \in \Lambda(n,r)} V^\lambda$. *Similarly for right modules.*

Proof Using the idempotent decomposition of the identity, it is clearly enough to demonstrate that $V^\lambda = \xi_\lambda V$ for any $\lambda \in \Lambda(n,r)$.

For d a diagonal element as in (1.8) and $\mu \in \Lambda(n,r)$, $T_{n,r}(d)\xi_\mu = \xi_\mu T_{n,r}(d) = \chi_\mu(d)\xi_\mu$. Then if $v \in \xi_\mu V^\lambda$, $dv = \chi_\lambda(d)v = \chi_\mu(d)v$ for all d. So if $\lambda \neq \mu$, $v = 0$, and so $V^\lambda = (\sum_\mu \xi_\mu)V^\lambda = \xi_\lambda V^\lambda \subseteq \xi_\lambda V$.

Finally, if $v \in \xi_\lambda V$ then $dv = \chi_\lambda(d)v$ and we obtain the reverse inclusion. □

Green's Product Rule

As anyone who has tried to use Rule (2.8) knows, this formula is rather unwieldy for use as a calculating tool. As an illustration of the problem, let us consider the case $n = r = 2$. Take $G = \{1, \pi\}$, which acts on $I = I(2,2)$ as $(i_1 i_2)\pi = (i_2 i_1)$, and therefore on I^2 as $((12),(21))\pi = ((21),(12))$. We want to construct a basis, so by 2.2.6 we need to understand the 10 orbits of G on I^2. This is best done pictorially as follows:

Each orbit gives rise to a corresponding basis element, for example:

$$\xi_{12,21} = \begin{bmatrix} 0 & 0 & 0 & 0 \\ 0 & 0 & 1 & 0 \\ 0 & 1 & 0 & 0 \\ 0 & 0 & 0 & 0 \end{bmatrix} \qquad \xi_{11,22} = \begin{bmatrix} 0 & 0 & 0 & 1 \\ 0 & 0 & 0 & 0 \\ 0 & 0 & 0 & 0 \\ 0 & 0 & 0 & 0 \end{bmatrix}$$

$$\xi_{12,12} = \begin{bmatrix} 0 & 0 & 0 & 0 \\ 0 & 1 & 0 & 0 \\ 0 & 0 & 1 & 0 \\ 0 & 0 & 0 & 0 \end{bmatrix}.$$

The Product Rule may now be verified: e.g. the product of the first two elements gives zero, while $\xi_{12,21}\xi_{12,12} = \xi_{12,21}$.

In Green [1990a] there is derived a more useful result which is more effective as a calculating tool. We describe this now. Let $\alpha_{i,j}$ be that K-endomorphism of $E^{\otimes r}$ whose matrix has a 1 in the (i,j)th place and zeros elsewhere, with respect to the basis $\{e_i\}_I$ of $E^{\otimes r}$. Thus $\xi_{i,j} = \sum_{(p,q)\in\omega} \alpha_{p,q}$, where ω is the G-orbit of (i,j) on I^2. Further, for $i, j \in I$ put $G_{i,j} = G_i \cap G_j$. Finally, put $G_{i,j,l} = G_{i,j} \cap G_l$. Applying the (relative) trace map to $\alpha_{i,j}$ yields

$$\mathrm{Tr}_{G_{i,j}}^G(\alpha_{i,j}) = \sum_{\sigma\in X} \alpha_{i,j}^\sigma = \xi_{i,j}$$

where X is a transversal of the set of left cosets of $G_{i,j}$ in G. In this equation, if $\alpha \in \mathrm{End}_K(E^{\otimes r})$ and $\sigma \in G$, we shall write α^σ for the map defined by $\alpha^\sigma(x) = \alpha(x\sigma^{-1})\sigma$, $x \in E^{\otimes r}$. Using the Mackey Formula, we have

$$\xi_{i,j}\xi_{h,l} = \mathrm{Tr}^G_{G_{i,j}}(\alpha_{i,j})\mathrm{Tr}^G_{G_{h,l}}(\alpha_{h,l})$$

$$= \sum_{G_{i,j}\sigma G_{h,l}} \mathrm{Tr}^G_{G^\sigma_{i,j}\cap G_{h,l}}(\alpha^\sigma_{i,j}\alpha_{h,l}) \qquad (2.9)$$

where summation is over a transversal $\{\sigma\}$ of double cosets $G_{i,j}\sigma G_{h,l}$ in G. Now $\alpha^\sigma_{i,j} = \alpha_{i\sigma,j\sigma}$, hence the product is zero unless $j \sim h$. In this case $(h,l) \sim (j,l')$ so $\xi_{h,l} = \xi_{j,l'}$ and hence without loss of generality we may assume $j = h$. But $G^\sigma_{i,j} = \sigma^{-1}G_{i,j}\sigma = G_{i\sigma,j\sigma}$ for all $i,j \in I$ and $\sigma \in G$. Also $\alpha^\sigma_{i,j} = \alpha_{i\sigma,j\sigma}$ and unless $j\sigma = j$, $\alpha^\sigma_{i,j}\alpha_{j,l} = 0$ (that is, unless $\sigma \in G_j$). So we can omit from (2.9) all terms with $\sigma \notin G_j$.

Assuming $\sigma \in G_j$, $\alpha^\sigma_{i,j}\alpha_{j,l} = \alpha_{i\sigma,j\sigma}\alpha_{j,l} = \alpha_{i\sigma,l}$ and $G^\sigma_{i,j} \cap G_{j,l} = G_{i\sigma,j} \cap G_{j,l} = G_{i\sigma} \cap G_j \cap G_l = G_{i\sigma,j,l}$. Then

$$\mathrm{Tr}^G_{G^\sigma_{i,j}\cap G_{j,l}}(\alpha^\sigma_{i,j}\alpha_{j,l}) = \mathrm{Tr}^G_{G_{i\sigma,j,l}}(\alpha_{i\sigma,l})$$

$$= [G_{i\sigma,l} : G_{i\sigma,j,l}]\mathrm{Tr}^G_{G_{i\sigma,l}}(\alpha_{i\sigma,l})$$

$$= [G_{i\sigma,l} : G_{i\sigma,j,l}]\xi_{i\sigma,l}.$$

The sum of (2.9) then becomes

Theorem 2.2.11 (Green) *For all* $i,j,l \in I$,

$$\xi_{i,j}\xi_{j,l} = \sum_\sigma ([G_{i\sigma,l} : G_{i\sigma,j,l}]1_K)\xi_{i\sigma,l} \qquad (2.10)$$

over a transversal $\{\sigma\}$ *of double cosets* $G_{i,j}\sigma G_{j,l}$ *in* G_j.

Remarks.
(1) In (2.10) we always assume that $\sigma = 1$ lies in the transversal.
(2) The elements $\xi_{i\sigma,l}$ are not necessarily distinct.

Computing the transversal. The cosets $G_{i,j}\sigma$ ($\sigma \in G_j$) are in one-to-one correspondence with elements $i\sigma$ of the G_j-orbit iG_j of i in I; hence double cosets $G_{i,j}\sigma G_{j,l}$ in G_j correspond to $G_{j,l}$-orbits of iG_j. Hence (2.10) may be re-written as

$$\xi_{i,j}\xi_{j,l} = \sum_p [G_{p,l} : G_{p,j,l}]\xi_{p,l}$$

over a transversal $\{p\}$ of the $G_{j,l}$-orbits in iG_j. Again we can choose this transversal so that it contains i.

2.3 The Schur algebra $S_K(n,r)$

The definition of S_r in 2.1.1 is different from the original formulation given in Green [1980]: there he starts off with the coalgebra A_r and defines $S_K(n,r)$ (which he calls the Schur algebra) to be the vector space dual A_r^*. The identification of S_r and $S_K(n,r)$ is made by first turning A_r into a S_r-bimodule by means of $\langle\ ,\ \rangle$. We recall from 1.3.5 that A_r has the structure of a (Γ,Γ)-bimodule; the (left and right) actions of Γ will then induce actions of $K\Gamma$ given by the following lemma.

Lemma 2.3.1 *For $\kappa \in K\Gamma$, $f \in A_r$ and $\xi \in S_r$,*

$$\langle \kappa \circ f, \xi \rangle = \langle f, \xi T_{n,r}(\kappa) \rangle$$
$$\langle f \circ \kappa, \xi \rangle = \langle f, T_{n,r}(\kappa)\xi \rangle.$$

Proof We simply put $\xi = T(\tau)$ for some $\tau \in K\Gamma$. Then

$$\langle \kappa \circ f, \xi \rangle = (\kappa \circ f)(\tau) = f(\tau\kappa) = \langle f, T(\tau\kappa) \rangle = \langle f, \xi T(\kappa) \rangle.$$

The second equality is proved similarly. $\square$

Corollary 2.3.2 *For $f \in A$ and $\tau \in \operatorname{Ker} T$, we have*

$$\tau \circ f = 0 = f \circ \tau.$$

Proof $\langle \tau \circ f, \xi \rangle = \langle f, \xi T(\tau) \rangle = 0$ for all $\xi \in S$, which implies $\tau \circ f = 0$. $\square$

Corollary 2.3.2 together with the criterion of 2.2.3 (with $V = A_r$) shows that $A_r \in \mathcal{P}_K(n,r)$ as a left $K\Gamma$-module and similarly that $A_r \in \mathcal{P}'_K(n,r)$ as a right $K\Gamma$-module. We conclude that A_r is an (S_r, S_r)-bimodule with S_r acting according to (2.6) and its right analogue. To be explicit, we have

Corollary 2.3.3 *A_r is a (S,S)-bimodule. The element ξ of S acts on A_r as follows:*

$$\xi \circ c_{i,j} = \sum_{t \in I} \langle c_{t,j}, \xi \rangle c_{i,t} \tag{2.11}$$

$$c_{i,j} \circ \xi = \sum_{s \in I} \langle c_{i,s}, \xi \rangle c_{s,j} \tag{2.12}$$

for all $i, j \in I$.

Proof Recalling (2.6), if $i,j \in I$, $\xi \circ c_{i,j} = \kappa \circ c_{i,j}$ where $\kappa \in K\Gamma$ and $\xi = T_{n,r}(\kappa)$. Choosing $\kappa = g$ yields

$$\xi \circ c_{i,j} = g \circ c_{i,j} = \sum_{t \in I} g_{t,j} c_{i,t} = \sum_{t \in I} \langle c_{t,j}, \xi \rangle c_{i,t}, \qquad (2.13)$$

since from the proof of 2.2.5(i) we know that $\langle c_{t,j}, T_{n,r}(g) \rangle = g_{t,j}$. By linear extension this remains true for all $\kappa \in K\Gamma$. Equation (2.12) follows similarly. $\qquad\square$

This allows us finally to reconcile the two definitions of the Schur algebra. We note that

Lemma 2.3.4 $\langle\ ,\ \rangle$ *is (S_r, S_r)-invariant.*

Proof $\xi, \eta \in S_r$ and $f \in A_r$ imply that $\langle \xi \circ f, \eta \rangle = \langle f, \eta\xi \rangle$ and $\langle f \circ \xi, \eta \rangle = \langle f, \xi\eta \rangle$. $\qquad\square$

The main theorem now follows: let $S_K(n,r) = A_K(n,r)^*$.

Theorem 2.3.5 *The map $\psi : S_r(\Gamma) \to S_K(n,r)$ defined by $\psi(\xi)(f) = \langle f, \xi \rangle$ is an (S_r, S_r)-isomorphism.*

The algebras $S_K(n,r)$ and $S_r(\Gamma)$ may now be identified. For, given $c \in A_r$, suppose $\Delta(c) = \sum_i c_i \otimes c_i'$. Hence multiplication in the algebra A_r^* is defined by

$$(\alpha\beta)(c) = \sum \alpha(c_i)\beta(c_i') \qquad (2.14)$$

for all $\alpha, \beta \in A_r^*$. By the definition of ψ, and 2.2.5(ii)

$$\psi(\xi\eta)(c_{i,j}) = \sum_h \psi(\xi)(c_{i,h})\psi(\eta)(c_{h,j})$$

for all $\xi, \eta \in S_r(\Gamma)$, $i, j \in I$. Since the $c_{i,j}$ span A_r it follows that

$$\psi(\xi\eta)(c) = \sum_i \psi(\xi)(c_i)\psi(\eta)(c_i') \qquad (2.15)$$

for all $c \in A_r$. Putting $\alpha = \psi(\xi)$ and $\beta = \psi(\eta)$ into (2.14), and comparing with (2.15), we deduce that $\psi(\xi)\psi(\eta) = \psi(\xi\eta)$. Since also $\psi(1_{S_r(\Gamma)}) = \varepsilon$ we have an algebra map.

Remarks.

(1) My policy is to use the equivalent terms S_r, $S_r(\Gamma)$, $S_K(n,r)$ and

$S(n, r)$ at various points in the sequel, as and when I want to emphasise different parameters.

(2) In characteristic zero Schur exploited the category equivalence between $\mathcal{P}_\mathbf{C}(n, r)$ and $_{A_\mathbf{C}(n,r)}\cdot\mathbf{mod}$. Green [1980] emphasised the importance of the K-algebra $A_K(n, r)^*$, but it is worth drawing attention to the fact that $A_K(n, r)^*$ is only one of several possible choices of suitable subalgebra of the dual of the coordinate ring of Γ that merit study. We might, for example, wish to study some 'dense' subalgebra S of $K[\Gamma]^*$ (S is *dense* if for all $f \in K[\Gamma]$ there is an $s \in S$ such that $s(f) \neq 0$). More generally, in studying semisimple algebraic groups $\mathbf{G}$ over algebraically closed fields several authors have taken S to be the *hyperalgebra* (or the *algebra of distributions*), $\mathcal{U}$ of $\mathbf{G}$; see Jantzen [1987, p. 116], [1980], Sullivan [1977] or Cline, Parshall and Scott [1980] (henceforth this latter triumvirate is denoted by 'CPS'). In the case where $\mathbf{G}$ is semisimple and simply connected we have *Verma's Conjecture*: the category of finite-dimensional rational modules for $\mathbf{G}$ is equivalent to the category of finite-dimensional $\mathcal{U}$-modules (actually this 'conjecture' is now a 'theorem'). The hyperalgebra may be identified with the algebra $\mathcal{U}_\mathbf{Z} \otimes_\mathbf{Z} K$ where $\mathcal{U}_\mathbf{Z}$ is the Kostant $\mathbf{Z}$-form of the universal enveloping algebra of the Lie algebra of the same type as $\mathbf{G}$. One then recovers the coordinate ring $K[\mathbf{G}]$ by taking the restricted dual of $\mathcal{U}$. A strong reason for working with $\mathcal{U}$-modules is to avoid the sophisticated geometric approach in the first instance; a perfect example of this is contained in the paper of Carter and Lusztig [1974] who employed $\mathcal{U}$ to great effect in their study of rational representations of $\mathrm{GL}_n(K)$.

2.4 Bideterminants and codeterminants

Henceforth, we fix $n, r \in \mathbf{N}$, $\lambda \in \Lambda^+(r)$ and choose (and fix) an infinite field K and a basic λ-tableau T^λ.

Bideterminants

The building blocks of the representation theory of the Schur algebra are certain elements of $A_K(n, r)$ called *bideterminants*; the name 'bideterminant' is intended to suggest that this object is a product of certain determinants depending on two indices $i, j \in I(n, r)$.

Definition 2.4.1 Given $i, j \in I(n, r)$, define the element

$$T^\lambda(i:j) = \sum_{\pi \in C(T^\lambda)} (-1)^\pi c_{i\pi,j} = \sum_{\pi \in C(T^\lambda)} (-1)^\pi c_{i,j\pi}$$

of A_r. This element is called a *bideterminant of shape* λ.

Here, I have written $(-1)^\pi$ for the sign of the permutation π. Sometimes one writes T for T^λ if the shape is understood.

Remarks.

(1) One can trace this definition back to Rota and his collaborators (see Rota [1976/77, p. 9], Désarménien *et al.* [1978, p. 67], DeConcini *et al.* [1980]). In the case $K = \mathbf{C}$, they appear in Deruyts [1892] (see Green [1991b, p. 60]) as 'semi-invariant functions'. Young (cf. Robinson [1977]) had also considered them in his work on representations of the octahedral group. The nomenclature, of course, varies considerably: variants of 2.4.1 appear in the guise of 'bitableaux' and 'double tableaux' in more contemporary literature.

(2) The second equality holds because $c_{i,j\pi} = c_{i\pi^{-1},j}$, for any $\pi \in G$.

(3) Let $\mu = (\mu_1, \ldots, \mu_r)$ be the conjugate partition λ'. Then $T^\lambda(i:j)$ is the product over all columns t of $[\lambda]$ of the $\mu_t \times \mu_t$ determinants

$$\det(c_{i_{a(s,t)}, j_{a(s',t)}})_{s,s'=1,\ldots,\mu_t},$$

with the convention that the determinant is defined to be 1 if $\mu_t = 0$. This is a consequence of the definitions.

(4) By (3), $T(i:j) = 0$ if and only if there are equal entries at two distinct places in the same column of T_i or of T_j.

(5) If $\pi \in C(T^\lambda)$ then $T^\lambda(i\pi:j) = (-1)^\pi T^\lambda(i:j) = T^\lambda(i:j\pi)$.

(6) By 1.3.5 and the definition of a bideterminant, for $g, h \in \Gamma$,

$$g \circ T^\lambda(i:j) = \sum_{t \in I} g_{t,j} T^\lambda(i:t) \tag{2.16}$$

$$T^\lambda(i:j) \circ h = \sum_{u \in I} h_{i,u} T^\lambda(u:j). \tag{2.17}$$

This defines actions of S_r on bideterminants, for example

$$\xi \circ T^\lambda(i:j) = \sum_{u \in I} \langle c_{u,j}, \xi \rangle T^\lambda(i:u) \tag{2.18}$$

for all $\xi \in S_r$. Recalling 1.4.6 we make the following crucial definition.

Definition 2.4.2 Let $\lambda \in \Lambda^+(r)$. We call $T^\lambda(i:j)$ a *standard bideterminant* if $i, j \in I^\lambda$.

Theorem 2.4.3 *If $\ell \in I(n,r)$ has weight λ and if T_ℓ^λ is standard, then T_ℓ^λ is given by*

$$
T_\ell^\lambda =
\begin{array}{ccccc}
1 & 1 & \cdots & \cdots & 1 \\
2 & 2 & \cdots & 2 & \\
3 & 3 & \cdots & 3 & \\
\vdots & \vdots & & & \\
n & \cdots & n & &
\end{array}
$$

So ℓ is uniquely determined by the basic tableau T^λ. We call T_ℓ^λ the canonical λ-tableau.

Proof In the proof of 1.4.7, $\mu_1 = \lambda_1$ so all entries in $R_1(T)$ are 1s; then $\mu_1 + \mu_2 = \lambda_1 + \lambda_2$, so all entries in $R_2(T)$ are 2s. We continue in this manner. $\qquad\qquad\square$

We normally write $\ell = \ell(\lambda)$ for this element of $I = \mathbf{n}^\mathbf{r}$: note that it sends $1, 2, \ldots, \lambda_1$ to 1, $\lambda_1 + 1, \ldots, \lambda_1 + \lambda_2$ to 2, etc. Of course the G-stabiliser, G_ℓ, of ℓ is precisely Σ_λ.

Experience suggests that one cannot give too many examples of bideterminants; let us give two easy ones right away to familiarise oneself with these objects:

Example. Take $\lambda = (3, 2, 1)$. We have

$$
T_\ell^\lambda =
\begin{array}{ccc}
1 & 1 & 1 \\
2 & 2 & \\
3 & &
\end{array}
\qquad
T_i^\lambda =
\begin{array}{ccc}
u & v & w \\
x & y & \\
z & &
\end{array}.
$$

Using Remark (3) above we see that

$$
T^{(3,2,1)}(\ell : i) =
\begin{vmatrix}
c_{1u} & c_{1x} & c_{1z} \\
c_{2u} & c_{2x} & c_{2z} \\
c_{3u} & c_{3x} & c_{3z}
\end{vmatrix}
\begin{vmatrix}
c_{1v} & c_{1y} \\
c_{2v} & c_{2y}
\end{vmatrix}
c_{1w} \in A(3, 6).
$$

Example. It should be clear that $T^\lambda(\ell : \ell)$ is the product $\bar{\mu}_1 \bar{\mu}_2 \ldots$, where for any $0 \le m \le n$, $\bar{\mu}_m$ is the μ_mth leading minor of $[c_{\alpha\beta}]$.

The properties inherent in Remarks (4) and (5) above are special cases of well-known results on antisymmetrising functions. As we shall use these again in connection with Weyl modules, let us explore them in more detail. Suppose $f : I \to A$ is an arbitrary function into an additive abelian group, A; then one defines $f^\lambda : I \to A$ by

$$f^\lambda(i) = \sum_{\sigma \in C(T^\lambda)} (-1)^\sigma f(i\sigma)$$

for all $i \in I$. So $T^\lambda(i : j) = f^\lambda(i)$ where $f(i) = c_{i,j}$ for fixed j. A useful lemma, valid for an arbitrary $f : I \to A$ may be stated.

Lemma 2.4.4 *For any* $\lambda \in \Lambda^+(n,r)$, $i \in I$,

  (i) $f^\lambda(i\sigma) = (-1)^\sigma f^\lambda(i)$ *for all* $\sigma \in C(T^\lambda)$.
  (ii) *If* T_i^λ *has equal entries* $i_\rho = i_{\rho'}$ *at two distinct places* $\rho \neq \rho'$ *in the same column of* T^λ *then* $f^\lambda(i) = 0$.
  (iii) $f^\lambda(i) = 0$ *if either (a)* $i \in \mu$ *and* $\mu \ntrianglelefteq \lambda$ *or (b)* $i \in \lambda$ *and* $i = \ell(\lambda)\sigma$ *for some* $\sigma \in G \setminus R(T)C(T)$.
  (iv) *Suppose we have* $f, g : I \to A$ *with* $g(i) = 0$ *unless* $i \in \lambda$. *Then*

  $$\sum_{q \in I} g(q) f^\lambda(q) = g^\lambda(\ell) f^\lambda(\ell),$$

  where $\ell = \ell(\lambda)$.
  (v) **(Carter-Lusztig Lemma).** *Suppose that conditions (i) and (ii) hold with* f^λ *replaced by* f; *assume also that* $f^\lambda(i) = 0$ *for any* $i \in I$, *where this time the sum (in the equation preceding this lemma) is restricted to be over all* $\sigma \in G_{X,Y}$. *Then, for some basic* λ-*tableau* T, *and any* $\emptyset \neq X \subseteq C_{t+1}(T)$ *for* $1 \leq t < r$, $G_{X,Y}$ *is a transversal of cosets* $\{\sigma V : \sigma \in Y\}$ *with* Y *the subgroup of* G *consisting of all* $\pi \in G$ *that fix every element outside* $C_t(T) \cup X$ *and* $V = C(T) \cap Y$. *Then*

  $$\mathrm{Im}\, f \leq \langle f(i) : i \in I^\lambda \rangle \leq A.$$

Proof (i) and (ii) are straightforward. Part (iii) follows from (ii) using a couple of combinatorial lemmas from James and Kerber [1981, 1.5.17, 1.5.8]. For (iv) we notice that the stated sum is over all $q \in \lambda$, i.e. over all q of the form $q = \ell\pi$ ($\pi \in G$). If $q = \ell\sigma$ with $\sigma \notin R(T)C(T)$, then $f^\lambda(i) = 0$ by (iiib). If $\sigma \in R(T)C(T)$, then $\ell\sigma = \ell\pi$ where $\pi \in C(T)$ since $R(T) = G_\ell$. Also, if $\alpha, \beta \in C(T)$ then $\ell\alpha = \ell\beta$ if and only if $\alpha = \beta$, since $R(T)$ and $C(T)$ intersect trivially. Hence (iv) may be written

$$\sum_{\sigma \in C(T)} g(\ell\sigma) f^\lambda(\ell\sigma) = \sum_{\sigma \in C(T)} (-1)^\sigma g(\ell\sigma) f^\lambda(\ell) = g^\lambda(\ell) f^\lambda(\ell),$$

using (ii).

The first condition in (v) guarantees that the sum is independent of the choice of transversal $G_{X,Y}$. The final condition in (v) is equivalent to condition (37) in Carter and Lusztig [1974, pp. 212–214]. It is also provable by mimicking the proof of the usual Garnir relations in James [1978a, p. 29]. The reader is left to chase these up.   □

Remark. The Carter-Lusztig Lemma is an analogue of the *Garnir relations* considered by James [1978a, §7].

Examples.
(1) Let $\lambda = (3,2,1)$. Defining i, $j \in I(n,6)$ as in the example after (1.7), we have

$$
T_i^\lambda = \begin{matrix} i_1 & i_2 & i_4 \\ i_3 & i_5 \\ i_6 \end{matrix}
\qquad
T_j^\lambda = \begin{matrix} j_1 & j_2 & j_4 \\ j_3 & j_5 \\ j_6 \end{matrix}.
$$

Thus

$$
T^\lambda(i:j) = \begin{vmatrix} c_{i_1 j_1} & c_{i_1 j_3} & c_{i_1 j_6} \\ c_{i_3 j_1} & c_{i_3 j_3} & c_{i_3 j_6} \\ c_{i_6 j_1} & c_{i_6 j_3} & c_{i_6 j_6} \end{vmatrix} \begin{vmatrix} c_{i_2 j_2} & c_{i_2 j_5} \\ c_{i_5 j_2} & c_{i_5 j_5} \end{vmatrix} c_{i_4 j_4} \in A(n,6).
$$

(2) With $r \le n$, let $\omega = (1,1,\ldots,1,0,0,\ldots,0) \in \Lambda^+(n,r)$ with r 1s.

Take the basic ω-tableau as $\begin{matrix} 1 \\ 2 \\ \vdots \\ r \end{matrix}$. Then, for $i \in I$, $T^\omega(\ell : i)$ is the

r-rowed determinant

$$
T^\omega(\ell : i) = \begin{vmatrix} c_{1 i_1} & c_{1 i_2} & \cdots & c_{1 i_r} \\ c_{2 i_1} & c_{2 i_2} & \cdots & c_{2 i_r} \\ \vdots & \vdots & & \vdots \\ c_{r i_1} & c_{r i_2} & \cdots & c_{r i_r} \end{vmatrix}.
$$

(3) Take $\lambda = (r,0,\ldots,0)$, in which case we may take $T^{(r)} = 1\, 2 \cdots r$. Then for $i, j \in I$,

$$
T^{(r)}(i:j) = c_{i_1 j_1} \cdots c_{i_r j_r} = c_{i,j}
$$

which is standard if and only if $i_1 \le \cdots \le i_r$ and $j_1 \le \cdots \le j_r$. This also demonstrates that any monomial $c_{i,j}$ is itself a bideterminant of shape (r).

(4) *The weight of a bideterminant.* Let $i, j \in I$ have weights $\lambda, \mu \in$

$\Lambda(n,r)$ respectively. From the example following 1.5.1, or directly from (2.16) and (2.17), the bideterminant $T^\lambda(i:j) = \sum_{C(T^\lambda)} (-1)^\pi c_{i\pi,j}$ is a left (right) weight element of the (Γ, Γ)-bimodule A, of right (left) weight λ (μ).

In fact, λ is the *highest* weight of any standard bideterminant of shape λ:

Lemma 2.4.5 *Let $\lambda \in \Lambda^+(n,r)$ and suppose that $i, j \in I$ are such that $T^\lambda(i:j)$ is standard.*

(i) *Then the left and right weights of $T^\lambda(i:j)$ are both dominated by λ.*

(ii) *Suppose that $\mu \trianglelefteq \lambda$, and $T^\mu(i:j)$ is standard. Then*

$$T^\mu(i:j) \text{ has left (right) weight } \lambda \Leftrightarrow \mu = \lambda \text{ and } j = \ell \ (i = \ell).$$

Proof For (i) apply 1.4.7, (2.16) and (2.17). For (ii), use also 2.4.3. $\square$

Codeterminants

We turn now to the dual situation.

Definition 2.4.6 Given $i, j \in I$ and $\lambda \in \Lambda(n,r)$, define the element

$$C^\lambda(i:j) = \xi_{i,\ell(\lambda)}\xi_{\ell(\lambda),j}$$

of S_r. This element $C^\lambda(i:j)$ is called the *codeterminant of shape λ*. The codeterminant is *standard* if $i, j \in I^\lambda$ (cf. 2.4.2).

Remark. Any non-zero product, ξ, of a pair of elements in the basis 2.2.6 produces a codeterminant, since $\xi = \xi_{i,j}\xi_{k,l}$ with $i, j, k, l \in I$ and $j \sim k$ from 2.2.9. If $j \in \lambda$ then $j \sim \ell(\lambda) \sim k$ and so $\xi_{i,j} = \xi_{p,\ell(\lambda)}$ and $\xi_{k,l} \sim \xi_{\ell(\lambda),q}$ for some $p, q \in I$. Thus $\xi = C^\lambda(p:q)$.

Exercises.

(1) The alert reader will point out that 2.4.6 depends on an initial choice of basic λ-tableau since $\ell(\lambda)$ does. Check that if we define $\hat{C}^\lambda(i:j)$ using a different basic λ-tableaux then there exists $\pi \in G$ such that $C^\lambda(i:j) = \hat{C}^\lambda(i\pi : j\pi)$ for all $i, j \in I$.

(2) Formally, the above definition does not look very similar to that of the bideterminant in 2.4.1. Show, using (2.10), that

$$C^\lambda(i:j) = \sum_{\sigma \in R(T^\lambda)} s(i\sigma, j)\xi_{i\sigma,j},$$

where $s(i\sigma, j) = |G_{i\sigma,j}|/|G_{i,\ell}||G_{\ell,j}|$. This is somewhat like 2.4.1. The result in 2.2.11 can be used to prove the next two statements.

(3) Refer to Example (2) after 2.4.4. We take $\ell(\omega) = (1, 2, \ldots, r) \in I$, and then it follows that $C^\omega(i:j) = |G_{i,j}|\xi_{i,j}$.

(4) The other extremal case (Example (3) after 2.4.4) gives, for any choice of basic (r)-tableau, $\ell(r) = (1, 1, \ldots, 1) \in I$. Then, we have

$$C^{(r)}(i:j) = \sum_{\sigma} \xi_{i\sigma,j}$$

over a transversal of double cosets $G_i \sigma G_j$ in G.

Immediate aims

The reader is perhaps familiar with the theorem from the representation theory of G that states that bijective standard λ-tableaux parametrise a basis for the Specht modules $S_\mathbf{Q}^\lambda$ (see James [1978a, §8]). The final two sections of this chapter are devoted to finding a 'similar' indexing set for a K-basis of A_r and S_r.

Recalling the definition of I^λ in 1.4.6, let (Ξ) denote the set

$$\{(\lambda, i, j) : \lambda \in \Lambda^+(n, r), \ i, j \in I^\lambda\}.$$

The claim is now that (Ξ) is an indexing set for a basis of A_r:

Theorem 2.4.7 *Let n, r and K be given as above. Then A_r has K-basis*

$$\mathcal{L} = \{T^\lambda(i:j) : (\lambda, i, j) \in (\Xi)\}. \tag{2.19}$$

The parallel result for S_r is

Theorem 2.4.8 *The set*

$$\mathcal{M} = \{C^\lambda(i:j) : (\lambda, i, j) \in (\Xi)\} \tag{2.20}$$

is a K-basis for S_r.

More generally over a commutative ring R, $\mathcal{L}$ is a free R-basis for A_r and $\mathcal{M}$ is a free R-basis for S_r. The proofs given below over K are valid also over R.

At this point we have a choice—either to work with the putative basis $\mathcal{L}$ or with $\mathcal{M}$. Traditionally, the approach has been to tackle the former and deduce corresponding properties about (2.20) 'by duality'. This is the line I take, but one should be aware, thanks to an elegant combinatorial result of Woodcock [1993], that it is possible to deal directly with $\mathcal{M}$.

2.5 The Straightening Formula

In this section we prove that the bideterminants appearing in (2.19) K-span the space A_r: we will in fact show that every bideterminant is expressible as a $\mathbf{Z}$-linear combination of elements appearing in $\mathcal{L}$. Here one regards A_r as a $\mathbf{Z}$-module via $zc = (z1_K)c$, for all $z \in \mathbf{Z}$ and $c \in A_r$ (but we are not saying that A_r is $\mathbf{Z}$-free!). This result is famous as the *Straightening Formula*. It boasts a long history, and is described in the wonderfully candid introduction to Désarménien *et al.* [1978] as

(probably) one of the fundamental algorithms of multilinear algebra. [It] is the culmination of a trend of thought that can be traced back to Capelli, and was developed most notably by Alfred Young and the Scottish invariant theorists.

The first modern proof appears in Doubilet *et al.*, though our version is inspired by DeConcini *et al.* [1980, pp. 140–145]: the latter realised that Hodge's straightening rule for the homogeneous coordinate ring of a Grassmann variety would specialise to polynomial rings, thus reducing the language to that of linear algebra. Other formulations appear in Désarménien *et al.* [1978], Doubilet *et al.* [1974], Clausen [1979, 4.4] and Akin *et al.* [1982].

The basic method is simple: induction on the linear ordering of the tableaux (see 1.4.9) is used. The induction step expresses a given nonstandard λ-bideterminant as a $\mathbf{Z}$-linear combination of smaller bideterminants of the same shape λ, together with those having shapes dominated by λ. In the above references, and in the original work of Mead [1972], the identity is proved using the Laplace expansion of a suitable determinant. We give a simplified version of this result (due to Clausen, see Green [1991a]) which is sufficient for our present purposes.

Theorem 2.5.1 ('Laplace Duality') *Let $\lambda, \mu \in \Lambda^+(n,r)$ and let T^λ, T^μ be basic. Let X be a transversal of left cosets of $C(T^\lambda) \cap C(T^\mu)$ in $C(T^\mu)$, and Y a transversal of right cosets of $C(T^\lambda) \cap C(T^\mu)$ in $C(T^\lambda)$. Then, for any $i, j \in I$,*

$$\sum_{\pi \in X} (-1)^\pi T^\lambda(i\pi : j) = \sum_{\sigma \in Y} (-1)^\sigma T^\mu(i : j\sigma^{-1}).$$

Proof One can express the set $Z = C(T^\mu)C(T^\lambda)$ as a disjoint union of cosets in two ways:

$$Z = \bigcup_{\pi \in X} \pi C(T^\lambda) = \bigcup_{\sigma \in Y} C(T^\mu)\sigma.$$

Then

$$
\begin{aligned}
\sum_{\pi \in X} (-1)^\pi T^\lambda(i\pi : j) &= \sum_{\pi \in X} \sum_{\rho \in C(T^\lambda)} (-1)^{\pi\rho} c_{i\pi\rho,j} = \sum_{z \in Z} (-1)^z c_{iz,j} \\
&= \sum_{\sigma \in Y} \sum_{\phi \in C(T^\mu)} (-1)^{\phi\sigma} c_{i\phi\sigma,j} \\
&= \sum_{\sigma \in Y} \sum_{\phi \in C(T^\mu)} (-1)^{\phi\sigma} c_{i\phi,j\sigma^{-1}} \\
&= \sum_{\sigma \in Y} (-1)^\sigma T^\mu(i : j\sigma^{-1}).
\end{aligned}
$$

$\square$

A filtration of A_r

For the inductive proof of Straightening, we filter A_r by a series of subspaces, which are introduced now. First, we use (2.16) and (2.17) (or (2.18)) to check that

Proposition 2.5.2 *Given $\lambda \in \Lambda(r)$, let a^λ be the K-space spanned by all bideterminants $T^\lambda(i : j)$ for all $i, j \in I$. Then a^λ is independent of the choice of basic λ-tableau and is a (Γ, Γ)-sub-bimodule of A_r. Equivalently, it is an (S_r, S_r)-sub-bimodule of A_r, cf. Corollary 2.3.3.*

Definition 2.5.3 Define the subspace

$$A(\leq \lambda) = \sum_{\mu \leq \lambda} a^\mu = \sum_{\mu \leq \lambda} \sum_{i,j \in I} K T^\mu(i : j) \tag{2.21}$$

of A_r.

Here $\leq$ is given by

Definition 2.5.4 Suppose we list the dominant weights in decreasing order:

$$\lambda_1 = (r) > \lambda_2 > \cdots > \lambda_t,$$

where $t = |\Lambda^+(n, r)|$. Here $\lambda > \mu$ means that λ' comes before μ' in the lexicographic order. The order so defined is termed the *reverse lexicographic order on conjugates*.

This defines a chain of sub-bimodules

$$A_r = A(\leq \lambda_1) \geq A(\leq \lambda_2) \geq \cdots \geq A(\leq \lambda_t) \geq \{0\},$$

the *fundamental filtration* of A_r. We may collect together the basic properties of $A(\leq \lambda)$.

Proposition 2.5.5 *For $\lambda \in \Lambda(r)$,*

  (i) $A(\leq \lambda)$ *is a sub-(S_r, S_r)-bimodule of A_r.*
 (ii) $\lambda \geq \mu \Rightarrow A(\leq \lambda) \geq A(\leq \mu)$;
(iii) $A(\leq (r)) = A_r$.
 (iv) $A(\leq \lambda)$ *has K-basis $\{T^\mu(i:j): \mu \leq \lambda \ i, j \in I^\mu\}$.*
  (v) $A(\leq \lambda) \neq 0 \Leftrightarrow \lambda \in \Lambda^+(n, r)$.

Proof For (i) and (ii) see 2.5.2 and 2.5.3. For (iii) refer to Example (3) preceding 2.4.5. (iv) will be proved in the course of 3.1 (we do not need it yet). (v) is a simple corollary of (iv) as follows:

(v)($\Rightarrow$). If $A(\leq \lambda) \neq 0$, then find $i, j \in I^\mu$ for some $\mu \leq \lambda$. Now, T_i^λ is standard, so we may deduce that $\mu'_1 \leq n$. For, if $\mu'_1 > n$ we know that the longest column of T_i^μ has more than n entries; at least two of these entries coincide, which contradicts the definition of standardness of T_i^μ. But $\mu \leq \lambda$ so $\mu'_1 \geq \lambda'_1$ by 2.5.4, and so $\lambda'_1 \leq n$. We have now shown that $\lambda \in \Lambda^+(n, r)$.

(v)($\Leftarrow$). Let $d = \lambda'_1 \leq n$. Then $\ell \in I(d, r)$ defines the canonical λ-tableau T_ℓ^λ. This is standard, and so the basis in (iv) contains the standard bideterminant $T^\lambda(\ell : \ell)$, which proves that $A(\leq \lambda) \neq 0$. $\square$

The filtration defined above plays a fundamental rôle, so it is as well to develop an understanding of it now. We shall therefore work through an extended calculation.

Example

Consider the special case $n = r = 2$ over a field of characteristic two. Let $A = A_K(2,2)$ be the 10-dimensional Schur coalgebra of polynomial functions of the form $c_{i,j} = c_{i_1 i_2, j_1 j_2}$ for $i, j \in I = I(2,2)$. The dominant weights form the set $\Lambda^+(2,2) = \{(2), (1^2)\}$, and so we just have a sequence with three terms:

$$A = A(\leq (2)) \geq A(\leq (1^2)) \geq \{0\}.$$

To construct $A(\leq (1^2))$, consider the determinant

$$\delta = \begin{vmatrix} c_{11} & c_{12} \\ c_{21} & c_{22} \end{vmatrix} = c_{11}c_{22} - c_{12}c_{21} = c_{12,12} - c_{12,21} \in A.$$

It is clear that $g \circ \delta = \delta \circ g = (\det g)\delta$, which means that $K\delta$ is a (Γ, Γ)-bimodule. We have $A(\leq (1^2)) = K\delta$.

Now $T_i^{(2)}$ is standard if and only if $i_1 \leq i_2$, and $c_{i',j'}$ is standard if and only if $c_{i',j'} = c_{i,j}$ with $i_1 \leq i_2$, $j_1 \leq j_2$. So the standard indices in I comprise the set $S = \{t = (1,1), u = (1,2), v = (2,2)\}$. There is only one non-standard index, namely $u' = (2,1)$. Consider first the standard bideterminants:

	t	u	v
t	$c_{t,t}$	$c_{t,u}$	$c_{t,v}$
u	$c_{u,t}$	$c_{u,u}$	$c_{u,v}$
v	$c_{v,t}$	$c_{v,u}$	$c_{v,v}$

For the other cases, notice that we have

$$\begin{aligned} c_{u,u'} &\equiv c_{t,v} \bmod A(\leq (1^2)) \\ c_{u,u'} &\equiv c_{u,u} \bmod A(\leq (1^2)) \\ c_{t,u'} &= c_{t,t}, \quad c_{u,u'} = c_{u,t}. \end{aligned} \qquad (2.22)$$

To proceed, we take the 'span of successive rows'.

(1) *Define* $R(z) = K\langle c_{z,11}, c_{z,12}, c_{z,22} \rangle$ for $z \in S$. We claim that this is a left $\mathrm{GL}_2(K)$-submodule (modulo $A(\leq (1^2))$). At least we have a left

Γ-action, for

$$g \circ c_{t,i} = \sum_{s \in I} g_{s,i} c_{t,s} \in R(t) \text{ as } c_{11,21} = c_{11,12}$$

$$g \circ c_{u,i} = \sum_{s \in I} g_{s,i} c_{u,s} \in R(u)$$

$$g \circ c_{v,i} = \sum_{s \in I} g_{s,i} c_{v,s} \in R(v) \text{ as } c_{22,21} = c_{22,12}.$$

The second of these is true because $c_{12,21} \equiv c_{12,12}$ (modulo $A(\leq (1^2))$), by (2.22). From these equations we have Γ-isomorphisms $R(t) \cong R(u) \cong R(v)$ (mod $A(\leq (1^2))$) and so we write $M(2,0)$ for one of these.

Similarly we take the span of successive columns:

(2) *Define* $C(z) = K \langle c_{11,z}, c_{12,z}, c_{22,z} \rangle$ *for* $z \in S$. These are isomorphic right Γ-modules modulo $A(\leq (1^2))$ which we write as $N(2,0)$.

Finally, suppose we write $M(1^2)$ for $A(\leq (1^2))$ considered as a left Γ-module, and $N(1^2)$ for $A(\leq (1^2))$ as a right Γ-module. We now have complete information about our filtration in this special case:

Proposition 2.5.6 *The filtration for* $A_K(2,2)$ *is a series*

$$A_K(2,2) = A(\leq (2,0)) > A(\leq (1^2)) > \{0\}$$

with factors

(i) $A(\leq (2,0))/A(\leq (1^2)) \cong M(2,0) \otimes N(2,0)$, *an isomorphism of 9-dimensional* (Γ, Γ)-*bimodules;*

(ii) $A(\leq (1^2)) \cong M(1^2) \otimes N(1^2)$.

Actually, we have encountered these modules before:

Exercise. Check that if char $K = 0$, $M(2,0)$ and $M(1^2)$ are the modules $T_{(2,0)}$ and $T_{(1,1)}$ appearing at the end of 1.2.

Basic combinatorial lemma

Let $\lambda \in \Lambda^+(r)$, $i \in I$ and choose a basic λ-tableau T^λ. One should at this point recall the ordering on the T_i^λ from 1.4.9. Our first move is:

Lemma 2.5.7 *If* T_i^λ *is not standard, there exist* $z_{i,k} \in \mathbf{Z}$ *such that for all* $j \in I$ *there holds*

$$T^\lambda(i:j) \equiv \sum_k z_{i,k} T^\lambda(k:j) \mod A(\leq \lambda)^*$$

where the sum is over all $k \in I$ with $T_k^\lambda < T_i^\lambda$. Here

$$A(\leq\lambda)^* = \begin{cases} A(\leq\lambda_{p+1}) & \text{if } \lambda = \lambda_p, \ 1 \leq p \leq t-1 \\ 0 & \lambda = \lambda_t \end{cases}.$$

Proof of Lemma 2.5.7 and Corollary

The procedure strikes one as reminiscent of Garnir's Theorem in James [1978a, §7]. Each numbered item below is accompanied by a discussion or a proof.

We assume that T_i^λ is non-standard.

(1) *We may assume that the entries in each column of* T_i^λ *are distinct, and increase strictly downwards.* If any column has a repeat, then $T^\lambda(i : j) = 0$, for all j. Further, if the entries are not decreasing, one can find $\sigma \in C(T^\lambda)$ such that $T_{i\sigma}^\lambda$ has entries in each column strictly increasing down columns. Then one can replace T_i by $T_{i\sigma}$ because of the property of bideterminants stated in Remark (5) after 2.4.1.

We know, by the assumption of (1) and the fact that T^λ is supposed to be non-standard, that there must exist a first (in the T^λ-order of 1.4.8) entry $\alpha \in \mathbf{r}$, say in column a of T_i^λ whose right-hand neighbour α' in column $a+1$ is such that $i_\alpha > i_{\alpha'}$. We may picture this as follows:

	C_a	C_{a+1}		C_a	C_{a+1}
		$*$			$*$
T^λ		ε'	T_i^λ		$i_{\varepsilon'}$
	α	α'		$i_\alpha \ >$	$i_{\alpha'}$
	β	β'		i_β	$i_{\beta'}$
	γ	γ'		i_γ	$i_{\gamma'}$
	$*$	$*$		$*$	$*$

Let $\mathcal{S} = \{\cdots, \gamma, \beta, \alpha, \alpha', \varepsilon', \ldots\} \subseteq C_a \cup C_{a+1}$ as shown.

(2) $\cdots > i_\gamma > i_\beta > i_\alpha > i_{\alpha'} > i_{\varepsilon'} > \cdots$. This is a by-product of the construction.

Let $\sigma' \in \Sigma_\mathcal{S}$.

(3) *There is a unique* σ *in the coset* $\sigma'(\Sigma_\mathcal{S} \cap C(T^\lambda))$ *such that* $\cdots > i_{\sigma(\gamma)} > i_{\sigma(\beta)} > i_{\sigma(\alpha)}$ *and* $i_{\sigma(\alpha')} > i_{\sigma(\varepsilon')} > \cdots$; *moreover* $\sigma(\rho) = \rho$ *for all* $\rho \in C_1, \ldots, C_{a-1}, C_a \setminus \mathcal{S}, C_{a+1} \setminus \mathcal{S}, C_{a+2}, \ldots$.

Now let X be the set of elements of $\Sigma_\mathcal{S}$ satisfying these inequalities. Note that X is a transversal of left cosets of $\Sigma_\mathcal{S} \cap C(T^\lambda)$ in $\Sigma_\mathcal{S}$. Also, we now prove

(4) $1 \neq \sigma \in X \Rightarrow T_{i\sigma}^{\lambda} < T_i^{\lambda}$. We assert that the set $\mathcal{S}' = \{\ldots, i_{\sigma(\gamma)},$ $i_{\sigma(\beta)}, i_{\sigma(\alpha)}\}$ contains at least one element of $\mathcal{S}'' = \{\ldots, i_{\alpha'}, i_{\varepsilon'}, \ldots\}$. For if not then σ is a permutation of $\{\ldots, \gamma, \beta, \alpha\}$, and hence lies in $C(T^{\lambda}) \cap \Sigma_{\mathcal{S}}$, and then by (3) above σ is the identity. Now we find the least member i_w of $\mathcal{S}''$ that also lies in $\mathcal{S}'$. By (2), i_w is less than any of $i_{\alpha}, i_{\beta}, i_{\gamma}, \ldots$.

$$
\begin{array}{ccc}
 & * & * \\
T_{i(\sigma)}^{\lambda} & & i_{\sigma(\varepsilon')} \\
 & i_{\sigma(\alpha)} & i_{\sigma(\alpha')} \\
 & i_{\sigma(\beta)} & i_{\sigma(\beta')} \\
 & i_{\sigma(\gamma)} & i_{\sigma(\gamma')} \\
 & * & *
\end{array}
$$

Hence i_w is the least member of $\mathcal{S}'$, so $i_w = i_{\sigma(\alpha)}$ by the result (3). Hence $i_{\sigma(\alpha)} < i_{\alpha}$. Since also $\sigma \in \Sigma_{\mathcal{S}}$, $i_w = i_{\sigma(\nu)}$ for all $\nu < \alpha$ in the T^{λ}-order, the result (4) follows.

(5) *The definition of* ζ. Consider the columns $C_1, C_2, \ldots$ of T^{λ} having lengths $\lambda_1' \geq \lambda_2' \geq \cdots$. Replace C_a and C_{a+1} by the three (possibly empty) sets $C_a \setminus \mathcal{S}, C_{a+1} \setminus \mathcal{S}$ and $\mathcal{S}$, and leave the other columns alone. Let $C_1', C_2', \ldots$ be the new list, in descending order of size. Define ζ to be that partition with conjugate

$$\zeta' = (|C_1'|, |C_2'|, \ldots).$$

(6) $\zeta' > \lambda'$ *in the lexicographic order.* If we can show this, then it will follow that $\zeta < \lambda$ as required, (cf. 2.5.4). Now $|\mathcal{S}| = |C_a| + 1$, and hence $|\mathcal{S}|$ exceeds each of $|C_a|$; $|C_{a+1}|, |C_{a+2}|, |C_a \setminus \mathcal{S}|, |C_{a+1} \setminus \mathcal{S}| < |C_a|$, so we may list the sets in (5) in such a way that

$$C_1' = C_1, \ldots, C_b' = C_b, C_{b+1}' = \mathcal{S}$$

where $C_1, \ldots, C_b$ are the columns of T^{λ} of cardinality exceeding $|C_a|$. So $|C_{b+1}'| > |C_{b+1}|$, and result (6) follows.

Now we choose a basic ζ-tableau T^{ζ} having columns $C_1', C_2', \ldots$ as in (5).

(7) X *is a transversal of left cosets of* $C(T^{\zeta}) \cap C(T^{\lambda})$ *in* $C(T^{\zeta})$. But

$$C(T^{\zeta}) \quad = \quad \Sigma_{C_1'} \times \Sigma_{C_2'} \times \cdots$$

$$= \left(\prod_{f \neq a, a+1} \Sigma_{C_f} \right) \times \Sigma_{C_a \backslash \mathcal{S}} \times \Sigma_{C_{a+1} \backslash \mathcal{S}} \times \Sigma_{\mathcal{S}}$$

Of course $\Sigma_{C_a \backslash \mathcal{S}} \leq \Sigma_{C_a}$ and $\Sigma_{C_{a+1} \backslash \mathcal{S}} \leq \Sigma_{C_{a+1}}$ whence $C(T^\varsigma)C(T^\lambda) = \Sigma_{\mathcal{S}} C(T^\lambda)$, and result (7) follows.

(8) *Completion of proof.* Let Y be any right transversal of cosets $C(T^\varsigma) \cap C(T^\lambda)$ in $C(T^\lambda)$. By 'Laplace Duality', for all j

$$T^\lambda(i:j) + \sum_{1 \neq \pi \in X} (-1)^\pi T^\lambda(i\pi : j) = \sum_{\sigma \in Y} (-1)^\sigma T^\varsigma(i : j\sigma^{-1}) \in A(\leq \lambda)^*,$$

because $\varsigma < \lambda$ by (6). Rewriting this we have

$$T^\lambda(i:j) \equiv - \sum_{1 \neq \pi \in X} T^\lambda(i\pi : j) \bmod A(\leq \lambda)^*$$

which is of the required form, since $T^\lambda_{i\pi} < T^\lambda_j$ for all $1 \neq \pi \in X$, by (4). At last the proof of the lemma is complete.

An easy induction now gives us an important corollary.

Corollary 2.5.8

(i) *With λ, T^λ and i fixed and for $h \in I^\lambda$, there exist integers $w_{i,h}$ such that, for all $j \in I$,*

$$T^\lambda(i:j) \equiv \sum_{T^\lambda_h \text{ standard}} w_{i,h} T^\lambda(h:j) \bmod A(\leq \lambda)^*.$$

(ii) *With λ, T^λ and j fixed, and for $k \in I^\lambda$, there exist integers $w'_{i,h}$ such that, for all $i \in I$,*

$$T^\lambda(i:j) \equiv \sum_{T^\lambda_k \text{ standard}} w'_{k,j} T^\lambda(i:k) \bmod A(\leq \lambda)^*.$$

Remark. We record for later that the integers $w_{i,h}$ are *independent of j and of K*.

Proof If (i) fails to hold, we find the least T^λ_i providing a counterexample. T^λ_i is non-standard since if it were standard, we could take $w_{i,h}$ to be 1 or 0 according to whether $i = h$ or $i \neq h$, and we would have equality. Hence by the Basic Combinatorial Lemma, there is an expansion of the form given in 2.5.7. Since (i) holds for each $T^\lambda_p < T^\lambda_i$ we have by the Basic Combinatorial Lemma that (i) holds also for T^λ_i, contrary to our assumption.

To get (ii), which is the right-hand analogue of (i), without any effort, use the K-algebra isomorphism induced by $c_{\alpha\beta} \mapsto c_{\beta\alpha}$ ($\alpha, \beta \in \mathbf{n}$) of A_r. $\square$

Proof of the Straightening Formula: the final step. It is easy enough to show by induction on λ that $\{T^\lambda(h:k) : h, k \in I^\lambda\}$ spans $A(\leq \lambda)$ modulo $A(\leq \lambda)^*$. But 2.5.8(i) and (ii) show that for fixed j we have, for all i,

$$T^\lambda(i:j) \equiv \sum_{h,k} w_{i,h} w'_{k,j} T^\lambda(h:k) \bmod A(\leq \lambda)^*$$

summed over all $h, k \in I^\lambda$. The proof of the Straightening Formula is now complete.

2.6 The Désarménien matrix and independence

It remains to establish independence of the the set $\mathcal{L}$ in (2.19). Several approaches are now open to us.

(1) We know that for $p \in \mathbf{t}$, $A(\leq \lambda_p)$ is spanned modulo $A(\leq \lambda_{p+1})$ by the set $\{T^{\lambda_p}(i:j) : i, j \in I^{\lambda_p}\}$. Each of these sets has $d_p{}^2$ elements, where $d_p = |I^{\lambda_p}|$, and so to prove that the union of these two sets is a basis we might try to establish that

$$\sum_{p=1}^{t} d_p{}^2 = \dim A_r. \tag{2.23}$$

This latter fact has a combinatorial proof due to Stanley [1971, §§5–10], which is amusing to recall: there is a formal identity involving Schur functions which states that

$$\prod_{a,b=1}^{n} \frac{1}{1 - X_a Y_b} = \sum_{\lambda \in \Lambda^+(r)} s_\lambda(\mathbf{X}) s_\lambda(\mathbf{Y}),$$

where the Ys are independent of the Xs, (Macdonald [1979, p. 33]). Considering only the part of this in given degree r in both the X and Y, and observing that $s_\lambda(\mathbf{X}) = 0$ if λ has length exceeding n, yields

$$\sum_{\lambda \in \Lambda^+(n,r)} s_\lambda(\mathbf{X}) s_\lambda(\mathbf{Y}) = H_r(\mathbf{X}, \mathbf{Y})$$

where the function on the right is the sum of all monomials of degree r in the n^2 terms $X_a Y_b$. In particular, substituting $X_a = Y_b = 1$ ($a, b \in \mathbf{n}$),

$H_r(1,1) = \begin{pmatrix} n^2 - r + 1 \\ r \end{pmatrix}$ and $s_\lambda(1)$ is the number of standard T_i^λ by
Macdonald [1979, (5.12)], which is what we called d_λ. This gives (2.23).

(2) Following Green [1991a], who in turn bases his arguments on those
of Mead [1972] and DeConcini *et al.* [1980], one defines a certain (non-
obvious) filtration on I^λ and applies induction. The method involves
studying the right action of the subgroup of lower unitriangular matrices,
U, in Γ on A_r. In fact one can in the process obtain a basis of the set
A_r^U, the semi-invariant functions of A_r.

(3) I would also like to mention the paper of Désarménien *et al.* [1978];
they give a self-contained account of the Basis Theorem and get independence using so-called *Capelli operators*. As such their work is seen as
the characteristic-free version of the classical work of Capelli and Gor-
dan. By splitting the Capelli operator into a letter- and a place-part,
Clausen [1979, 1980] gave a refinement of this approach.

(4) One could use the Carter-Lusztig Lemma by checking that $f(i) =
T^\lambda(\ell : i)$ satisfies all the conditions of 2.4.4(v).

(5) We employ a new and direct route, due to Green [1992], which
analyses the dual properties of bideterminants and codeterminants. This
will also be useful when we come to define the Schur modules in Chapter
3.

The Désarménien matrix

We fix $\lambda, \mu \in \Lambda^+(n,r)$, and indices $i, j, k, l \in I$ and recall the pairing
$\langle\ ,\ \rangle$ from 2.2.4.

Proposition 2.6.1 *Writing* $\ell = \ell(\lambda)$ *we have*

(i) $\langle T^\lambda(i : j), C^\mu(k : l) \rangle \neq 0 \Rightarrow \mu \trianglelefteq \lambda$.

(ii) $\langle T^\lambda(i : j), C^\lambda(k : l) \rangle = \langle T^\lambda(i : \ell), \xi_{k,\ell} \rangle \langle T^\lambda(\ell : j), \xi_{\ell,l} \rangle$.

(iii) $\langle T^\lambda(i : \ell), \xi_{k,\ell} \rangle = \Omega(i,k)$ *and* $\langle T^\lambda(\ell, j), \xi_{\ell,m} \rangle = \Omega(j,m)$, *where,*
for $p, q \in I$

$$\Omega(p,q) = \sum_\sigma (-1)^\sigma,$$

summed over all $\sigma \in C(T)$ *such that* $p\sigma$ *and* q *are in the same*
$R(T)$*-orbit of* I.

Proof First, we note that $\langle T^\lambda(i : j), \xi_{k,l} \rangle = 0$ unless $i \sim k$ and $j \sim l$
from 2.2.9. Now suppose that $\mu \ntrianglelefteq \lambda$. Consider $f^\lambda(j) = \langle T^\lambda(i : j), \xi_{k,\ell(\mu)} \rangle$

where $f(j) = \langle c_{i,j}, \xi_{k,\ell(\mu)} \rangle$. By 2.4.4(iii) $f^\lambda(j) = 0$ if $j \sim \ell(\mu)$. Also if $j \nsim \ell(\mu)$ it is zero by the first sentence. Thus $f^\lambda(j) = 0$ identically for all $j \in I$.

For (i), from 2.2.5 we have

$$\langle T^\lambda(i:j), \xi_{k,\ell(\mu)} \xi_{\ell(\mu),l} \rangle = \sum_{q \in I} \langle c_{i,q}, \xi_{k,\ell(\mu)} \rangle \langle c_{q,j}, \xi_{\ell(\mu),l} \rangle. \qquad (2.24)$$

Now (i) follows from the first paragraph.

For (ii), taking $g(q) = \langle c_{q,j}, \xi_{\ell(\lambda),l} \rangle$ and $f(q) = \langle c_{i,q}, \xi_{k,\ell(\lambda)} \rangle$ in 2.4.4(iv) and using (2.5) we see that $g(q) = 0$ unless $q \in \lambda$. Now $f^\lambda(q) = \langle T^\lambda_{i,q}, \xi_{k,\ell(\lambda)} \rangle$ and so the left-hand side of the sum appearing in 2.4.4(iv) equals $\langle T^\lambda(i:j), C^\lambda(k:l) \rangle$ by (2.24). But the right-hand side of this sum is the right-hand side of (ii), hence we're home.

For (iii), $\langle T^\lambda(i:\ell), \xi_{k,\ell} \rangle = \sum_\sigma (-1)^\sigma \langle c_{i\sigma,\ell}, \xi_{k,\ell} \rangle$. By (2.5) $\langle c_{i\sigma,\ell}, \xi_{k,\ell} \rangle = 1$ if $(i\sigma, \ell) \sim (k, \ell)$, otherwise it is zero. But $(i\sigma, \ell) \sim (k, \ell)$ if and only if there exists $\pi \in G_\ell = R(T^\lambda)$ such that $i\sigma\pi = k$. This proves the first half of (iii); proof of the other half is similar. □

Ordering I^λ

The final step is the definition of a partial ordering on I^λ introduced by Clausen [1991, pp. 492–494].

Definition 2.6.2 If $a, b \in \mathbf{N}$ and $i \in I$, let $r_{ab}(i)$ be the number of entries $i_\rho \le b$ in the first a rows of T^λ_i. If $i, j \in I$, write $i \sim_R j$, or respectively $i \sim_C j$, if i, j are in the same $R(T^\lambda)$-orbit, or respectively the same $C(T^\lambda)$-orbit, of I.

In Clausen [1991, p. 493] it is shown that if $i, j \in I^\lambda$ and $p \in I$ are such that $i \sim_R p \sim_C j$ then $r_{ab}(i) = r_{ab}(p) \le r_{ab}(j)$ for all a and b.

Definition 2.6.3 (Clausen ordering) Define a relation $\preceq$ on I^λ by the rule: $i \preceq j$ if there are elements $i = j_0, j_1, \ldots, j_m = j \in I^\lambda$ and elements $k_1, \ldots, k_m \in I$ such that

$$j_0 \sim_R k_1 \sim_C j_1 \sim_R k_2 \sim_C j_2 \cdots \sim_R k_m \sim_C j_m.$$

Reflexivity and transitivity are immediate; antisymmetry comes from the observation following 2.6.2, because $r_{ab}(i) = r_{ab}(j)$ (for all a, b) means $i = j$, since T^λ_i and T^λ_j are both standard.

The next result is innocuous enough, yet it proves crucial; see Désarménien's appendix in Rota [1976/77, Proposition 2], or Désarménien [1980].

Lemma 2.6.4 *For* $p, q \in I^\lambda$, $\Omega(p, q) \neq 0 \Rightarrow q \preceq p$. *Also* $\Omega(p, p) = 1$.

Proof In the case $\Omega(p, q) \neq 0$, we can find $\sigma \in C(T)$ such that $p\sigma \sim_R q$, by 2.6.1(iii). So $q \sim_R p\sigma \sim_C p$ which implies $q \preceq p$.

For the second assertion, we begin by showing that $p\sigma \sim_R p$ ($\sigma \in C(T)$) implies $\sigma = 1$. For if not, we can find $a, b \in \mathbf{N}$ such that $r_{ab}(p) > r_{ab}(p\sigma)$. This is impossible, since $p\sigma \sim_R p$ would imply $r_{ab}(p) = r_{ab}(p\sigma)$ for all a, b.   □

Now we introduce an order on (Ξ):

Definition 2.6.5 If $(\lambda, i, j), (\mu, p, q) \in (\Xi)$ we say that $(\mu, p, q) \leq (\lambda, i, j)$ if either $\mu \lhd \lambda$, or $\mu = \lambda$ and $p \preceq i$ and $q \preceq j$.

Proposition 2.6.6 *Given* $(\lambda, i, j), (\mu, p, q) \in (\Xi)$,

  (i) $\langle T^\lambda(i : j), C^\mu(p, q) \rangle \neq 0 \Rightarrow (\mu, p, q) \leq (\lambda, i, j)$.

  (ii) $\langle T^\lambda(i : j), C^\lambda(i : j) \rangle = 1$.

Proof Proposition 2.6.1 and Lemma 2.6.4 are used.   □

Definition 2.6.7 The matrix $D = [\langle T^\lambda(i : j), C^\mu(p : q) \rangle]$ whose rows and columns are indexed by (Ξ) will be called the *Désarménien matrix*.

The next result is due to Désarménien [1980].

Theorem 2.6.8 *The matrix D is unimodular, hence D is also invertible.*

Proof Suppose we arrange the rows and columns of D subject to a total order on (Ξ) that contains the order of 2.6.5. The force of 2.6.6 is that D is then upper triangular with diagonal entries all equal to 1. Thus D is unimodular.   □

Independence

The independence of (Ξ) is now a simple corollary of 2.6.8. For, assume that there is a relation

$$\sum k_{i,j}^{\lambda} T^{\lambda}(i:j) = 0$$

summed over all $(\lambda, i, j) \in (\Xi)$ and where $k_{i,j}^{\lambda} \in K$. Take the inner product with $C^{\mu}(p:q)$ to obtain an expression

$$\sum k_{i,j}^{\lambda} \langle T^{\lambda}(i:j), C^{\mu}(p:q) \rangle = 0,$$

for all $(\mu, p, q) \in (\Xi)$. Since D is nonsingular, $k_{i,j}^{\lambda} = 0$ for all (λ, i, j) and the independence of (Ξ) is shown.

Turning now to 2.4.8, let $\{Z^{\lambda}(i:j) : (\lambda, i, j) \in (\Xi)\}$ be the basis of S_r dual to $\mathcal{L}$ with respect to $\langle \, , \, \rangle$. Thus $C^{\mu}(p:q) = \sum \langle T^{\lambda}(i:j), C^{\mu}(p:q) \rangle Z^{\lambda}(i:j)$ for all $(\mu, p, q) \in (\Xi)$. Unimodularity of D now shows that $\mathcal{M}$ and the $Z^{\lambda}(i:j)$ form a basis of S_r.

Finally we strengthen the Straightening result as follows:

Proposition 2.6.9 *Given $\lambda \in \Lambda^{+}(n, r)$ and $h, k \in I$,*

(i) *Suppose that $T^{\lambda}(h:k) = \sum k_{i,j}^{\nu} T^{\nu}(i:j)$ where $k_{i,j}^{\nu} \in K$. Then for any $(\nu, i, j) \in (\Xi)$, $k_{i,j}^{\nu} \neq 0 \Rightarrow \nu \trianglelefteq \lambda$.*

(ii) *Suppose that $C^{\lambda}(h:k) = \sum c_{i,j}^{\nu} C^{\nu}(i:j)$ where $c_{i,j}^{\nu} \in K$. Then for any $(\nu, i, j) \in (\Xi)$, $c_{i,j}^{\nu} \neq 0 \Rightarrow \nu \trianglerighteq \lambda$.*

Proof Clearly these statements are dual, so we just prove (i).

Without loss, $T^{\lambda}(h:k) \neq 0$. Choose (μ, p, q) maximal in $X = \{(\nu, i, j) : k_{i,j}^{\nu} \neq 0\}$ with respect to the order of 2.6.5. All members of X other than this chosen one satisfy $(\nu, i, j) \not\geq (\mu, p, q)$. By 2.6.6(i) $\langle T^{\nu}(i:j), C^{\mu}(p, q) \rangle = 0$. So $\langle T^{\lambda}(h:k), C^{\mu}(p:q) \rangle = k_{p,q}^{\mu} \langle T^{\mu}(p:q), C^{\mu}(p:q) \rangle = k_{p,q}^{\mu} \neq 0$, whence $\mu \trianglelefteq \lambda$ by 2.6.1(i). But given any element $(\nu, i, j) \in X$ there is some maximal $(\mu, p, q) \in X$ satisfying $(\nu, i, j) \leq (\mu, p, q)$. Thus $\nu \trianglelefteq \mu$ (by 2.6.5) and we've just shown that $\mu \trianglelefteq \lambda$ so we have $\nu \trianglelefteq \lambda$, as required. $\qquad\square$

Remark. That any bideterminant of shape λ is expressible as a linear combination of standard bideterminants whose shapes are dominated by λ is due to DeConcini *et al.* [1980] as mentioned above; another version is due to Clausen [1991]. But at the same time we find that any codeterminant of shape λ is expressible as a linear combination of codeterminants whose shapes dominate λ. This is a 'straightening rule' for codeterminants, which is highly non-constructive! Woodcock's [1993] codeterminantal straightening procedure gives a method for proving that the codeterminants form a K-basis without first using the bideterminant Straightening Formula.

3

Representation theory of the Schur algebra

The aim of this chapter is to discuss the representation-theoretic aspects of the Schur algebra, S_r. Fulfilling a promise made in 1.6, we construct explicitly a complete set of irreducible S_r-modules indexed by the dominant weights $\Lambda^+(n,r)$. After our study of bideterminants and codeterminants in the previous chapter, most of the work has already been done. We define left and right Schur and Weyl modules as certain S_r-modules having bases consisting of standard bideterminants or codeterminants with shapes in $\Lambda^+(n,r)$.

In 3.3 we prove the fundamental fact that S_r is a quasi-hereditary algebra, which means roughly that S_r can be filtered by certain (S_r, S_r)-bimodules $V = S(\lambda)$ indexed by dominant weights and that for each $\lambda \in \Lambda^+(n,r)$ there is a factorisation $V \cong V^\lambda \otimes {}^\lambda V$ as (S_r, S_r)-bimodules into right and left weight spaces of V. The modules V^λ and ${}^\lambda V$ are precisely the left and right Weyl modules, and constitute the standard modules in the highest weight category $\mathcal{P}_K(n,r)$ (see CPS [1988b]). Taking duals we re-discover the fundamental filtration appearing in 2.5, which had been obtained by DeConcini *et al.* some dozen years before anyone had thought about such categories. The sections of this filtration are then seen to be the Schur modules i.e. the costandard modules in the highest weight category $\mathcal{P}_K(n,r)$.

Schur modules have a long history: they were studied as 'primary covariants' by Deruyts as long ago as 1892 when $K = \mathbf{C}$ (Deruyts [1892], Green [1991b]). Descriptions for arbitrary characteristic were given by Clausen [1979, 1980] who called such modules 'Weyl modules of type 1', and by James [1978a] who proved that Schur modules were induced modules in the sense of algebraic group theory: we derive this in 3.2.6 quite formally. Schur modules are called 'dual Weyl modules' in Green's monograph [1980] For more on the historical development of the subject see Towber [1979].

In 3.4 we present structure theorems cataloguing basic properties of both Schur modules and Weyl modules. As a corollary we show that the socles of the Schur modules or the heads of the Weyl modules are simple and provide us with complete sets of simple objects in $_{S_r}$**mod**. A general discussion of contravariant duality appears in 3.5 and as an example the contravariant duals to the Schur modules are seen to be none other than the Weyl modules. Finally, we discuss a procedure whereby a 'modular representation theory' for S_r and Γ may be sensibly constructed, and we settle in the affirmative the question of whether the Schur and Weyl modules have integral forms. I should also say again that the pioneering work of Carter and Lusztig [1974] was the first to address this question successfully.

3.1 Modules for A_r and S_r

Let $\lambda \in \Lambda^+(n,r)$, and recall from 2.5.2 the definition of the (S_r, S_r)-sub-bimodule a^λ of A_r. Analogously for S_r we are led to consider

Proposition 3.1.1 *Let s^λ be the K-space spanned by all codeterminants $C^\lambda(i:j)$ for all $i,j \in I$. Then s^λ is independent of the choice of basic λ-tableaux and is an (S_r, S_r)-sub-bimodule of S_r.*

Proof The space s^λ is well defined because of Exercise (1) after 2.4.6. Also, 2.2.9 proves that $s^\lambda = S_r \xi_\lambda S_r$ where $\xi_\lambda = \xi_{\ell(\lambda),\ell(\lambda)}$. □

Definition 3.1.2 Let $\pi \subseteq \Lambda^+(n,r)$. Define the spaces

$$A(\pi) = \sum_{\lambda \in \pi} a^\lambda \qquad S(\pi) = \sum_{\lambda \in \pi} s^\lambda.$$

Now 3.1.1 and 2.5.2 imply that we actually have (S_r, S_r)-sub-bimodules of A_r and S_r respectively. Using results in 2.6 we can easily obtain bases for these spaces. First, we mention some notation culled from the representation theory of algebraic groups: we shall call $\pi \subseteq \Lambda^+(n,r)$ a *decreasing saturation* if

$$\mu \trianglelefteq \lambda \Rightarrow \mu \in \pi, \text{ for any } \mu \in \Lambda^+(n,r), \ \lambda \in \pi.$$

Similarly an *increasing saturation* is a subset $\pi' \subseteq \Lambda^+(n,r)$ such that

$$\mu \trianglerighteq \lambda \Rightarrow \mu \in \pi', \text{ for any } \mu \in \Lambda^+(n,r),\ \lambda \in \pi'.$$

Proposition 3.1.3

(i) *If π is a decreasing saturation then $A(\pi)$ has as K-basis the set*

$$T(\pi) = \{T^\lambda(i:j) : (\lambda, i, j) \in (\Xi),\ \lambda \in \pi\}.$$

(ii) *If π is a increasing saturation then $S(\pi)$ has as K-basis the set*

$$C(\pi) = \{C^\lambda(i:j) : (\lambda, i, j) \in (\Xi),\ \lambda \in \pi\}.$$

Proof By the usual duality we need only prove (ii). Now $C(\pi) \subseteq S(\pi)$ and is a linearly independent set being a subset of $\mathcal{M}$; see (2.20). That $S(\pi)$ lies in the span of $C(\pi)$ follows from 2.6.9(ii) plus the given saturation property. $\qquad\qquad\square$

We refine these notions. Take any partial ordering, $\geq$, on $\Lambda^+(n,r)$ containing the dominance ordering. Thus, if π is an increasing saturation with respect to $\geq$ then it is an increasing saturation with respect to $\trianglerighteq$. Choose and fix $\lambda \in \Lambda^+(n,r)$ and take $\pi = \{\mu \in \Lambda^+(n,r) : \mu \geq \lambda\}$, and $\pi' = \{\mu \in \Lambda^+(n,r) : \mu > \lambda\}$.

Definition 3.1.4 Let $S(\geq \lambda) = S(\pi)$ and $S(> \lambda) = S(\pi')$. Since $S(\geq \lambda) \supseteq S(> \lambda)$ we may form $V = S(\lambda) = S(\geq \lambda)/S(> \lambda)$. Also let $\bar{S} = S/S(>\lambda)$, and write $\bar{\xi}$ for the image of ξ in $\bar{S}$.

Notice that one may think of $\bar{S}$ as either a quotient K-algebra or as a quotient (S, S)-bimodule, so it makes sense to write $\overline{\xi\eta} = \bar{\xi}\bar{\eta} = \bar{\xi}\eta = \xi\bar{\eta}$, for any $\xi, \eta \in S$. I should point out that the isomorphism type of the (S, S)-bimodule $V = S(\lambda)$ does not depend on the order $\leq$, provided it refines dominance. The next exercise should convince the reader of this:

Exercise. Prove that $S(> \lambda) + S(\trianglerighteq \lambda) = S(\geq \lambda)$ and that $S(> \lambda) \cap S(\trianglerighteq \lambda) = S(\triangleright \lambda)$. This implies $S(\geq \lambda)/S(> \lambda) \cong S(\trianglerighteq \lambda)/S(\triangleright \lambda)$.

Now π and π' as defined above are increasing saturations of $\Lambda^+(n,r)$, so by 3.1.3 there are R-bases $C(\pi)$ and $C(\pi')$ of $S(\geq \lambda)$ and $S(> \pi)$ respectively. Using the basis $\mathcal{M} = C(\Lambda^+(n,r))$ of S_r from (2.20) we may deduce the existence of bases

$$\mathcal{M}' = \{\bar{\xi}_{i,\ell(\mu)}\bar{\xi}_{\ell(\mu),j} : \mu \not> \lambda,\ i, j \in I^\mu\}$$

of $\bar{S}$ and

$$\mathcal{M}'' = \{\bar{\xi}_{i,\ell(\lambda)}\bar{\xi}_{\ell(\lambda),j} : i, j \in I^\lambda\}$$

of $V = S(\lambda)$. We reserve the notation $V(\lambda)$ for the particular weight space $V^\lambda = V\xi_\lambda$, because later on we shall want to identify this as a Weyl module in the algebraic group sense.

Theorem 3.1.5 *Let $\leq$ be a partial order containing dominance on $\Lambda^+(n,r)$. For $\lambda \in \Lambda^+(n,r)$, write $\ell = \ell(\lambda)$. Then*

(i) $V = \bar{S}\bar{\xi}_\lambda\bar{S}$.

(ii) $V(\lambda) = V^\lambda = \bar{S}\bar{\xi}_\lambda$ *has K-basis* $\{\bar{\xi}_{i,\ell} : i \in I^\lambda\}$, *and* $^\lambda V = \bar{\xi}_\lambda\bar{S}$ *has K-basis* $\{\bar{\xi}_{\ell,j} : j \in I^\lambda\}$.

(iii) $\bar{\xi}_\lambda\bar{S}\bar{\xi}_\lambda = K\bar{\xi}_\lambda$.

(iv) *The (S,S)-map $\varphi : V^\lambda \otimes {}^\lambda V \to V$ given by $\varphi(\bar{\xi} \otimes \bar{\eta}) = \bar{\xi}\bar{\eta}$ is an (S,S)-isomorphism.*

Proof For (i), $S(\geq \lambda) = s^\lambda + S(> \lambda)$; we now use the fact that $s^\lambda = S\xi_\lambda S$ (see the proof of 3.1.1).

For (ii), consider the basis element $\bar{\xi}_{i,\ell(\mu)}\bar{\xi}_{\ell(\mu),j}$ in $\mathcal{M}'$. If $j \in \alpha$ then this element lies in the right α-weight space of $\bar{S}$. Now if $\alpha = \lambda$ we would have $\lambda \trianglelefteq \mu$ since $j \in I^\mu$ (cf. 1.4.7). Since also μ cannot be greater than λ we conclude that $\lambda = \mu$. The only $j \in I^\lambda$ of weight λ is $j = \ell(\lambda)$, so the only elements of $\mathcal{M}'$ of right weight λ are $\bar{\xi}_{i,\ell(\lambda)}\bar{\xi}_{\ell(\lambda),\ell(\lambda)} = \bar{\xi}_{i,\ell}$ ($i \in I^\lambda$) and this gives the first statement. A similar proof deals with the second statement.

For (iii), we observe that $\bar{\xi}_\lambda\bar{S}\bar{\xi}_\lambda = V^\lambda \cap {}^\lambda V$ and use (ii).

For (iv), φ maps a basis $\{\bar{\xi}_{i,\ell} \otimes \bar{\xi}_{\ell,j} : i, j \in I^\lambda\}$ of $V^\lambda \otimes {}^\lambda V$ to a basis $\mathcal{M}''$ of $V = S(\lambda)$, so is bijective. $\square$

To finish this section, let us consider the dual situation in A_r. We have already defined, in 2.5.3, the module $A(\leq \lambda)$ with basis $\mathcal{L}'$ from 2.5.5(iv). We introduce the (S,S)-sub-bimodule $A(< \lambda)$ by analogy to $S(> \lambda)$: let $M = A(\lambda) = A(\leq \lambda)/A(< \lambda)$. M has basis $\mathcal{L}'' = \{\bar{T}^\lambda(i:j) : i, j \in I^\lambda\}$. Write $\bar{c} = c + A(< \lambda)$ for $c \in A_r$. The dual theorem to 3.1.5 was proved some dozen years before it (!), by DeConcini *et al.* [1980, p. 147] and only a sketch proof is offered here.

Theorem 3.1.6 *Let $\leq$ be a partial order containing dominance on $\Lambda^+(n,r)$. For $\lambda \in \Lambda^+(n,r)$, write $\ell = \ell(\lambda)$. Then*

(i) $A(\leq \lambda)^{\lambda}$ has K-basis $\{T^{\lambda}(\ell : j) : j \in I^{\lambda}\}$, and $^{\lambda}M(\leq \lambda)$ has K-basis $\{T^{\lambda}(i : \ell) : i \in I^{\lambda}\}$.

(ii) $A(\leq\lambda)^{\lambda} \cap A(<\lambda) = \{0\} = {}^{\lambda}A(\leq\lambda) \cap A(<\lambda)$, and so the canonical epimorphism $c \mapsto \bar{c}$ induces isomorphisms $A(\leq \lambda)^{\lambda} \cong M^{\lambda}$ and $^{\lambda}A(\leq\lambda) \cong {}^{\lambda}M$.

(iii) The (S,S)-map $\psi : M^{\lambda} \otimes {}^{\lambda}M \to M$ given by the mapping $\bar{T}^{\lambda}(\ell : j) \otimes \bar{T}^{\lambda}(i : \ell) \mapsto \bar{T}^{\lambda}(i : j)$ for all $i, j \in I^{\lambda}$ is an (S,S)-isomorphism.

Proof For (i), we alter the proof of 3.1.5(i): use 2.4.5(ii) to show that the only elements in $\mathcal{L}'$ with right/left weight λ are respectively the $T^{\lambda}(\ell : j)$ for $j \in I^{\lambda}$, or the $T^{\lambda}(i : \ell)$ for $i \in I^{\lambda}$.

For (ii), we use 3.1.3(i) and (i) above.

For (iii), the explicit construction of ψ is done in DeConcini *et al.* [1980] and we do not repeat the details here. One can deduce the factorisation $M \cong M^{\lambda} \otimes {}^{\lambda}M$ from the isomorphism $V \cong V^{\lambda} \otimes {}^{\lambda}V$ by means of the next result. $\qquad\square$

Proposition 3.1.7 *Recall the pairing of (2.5). There is a well-defined non-singular form* $(\, , \,) : M \times V \to K$ *defined by* $(\bar{c}, \bar{\xi}) = \langle c, \xi \rangle$ *for all* $c \in A(\leq \lambda)$ *and* $\xi \in S(\geq \lambda)$. *Then the map* $P : M \to V^{*}$ *defined by* $P(m)(v) = (m, v)$ *for all* $m \in M$, $v \in V$ *defines an* (S,S)-*isomorphism. Moreover* P *induces isomorphisms* $M^{\lambda} \cong ({}^{\lambda}V)^{*}$ *(as left* S-*modules) and* $^{\lambda}M \cong (V^{\lambda})^{*}$ *(as right* S-*modules).*

Proof From 2.6.1(i) $\langle A(\leq\lambda), S(>\lambda)\rangle = 0$ and $\langle A(<\lambda), S(\geq\lambda)\rangle = 0$, so $(\, , \,)$ is well defined. The matrix of the form with respect to the bases $\mathcal{L}''$ of M and $\mathcal{M}''$ of V is unimodular: it is some unitriangular submatrix of the Désarménien matrix D. Hence $(\, , \,)$ is non-singular. $\qquad\square$

3.2 Schur modules as induced modules

It is time to give names to some of the modules appearing in the last section.

Definition 3.2.1 Let $\lambda \in \Lambda^{+}(n, r)$. Recall that we have modules $M = A(\lambda)$ and $V = S(\lambda)$. We call $V(\lambda) = V^{\lambda}$ and $^{\lambda}V$ the *left and right Weyl modules* for S_r. We also call $M(\lambda) = M^{\lambda}$ and $^{\lambda}M$ the *left and right Schur modules* for S_r.

It is perhaps surprising, at first, to realise that the duality we have here is the natural one: the left Schur module is dual to the right Weyl module and the right Schur module is dual to the left Weyl module, by 3.1.7. It is therefore no loss to choose to deal with left modules from now on. Ideas on contravariant duality will be discussed in 3.4.

Recalling the explicit bases of the left Schur modules, namely that $M(\lambda)$ has basis

$$\{T^\lambda(\ell : j) : j \in I^\lambda\}, \tag{3.1}$$

it is immediate that $M(\lambda) \in \mathcal{P}_K(n, r)$. By comparing with Green [1980, p. 54] we have

Proposition 3.2.2 *The left Schur module $M(\lambda)$ is Green's 'dual Weyl module', $D_{\lambda,K}$.*

Borel subalgebras

In the light of 3.2.2 we should consider it feasible to provide alternative characterisations of the left Schur modules as induced modules. Actually we can treat Schur and Weyl modules together, but first we shall need something to induce from!

We recall the ordering on $I(n, r)$ defined after 1.4.4: this ordering contains dominance. The reader may refer to A.2 for the definitions of the Borel and unipotent subgroups of Γ. Let S^+ and S^- be the K-subspaces of S generated by all $\xi_{i,j}$ where $i, j \in I$ satisfy respectively $i \leq j$ and $i \geq j$. Actually these are subalgebras of S, indeed $S^+ = T_{n,r}(KB^+)$ and $S^- = T_{n,r}(KB^-)$ where B^+ and B^- are the upper and lower Borel subgroups of Γ; see Green [1990a, p. 270].

Definition 3.2.3 S^+ and S^- are called the *upper and lower Borel subalgebras* of S.

Remark. Green [1990a] intimates the importance of the Borel subalgebras; in particular he shows that $S = S^- S^+$, i.e. the Schur algebra is the linear closure of the set of all products $\xi\eta$ with $\xi \in S^-$, $\eta \in S^+$, proves that one (and hence both) subalgebras are quasi-hereditary, and finally proves a 'dominant weight decomposition' of S:

$$S = \sum_{\lambda \in \Lambda^+(n,r)} S^- \xi_\lambda S^+.$$

For each $\lambda \in \Lambda(n,r)$ define K-algebra maps $\chi_\lambda : KB^+ \to K$ and $\chi'_\lambda : S^+ \to K$ by $\chi_\lambda(b) = b_{11}^{\lambda_1} \ldots b_{nn}^{\lambda_n}$ for $b \in B^+$; $\chi'_\lambda(\xi_{i,j})$ equals 1 or 0 according to whether $i \sim j$ has weight λ or not, for $i \leq j$. Thus $\chi_\lambda(b) = \chi'_\lambda(T_{n,r}(b))$ for all $b \in B^+$. There are obvious 'lower' versions of these maps.

Definition 3.2.4 For each $\lambda \in \Lambda(n,r)$, ${}^+K_\lambda$, or respectively ${}^-K_\lambda$, is the field K regarded as a left module for S^+, or respectively for S^-, by the rule $\xi k = \chi'_\lambda(\xi)k$ for all $\xi \in S^+$, or respectively $\xi \in S^-$, and $k \in K$. One defines right modules K_λ^+ and K_λ^- similarly. Of course, it is convenient to view these as rational B^+- or B^--modules as appropriate.

In the next result one must regard S as a left or right S^+- or S^-- module, as appropriate, on one side and an S-module on the other side. We also need to use the 'adjoint isomorphism theorem', a very general homological fact, e.g. see Rotman [1979, (2.11)]. The statement is that if A, B are rings, $P \in {}_A\mathbf{mod}$, Q is a (B,A)-bimodule and $R \in {}_B\mathbf{mod}$, then there is a group isomorphism

$$\text{Hom}_B(Q \otimes_A P, R) \cong \text{Hom}_A(P, \text{Hom}_B(Q, R)).$$

Theorem 3.2.5 *There are isomorphisms*
  (i)

$$M^\lambda \cong \text{Hom}_{S^-}(K_\lambda^-, A_r) \cong \{c \in A_r : c \circ \xi = \chi'_\lambda(\xi)c, \ \forall \xi \in S^-\}$$

  and

$${}^\lambda M \cong \text{Hom}_{S^+}(S, K_\lambda^+) \cong \{c \in A_r : \xi \circ c = \chi'_\lambda(\xi)c, \ \forall \xi \in S^-\}.$$

  (ii) $V^\lambda \cong S \otimes_{S^+} {}^+K_\lambda$ *and* ${}^\lambda V \cong K_\lambda^- \otimes_{S^-} S$.

Proof (i) follows from (ii) by the adjoint isomorphism theorem. For example, suppose we know that ${}^\lambda V \cong K_\lambda^- \otimes_{S^-} S$. Then

$$\begin{aligned} M^\lambda &\cong ({}^\lambda V)^* \cong \text{Hom}_K(K_\lambda^- \otimes_{S^-} S, K) \\ &\cong \text{Hom}_{S^-}(K_\lambda^-, \text{Hom}_K(S, K)) \cong \text{Hom}_{S^-}(K_\lambda^-, A_r). \end{aligned}$$

The second isomorphism in (i) takes the S^--map $f : K_\lambda^- \to A_r$ to $f(1) \in A_r$.

  For (ii) we need only show that ${}^\lambda V \cong K_\lambda^- \otimes_{S^-} S$. Helpfully, both modules are cyclic: we notice that $X = K_\lambda^- \otimes_{S^-} S = \gamma S$ where $\gamma = 1 \otimes 1$, and $^\lambda V = \bar\xi_\lambda S$, so we need only show that there exist right S-maps $f : {}^\lambda V \to X$ and $g : X \to {}^\lambda V$ such that $f(\bar\xi_\lambda) = \gamma$ and $g(\gamma) = \bar\xi_\lambda$.

To show that f exists. Let $K = \mathrm{Ker}(\xi_\lambda S \to \bar\xi_\lambda \bar S)$. Now $\gamma\xi_\lambda = \gamma$, so we certainly have an S-map $F : \xi_\lambda S \to X$ such that $F(\xi_\lambda) = \gamma$. It is therefore enough to check that $K \subseteq \mathrm{Ker}F$.

By 3.1.3(ii) $S(> \lambda)$ has basis $C(> \lambda) = \{\xi_{i,\ell(\mu)}\xi_{\ell(\mu),j} : \mu > \lambda,\ i,j \in I^\mu\}$. So $K = \xi_\lambda S \cap S(> \lambda)$ is S-generated by the $\xi_{i,\ell(\mu)}$ where $\mu > \lambda$ and $i \in \lambda$ is in I^μ. The latter condition implies that $i \geq \ell(\mu)$ by 1.4.7, so $\xi_{i,\ell(\mu)} \in S^-$. Now $\mu \neq \lambda$, so $i \neq \ell(\mu)$ which forces $\chi'_\lambda(\xi_{i,\ell(\mu)}) = 0$. Then $F(\xi_{i,\ell(\mu)}) = 1\xi_{i,\ell(\mu)} \otimes 1 = 0$, hence $F(K) = 0$.

To show that g exists. Consider the following assertion:

$$\xi_\lambda \xi \equiv \chi'_\lambda(\xi) \pmod{S(> \lambda)} \quad \forall \xi \in S^-. \tag{3.2}$$

If this is valid then the assignation $g(k \otimes \xi) = k\bar\xi_\lambda\bar\xi$ $(k \in K,\ \xi \in S)$ is a well-defined map $X \to {}^\lambda V$ satisfying $\gamma \mapsto \bar\xi_\lambda$.

It is enough to check (3.2) with $\xi = \xi_{i,j}$ for $i \geq j$. Immediately, we notice that both sides of (3.2) are zero if $i \not\sim \ell(\lambda)$.

Suppose that $i = \ell(\lambda)$. If $j \in \mu$ then $i \geq j$ implies $\lambda \trianglelefteq \mu$. If $\lambda = \mu$, then $i \sim j$, and $i \geq j$ implies $i = j = \ell(\lambda)$, forcing $\xi = \xi_\lambda$. So (3.2) is also valid.

The final case occurs when $\xi = \xi_{\ell(\lambda),j}$ with $j \in \mu \triangleright \lambda$. Since $\ell(\lambda) > j$ implies $\chi'_\lambda(\xi) = 0$, we must show that $\xi \in S(> \lambda)$. First of all $j \sim \ell(\mu)$ so $\xi = \xi\xi_\mu \in s^\mu$. If $\mu \in \Lambda^+(n,r)$ then $s^\mu \subseteq S(> \lambda)$ and we're done. If not, we must find $w \in \Sigma_n$ such that $w(\mu) = \alpha \in \Lambda^+(n,r)$ (cf. the comments after 1.4.4). But $w(\mu) \trianglerighteq \mu$, so $\alpha \triangleright \lambda$ and $s^\alpha \subseteq S(> \lambda)$. Our final observation is that there exists an invertible $x_w \in S$ such that $x_w\xi_\mu x_w^{-1} = \xi_{w(\mu)} = \xi_\alpha$. Namely, define $x_w = \sum \xi_{w(i),i}$, where the sum is over a transversal of the G-orbits of I; notice that $x_w^{-1} = x_{w^{-1}}$. We conclude that $s^\alpha = S\xi_\alpha S = S\xi_\mu S = s^\mu$, i.e. $\xi \in S(> \lambda)$. $\qquad\square$

Existence of the second isomorphism in 3.2.5(i) implies that the left Schur module $M(\lambda)$ may be characterised as

$$M^\lambda = M(\lambda) = \{c \in A_r : c \circ b = \chi_\lambda(b)c,\ \forall\, b \in B^-\}.$$

But compare this with the usual definition of 'induced module' for algebraic groups: from A.7 or Jantzen [1987, p. 199] we know that $H^0(\lambda) = \{f \in K[\Gamma] : f(gb) = \chi_\lambda^{-1}(b)f(g),\ \forall g \in \Gamma,\ b \in B^-\}$. We thus obtain the following result of James [1978a] without effort:

Theorem 3.2.6 (James) $M(\lambda)$ *is the induced module*

$$H^0(\lambda) = \mathrm{Ind}_{B-}^{\Gamma}({}^-K_\lambda)$$

in the category $\mathcal{P}_K(n,r)$.

Remark. There exist dominant weights $\lambda \in X^+(T)$ that are not dominant polynomial such that $H^0(\lambda) \neq 0$ and for which the combinatorial description of 3.2.6 is false. For such λ, $\lambda_n < 0$ and $H^0(\lambda)$ is a tensor product of $M(\mu)$ with the λ_nth power of the determinant representation, where μ is the partition $(\lambda_1 - \lambda_n, \ldots, \lambda_{n-1} - \lambda_n, 0)$.

Let us now identify what we have called Weyl modules with certain modules in Jantzen's book [1987, p. 206]. He defines a 'Weyl module' for a reductive algebraic group to be $H^0(-\pi_0\lambda)^* = {}^\tau H^0(\lambda)$, where π_0 is the longest element of the Weyl group and τ is the anti-automorphism of order 2 appearing in Jantzen [1987, II.1.16] (for $GL_n(K)$, τ is merely matrix transposition). In our case, this module is just $({}^\lambda M)^*$ which, by 3.1.7, is isomorphic to $V^\lambda = V(\lambda)$. Other identifications will be made in 3.4.

3.3 Heredity chains

In their work on highest weight categories coming from representations of finite-dimensional semisimple complex Lie algebras, and also of algebraic groups, Cline, Parshall and Scott [1988a], [1988b], [1990] formed the notion of a quasi-hereditary algebra and studied its module category (which was termed a 'highest weight category'); see also the work of Dlab and Ringel [1989] and Scott [1987, p. 280]. I shall here set down a résumé of the relatively new area of quasi-hereditary algebras, and end with a two-line proof that the Schur algebras belong to this important class. The first explicit proof of quasi-hereditariness was by Donkin [1981], who showed that $_{S_r}\mathbf{mod}$ is a highest weight category (indeed he did this a little before the term was coined). Actually, a proof is implicit in the 1980 paper of DeConcini *et al.* [1980]: take the part of their 'fundamental filtration' of the coordinate ring that is homogeneous polynomial of degree r, and then dualise. Later Parshall [1989, 4.1], [1987] gave a proof depending heavily on methods from algebraic group theory. A simpler proof appears in some unpublished notes of Green [1990b], and this is essentially the approach we take here, though the work on weight spaces in 3.1 makes the final version quite transparent; see Green [1992].

Background

Consider the associative K-algebra S, say. Let $N = \operatorname{Rad} S$. All of what we say below remains true for algebras over commutative noetherian rings, though we shall limit ourselves to working over fields. For the next definition see CPS [1988b, pp. 92–93] and Parshall [1989, §3].

Definition 3.3.1 A non-zero (two-sided) ideal J of S is called a *heredity ideal* of S if the following three conditions all hold:

(i) J considered as a left S-module is projective;

(ii) $\operatorname{Hom}_S(J, S/J) = 0$;

(iii) $JNJ = 0$.

Conditions (ii) and (iii) guarantee that composition factors in $S/\operatorname{Rad}J$ do not occur in S/J and that $\operatorname{Rad}J = NJ$ is a module for S/J. In the presence of condition (i) and (iii) we can replace (ii) by

(ii′) J is idempotent, i.e. S possesses an idempotent e such that $J = SeS$.

In the presence of (iii) and (ii′) we can replace (i) by

(i′) The surjective multiplication map $Se \otimes_{eSe} eS \to J$ is bijective for any idempotent e satisfying $SeS = J$.

For these alternative conditions see Parshall [1989] or Parshall and Scott [1988]. The nomenclature is suggested by the observation that if S is already hereditary then $J = \operatorname{Soc}S$ satisfies the conditions of 3.3.1; the next definition is motivated by iterating this definition.

Definition 3.3.2 The finite-dimensional associative K-algebra (with 1) S is *quasi-hereditary* if there is a chain of (two-sided) ideals of S,

$$0 = J_0 < J_1 < \cdots < J_t = S,$$

such that for any $p \in \mathbf{t}$, J_p/J_{p-1} is a heredity ideal of S/J_{p-1}. We call such a chain of idempotent ideals a *heredity chain* or a *defining sequence* for S.

At this point we digress to link this definition with the notion of highest weight categories. The blueprint for this latter concept (e.g. see CPS [1988b, 3.1]) consists of

- $\mathcal{A}$: a K-finite abelian category. This guarantees that $\operatorname{Hom}(X, Y)$ is a finite-dimensional K-space $(X, Y \in \mathcal{A})$, composition is K-bilinear and all objects have composition series. The obvious example is $_S\mathbf{mod}$ where S is a finite-dimensional K-algebra;

• Λ: a poset (the 'weight poset'). This set will provide the indexing set for the simple $\mathcal{A}$-objects.

Definition 3.3.3 The category $\mathcal{A}$ is a *highest weight category* with *weight poset* Λ, if there exists a family $\{\Delta(\lambda) : \lambda \in \Lambda\}$ of objects of $\mathcal{A}$ (variously called the *Weyl objects*, the *standard objects* or the *Verma objects*) with the following properties:

(i) The head of $\Delta(\lambda)$ is simple; denoting this head by $L(\lambda)$ then $\{L(\lambda) : \lambda \in \Lambda\}$ is a complete set of simple objects in $\mathcal{A}$. For each $\lambda \in \Lambda$, the composition factors of $\mathrm{Ker}\{\Delta(\lambda) \to L(\lambda)\}$ are all of the form $L(\mu)$, for $\mu < \lambda$.

(ii) Each $L(\lambda)$ has a projective cover, $P(\lambda)$, in $\mathcal{C}$. There exists an epimorphism $P(\lambda) \to \Delta(\lambda)$ whose kernel is filtered by some $\Delta(\alpha)$ with $\alpha > \lambda$.

Dual statements exist about the costandard object $\nabla(\lambda)$ ($\lambda \in \Lambda$), its simple socle $L(\lambda)$ and the associated injective hull $I(\lambda)$ of $L(\lambda)$. My tendency is to use whichever concept is appropriate at the time. Other useful properties appear at the end of this section. First of all, a fundamental result of CPS [1988b, 3.4] shows the following.

Theorem 3.3.4 *The category $_S\mathbf{mod}$ together with $(\Lambda, \le)$ is a highest weight category if and only if S is quasi-hereditary.*

I now give an informal sketch indicating why this result is true. Somehow one has to manufacture standard objects for a given quasi-hereditary algebra S with a set of simple modules $\{L(\lambda)\}$ indexed by some finite set Λ of cardinality t. Let us begin by examining some general consequences of 3.3.2. Suppose we have some class $\mathcal{C} \subseteq {}_S\mathbf{mod}$ that is closed under isomorphisms. Denote by $\mathcal{F}(\mathcal{C})$ the class of $_S\mathbf{mod}$ consisting of all objects having filtrations of the form

$$0 = M_0 < M_1 < \cdots < M_m = M$$

where all filtered quotients M_i/M_{i-1} belong to $\mathcal{C}$ ($1 \le i \le m$). We call $\mathcal{F}(\mathcal{C})$ the class of $\mathcal{C}$-*good filtrations*. Take a complete collection $\{L(\lambda) : \lambda \in \Lambda\}$ of simple S-modules and suppose we have a partial order, $\le$ on the indexing set Λ. Denote by $P(\lambda)$ and $I(\lambda)$ respectively the projective cover and injective hull of $L(\lambda)$, for each $\lambda \in \Lambda$.

Definition 3.3.5 Define the *standard module* $\Delta(\lambda)$ as the largest quotient of $P(\lambda)$, all of whose composition factors have the form $L(\beta)$ for $\beta \leq \lambda$. The *costandard module* $\nabla(\lambda)$ is the largest submodule of $I(\lambda)$ all of whose composition factors have the form $L(\beta)$ for $\beta \leq \lambda$.

Remark. Usually, in applications we take $\mathcal{F}(\mathcal{C}) = \mathcal{F}(\Delta)$, the class of finite-dimensional S-modules that admit a Δ-good filtration, i.e. a filtration of the above type with M_i/M_{i-1} either zero or isomorphic to $\Delta(\lambda)$ for some $\lambda \in \Lambda$. Similarly we consider the class $\mathcal{F}(\nabla)$ of ∇-good filtrations. More will be said about this class in 4.5

Suppose we have chosen an ordering as above. Parts (i) and (ii) of the next result are sometimes used as an alternative to Definition 3.3.2. I believe that this formulation is due to Soergel [1990].

Proposition 3.3.6 *Assume that K is algebraically closed and use the notation set out in 3.3.5. Then S is quasi-hereditary if and only if (i) and (ii) below hold. Moreover if (i) holds, (ii) and (iii) are equivalent.*

(i) $\text{End}_S(\Delta(\lambda)) \cong K$ *i.e.* $L(\lambda)$ *occurs with multiplicity one in* $\Delta(\lambda)$.

(ii) $P(\lambda) \in \mathcal{F}(\Delta)$, *i.e.* $P(\lambda)$ *has a chain of submodules with quotients of the form* $\Delta(\beta)$ *with* $\beta \geq \lambda$ *and* $\Delta(\lambda)$ *has multiplicity one as a section.*

(iii) $\text{Ext}_S^2(\Delta(\lambda), \nabla(\mu)) = 0$ *for all* $\lambda, \mu \in \Lambda$.

If S is quasi-hereditary there are obvious dual statements valid for the costandard modules (it is immediate that $\text{End}_S(\Delta(\lambda)) \cong \text{End}_S(\nabla(\lambda))$ since they are both isomorphic to K).

Proof Note that if (i) holds then (iii) is equivalent to the assertion that $_SS \in \mathcal{F}(\Delta)$ and hence is equivalent to (ii). Let us quickly show that (i) and (iii) are equivalent to there being a defining sequence as in 3.3.2.

Given a heredity chain as in 3.3.2 we may always refine it in such a way that for all j the indecomposable summands of a fixed $_S(J_{j-1}/J_j)$ are all isomorphic (to some $\Delta(\gamma)$ say), hence giving us the standard modules. On the other hand, if (i) and (iii) hold then for all j one can find a maximal left ideal J_j of S that belongs to $\mathcal{F}(\{\Delta(j+1), \ldots, \Delta(t)\})$. It is clear that the J_j must be two-sided, so we have constructed a defining sequence $\{J_j\}_j$ for S as required. $\square$

It is clear that standard modules in the sense of 3.3.5 have a simple head, and by construction, the (S/J_j)-modules are precisely those whose composition factors are of the form $L(\lambda)$, $\lambda \leq j$. Conversely, let $P(\lambda) = Se_\lambda$

be a principal indecomposable module (PIM) and let $\Delta(\lambda)$ be standard in the sense of 3.3.3. Then if $U(\lambda) = \sum_{\mu > \lambda} Se_\mu Se_\lambda$ is the *good radical* of $P(\lambda)$ (i.e. it is the submodule of $P(\lambda)$ generated by all images of S-homomorphisms $P(\mu) \to P(\lambda)$ with $\mu > \lambda$) then $P(\lambda)/U(\lambda) \cong \Delta(\lambda)$ is the largest quotient of $P(\lambda)$ all of whose factors $L(\mu)$ satisfy $\mu \le \lambda$. Dually, construct $\nabla(\lambda)$ as the intersection of all kernels of maps $I(\lambda) \to I(\mu)$ for $\mu > \lambda$. Thus the terminology is consistent: (co)standard modules in the sense of 3.3.5 are (co)standard in the sense of 3.3.3 and vice versa.

Example. The prototype for a quasi-hereditary algebra comes from the 'category $\mathcal{O}$' defined by Bernstein, Gelfand and Gelfand [1976]. This category has a decomposition into blocks, $\mathcal{O}_\chi$, indexed by central characters, χ. $\mathcal{O}_\chi$ is equivalent to the category of finitely-generated S_χ-modules where S_χ is the endomorphism ring of a complete set of projectives.

These properties should remind us of Schur and Weyl modules. Now we specialise by taking $S = S_r$. we take any total order $\le$ on $\Lambda^+(n, r)$ containing dominance and enumerate $\Lambda^+(n, r)$ as $\lambda_1, \ldots, \lambda_t$ subject to the condition

$$\lambda_i \le \lambda_j \Rightarrow i \ge j. \tag{3.3}$$

Theorem 3.3.7 (Donkin, Parshall, Green) S_r *is a quasi-hereditary algebra; in fact*

$$\{0\} = J_0 \subseteq J_1 \subseteq \cdots \subseteq J_t = S_r \tag{3.4}$$

is a defining sequence for S_r where $J_p = S(\ge \lambda_p)$ $(p \in \mathbf{t})$.

Proof Observe that $V = S(\lambda)$ is a heredity ideal of $\bar{S}$ by virtue of 3.1.5. But $J_{p-1} = S(> \lambda_p)$ for any $p \in \mathbf{t}$, and hence (3.4) is a defining sequence as desired. $\square$

One often says that S_r is quasi-hereditary with respect to the ordering of (3.3). The next section shows that the Schur modules are the costandard modules and the Weyl modules are the standard modules in the highest weight category $S_{r(\Gamma)}\mathbf{mod}$.

Remark. That the Borel subalgebras S^+ and S^- are quasi-hereditary is shown in Green [1990a, §6].

Some cohomology theory

Quasi-hereditary algebras satisfy some quite restrictive conditions. In CPS [1988a, 3.1] (see also Parshall [1987, 2.1]) it is proved that if S is quasi-hereditary then the supremum of the projective dimensions of modules in $_S$mod is bounded. This supremum is called the *global dimension* of S. That the Schur algebra S_r has finite global dimension was deduced by Donkin [1986, 3.2d]: this was a special case of results for 'generalised Schur algebras' for arbitrary reductive groups; he also made frequent use of filtrations in $\mathcal{F}(\Delta)$. Independent proofs of this fact are given by Akin and Buchsbaum [1985] and Parshall [1989]. The official statement is:

Theorem 3.3.8 *The global dimension of S_r is at most $2|\Lambda^+(n,r)| - 2$.*

Another consequence is that S_r has Loewy length at most $2^{|\Lambda^+(n,r)|} - 1$.

Remark. In the light of 3.3.8 it is natural to ask for explicit projective resolutions of Weyl modules. The papers of Akin and Buchsbaum [1985] and Akin, Buchsbaum and Weyman [1982] give short exact sequences of Schur modules and Weyl modules associated to 'shapes'; in particular, they describe an inductive procedure for constructing finite projective resolutions of Weyl modules for $S_r(\Gamma_R) = A_R(n,r)^*$ in case $n = 2$, over a commutative domain R. Akin [1988], [1989] and Zelevinskii [1987] examine the characteristic 0 situation. Recently, Santana [1993] gave a construction of a complete minimal projective resolution in the case where the field has characteristic zero (for $n \leq 3$), and also obtained the first two terms in a minimal projective resolution in arbitrary characteristic, by purely combinatorial methods. The degree of complexity even at this stage is considerable. Santana's method is based on work of Woodcock [1992] who has shown that it is enough to construct projective resolutions for the simple modules for the Borel subalgebra S^+, since restriction induces an isomorphism

$$\mathrm{Ext}^i_{S_r(\Gamma)}(M, N) \cong \mathrm{Ext}^i_{S^+}(M, N)$$

for $i \geq 1$ and $M, N \in \mathcal{P}_K(n,r)$. This is reminiscent of the well-known fact that restriction to normalisers of T.I. Sylow subgroups of finite groups induces an isomorphism of Ext groups.

Some Ext groups are automatically zero as a result of the CPS investigations into highest weight categories; later we shall need one instance of this, namely that for $\lambda, \mu \in \Lambda$:

Proposition 3.3.9 *We recall the notation of 3.3.3. Suppose that $\lambda, \mu \in$
Λ. If either $\mathrm{Ext}^1_{\mathcal{A}}(\nabla(\mu), \nabla(\lambda)) \neq 0$ or $\mathrm{Ext}^1_{\mathcal{A}}(L(\mu), \nabla(\lambda)) \neq 0$, then
necessarily $\mu > \lambda$.*

Proof I should point out that $\mathrm{Ext}^1_{\mathcal{A}}(M, N)$ is finite-dimensional provided $M, N \in \mathcal{A}$ have finite length. Now if $\mathrm{Ext}^1_{\mathcal{A}}(L(\alpha), \nabla(\lambda)) \neq 0$
then $\mathrm{Hom}_{\mathcal{A}}(L(\alpha), I(\lambda)/\nabla(\lambda)) \neq 0$; taking a ∇-good filtration $\{F_j\}$ of
$I(\lambda)$, we see that $\mathrm{Hom}_{\mathcal{A}}(L(\alpha), F_j/F_1) \neq 0$ for some $j > 1$, so there
is an $i \geq 1$ such that the ith quotient $F_i/F_{i-1} \cong \nabla(\alpha)$. Also, if
$\mathrm{Ext}^1_{\mathcal{A}}(\nabla(\mu), \nabla(\lambda)) \neq 0$ then there is a composition factor $L(\alpha)$ of $\nabla(\mu)$
for which $\mathrm{Ext}^1_{\mathcal{A}}(L(\alpha), \nabla(\lambda)) \neq 0$. This implies $\mu \geq \alpha > \lambda$. $\qquad\square$

Exercise. Extend this proof to show that for $\lambda, \mu \in \Lambda$,

$$\mathrm{Ext}^1_{\mathcal{A}}(L(\lambda), L(\mu)) \neq 0$$

implies that λ and μ must be strictly comparable with respect to $>$
(meaning $\lambda > \mu$ or $\mu > \lambda$).

Remark. The results 3.3.9 show that if $(\mathcal{A}, \Lambda)$ is a highest weight
category then the ordering of simples can always be chosen to satisfy a
relation like (3.3).

3.4 Schur modules and Weyl modules

We recall from 3.2.1 the definition of the (left) Schur module, $M(\lambda)$, as
the left S_r-module with basis given in (3.1).

Examples.
(1) If $\lambda = (r, 0, \ldots, 0)$ then we have a map $T^{(r)}(\ell : i) \mapsto e_\alpha$ where $i \in \alpha$.
This induces a isomorphism $M(\lambda) \cong S^r(E)$.

More generally, we can exhibit the Schur module indexed by some
$\lambda \in \Lambda^+(r)$ as a submodule of $S^\lambda(E)$ (see Example (3) at the end of 1.5).
Here is one proof: another will be suggested in Exercise (4) in 4.2.

Two natural maps are important here; first we have a projection $\pi :$
$E^{\otimes r} \to S^\lambda(E)$. For example, take $\lambda = (3, 2)$ and send $e_{i_1} \otimes \cdots \otimes e_{i_5} \mapsto$
$\bar{e}_{1} \bar{e}_{2} \bar{e}_{3} \otimes \bar{e}_{4} \bar{e}_{5}$ (cf. James and Kerber [1981, 8.1]). This is a Γ-map.
Also, define an isomorphism $f : A_r \xi_\lambda = \sum_{i \in I} K c_{\ell,i} \to S^\lambda(E)$, again by
example: its action is as follows:

$$c_{\ell,i} = c_{11122, i_1 i_2 i_3 i_4 i_5} = c_{1 i_1} c_{1 i_2} c_{1 i_3} c_{2, 1_4} c_{2 i_5} \mapsto \bar{e}_{i_1} \bar{e}_{i_2} \bar{e}_{i_3} \otimes \bar{e}_{i_4} \bar{e}_{i_5}.$$

Let $\mathcal{V}^T$ be the signed column sum (the alternating sum of $C(T^\lambda)$—see the start of 4.2). Under f, $T^\lambda(\ell : i) = \sum_\sigma (-1)^\sigma c_{\ell,i\sigma}$ will be mapped to the image under π of $E^{\otimes r} \mathcal{V}^T$.

(2) If $\omega = (1, 1, \ldots, 1, 0, \ldots, 0)$ for $r \leq n$ then $T^\omega(\ell : i) = \det[c_{\ell,i}]$. Recall from 1.5 the rth exterior power $\Lambda^r(E)$, with basis consisting of all $e_{[i]} = e_{i_1} \wedge \ldots \wedge e_{i_r}$, $[i] \in P_r$. Now the map $M(\lambda) \to \Lambda^r(E)$ given by $\det[c_{\ell,i}] \mapsto e_{[i]}$ induces an isomorphism $\Lambda^r(E) \cong M(1^r)$.

(3) Call $\lambda \in \Lambda^+(n, r)$ a *Steinberg partition* of degree d if $\lambda_i - \lambda_{i+1} \equiv -1$ (mod p^d). If λ is a Steinberg partition then $M(\lambda)$ is called the dth *Steinberg module*, and is a simple projective object in $\mathcal{P}_K(n, r)$ (because λ is minimal in its block: see Chapter 5).

Remark. A forerunner of Example (1) appears in the characteristic zero work of Akin [1988], [1989] and Zelevinskii [1987] referred to at the end of the last section. For this case they exhibit the Weyl module $V(\lambda)$ as the cokernel of a map into $S^\lambda(E)$, so that taking the contravariant dual, $M(\lambda)$ is exhibited as the kernel of a map from $S^\lambda(E)$.

We now collect together the main properties of the left Schur module in a portmanteau theorem. The reader is left to formulate the analogue for right modules. The original proofs are due to many people, of which Carter and Lusztig [1974], Green [1980], [1991a], [1991b] and James [1978a, §24] deserve special mention.

Theorem 3.4.1 (Properties of Schur modules) *Let* $\lambda \in \Lambda^+(n, r)$ *be given.*

(i) *The Schur module* $M(\lambda)$ *is the induced module*

$$H^0(\lambda) = \mathrm{Ind}_{B^-}^\Gamma(^- K_\lambda)$$

in the category $\mathcal{P}_K(n, r)$. *Here the* KB^--*module* $^- K_\lambda$ *affords the character* χ_λ *of* B^- *as in 3.2.4. We may identify* $M(\lambda)$ *with the dual Weyl module,* $D_{\lambda,K}$, *of Green [1980, §4].*

(ii) $M(\lambda)$ *has a unique minimal submodule, namely the left* S_r-*socle* $L(\lambda) = S_r \omega_\lambda$, *where* $\omega_\lambda = T^\lambda(\ell : \ell)$. *Moreover* $L(\lambda)$ *is an irreducible* $K\Gamma$-*module.*

(iii) *If* $\mu \in \Lambda(n, r)$ *is a weight of* $M(\lambda)$ *then* $\lambda \trianglerighteq \mu$. *The left* λ-*weight space* $M(\lambda)^\lambda$ *of* $M(\lambda)$ *is 1-dimensional, generated by* ω_λ. *Hence* $M(\lambda)$ *is a highest weight module (see the discussion after (1.8)), with highest weight* λ.

(iv) *If* K *has characteristic zero,* $M(\lambda) = L(\lambda)$ *is an ordinary irreducible* $K\Gamma$-*module.*

(v) ch $M(\lambda)(\mathbf{X}) = s_\lambda(\mathbf{X})$.

Proof (i) is observed in 3.2.6.

(ii) will be seen to follow from 3.4.3(iii) by duality. However, we give a direct proof here as well. We recall that U^- and U^+ are respectively the groups of lower and upper unitriangular subgroups of Γ. The maximal torus T normalises these unipotent groups and the Borel subgroups are $B^- = U^- T = TU^-$ and $B^+ = U^+ T = TU^+$. Write U for U^- or U^+ and let $M = M(\lambda)$. Now U is unipotent, so it has only one simple module, namely K; hence regarding M as a module for U, it is immediate that the U-socle of M (i.e. the space of U-invariants, M^U) is non-zero, since $M \neq 0$.

Now any $f \in (H^0(\lambda))^{U^+}$ satisfies $f(utw) = \chi_\lambda(t)^{-1} f(1)$ where $u \in U^+$, $w \in U^-$, $t \in T$ (from the definition of the Γ-module $H^0(\lambda)$). So $f(1)$ determines $f|_{U^+ B}$. Since $U^+ B$ is dense in Γ (Jantzen [1987, II.1.9]), $f(1)$ determines f. The upshot is that

$$\dim H^0(\lambda)^{U^+} \leq 1;$$

from the start of the proof of (ii) the dimension is exactly 1.

Assume, for contradiction, that we have two non-isomorphic simple submodules L_1, L_2 of $H^0(\lambda)$. Then their direct sum is a submodule of $H^0(\lambda)$. Hence $L_1^{U^+} \oplus L_2^{U^+} \subseteq H^0(\lambda)^{U^+}$ implying that the latter space has dimension at least two. This contradicts the previous paragraph.

For (iii), apply 2.4.4.

For (iv), (v): using (iii) and the basis (3.1) of $M(\lambda)$, the μ-weight space of $M(\lambda)$ has dimension equal to $K_{\lambda\mu}$, the (λ, μ)th Kostka number, (Macdonald [1979, pp. 56–57]). From 1.5.3, ch $M(\lambda)(\mathbf{X}) = \sum_{\mu \in \Lambda(n,r)} K_{\lambda\mu} \mathbf{X}^\mu$.

By Macdonald [1979, p. 42] the function on the right-hand side is precisely $s_\lambda(\mathbf{X})$. Now since s_λ is the character of an irreducible module, in characteristic zero $M(\lambda)$ is simple as a left $K\Gamma$-module. So $M(\lambda) = L(\lambda)$ is irreducible. This also follows from the semisimplicity of S_r over $\mathbf{C}$, using (ii). $\qquad\square$

Remark. The proof of (ii) can be modified to show that

$$H^0(\lambda)^{U^+} = H^0(\lambda)^\lambda.$$

By (iii) we must prove that any $f \in M(\lambda)^{U^+}$ is a multiple of ω_λ. The evaluation $f \mapsto f(1)$ is a monomorphism between the B^--modules

$H^0(\lambda)$ and $^-K_\lambda$, which gives one inclusion, and hence equality by dimensions.

The irreducible objects of $\mathcal{P}_K(n,r)$ revisited

Now 3.4.1(ii) asserts that the socle of a Schur module is simple. We have in fact obtained a classification of all simple polynomial $K\Gamma$-modules, that is, we have a complete set of simple $S_r(\Gamma)$-modules.

Theorem 3.4.2 $\{L(\lambda) = \operatorname{Soc} M(\lambda) : \lambda \in \Lambda^+(n,r)\}$ *is a complete set of inequivalent absolutely irreducible $K\Gamma$-modules in $\mathcal{P}_K$. Moreover the formal character* ch $L(\lambda)$ *has leading term* $\mathbf{X}^\lambda$.

Proof By Remark (1) in 1.6, we need only show that the formal character ch $L(\lambda)$ has leading term $\mathbf{X}^\lambda$ (where the monomials are arranged with respect to the dictionary order).

Visibly, if $\mu \in \Lambda(n,r)$ is such that $\mu \trianglelefteq \lambda$, then $\mathbf{X}^\lambda$ precedes $\mathbf{X}^\mu$ in the dictionary order of monomials. By 3.4.1(v) ch $M(\lambda)$ has leading term $\mathbf{X}^\lambda$. Since $L(\lambda)$ is a $K\Gamma$-submodule, it is also a module for the toral group algebra KT, and we have that $L(\lambda)^\lambda = M(\lambda)^\lambda$, since the former weight space contains ω_λ. $\qquad\square$

Weyl modules

We recall from 3.1.5 that the left Weyl module $V(\lambda)$ has K-basis

$$\{\xi_{i,\ell(\lambda)} + S(>\lambda) : i \in I^\lambda\}. \tag{3.5}$$

$V(\lambda)$ is therefore some left $K\Gamma$-module lying in $\mathcal{P}_K(n,r)$. By the natural duality between Schur modules and Weyl modules, we can virtually read off the basic properties of the Weyl modules. First let me digress to mention another alternative definition.

Green's Weyl module. Given $\lambda \in \Lambda^+(n,r)$, recall Example (1) at the start of this section. The canonical map $E^{\otimes r} \to S^\lambda E$ was seen to induce an S_r-epimorphism $\varphi : E^{\otimes r} \to M(\lambda)$. Thus there is an short exact sequence in $\mathcal{P}_K(n,r)$

$$0 \to N \to E^{\otimes r} \xrightarrow{\varphi} M(\lambda) \to 0$$

where $N = \operatorname{Ker}\varphi$. Green [1980, 5.1a] called the orthogonal complement to N (relative to the usual form on $E^{\otimes r}$) a Weyl module and denoted it by $V_{\lambda,K}$.

In [1980, p. 67], $V_{\lambda,K}$ is identified with the Carter and Lusztig module $\bar{V}_\lambda$ of Carter and Lusztig [1974, p. 211]; the latter module had been defined using the Kostant $\mathbf{Z}$-form $\mathcal{U}_\mathbf{Z}$, (see James and Kerber [1981, 8.2.10]), of the universal enveloping algebra $\mathcal{U}$ of the Lie algebra $\mathrm{gl}_n(\mathbf{Q})$. It turns out that one can construct explicit polynomial generators for the centre of $\mathcal{U}$. As I mentioned in Remark (2) at the end of 2.3, using the hyperalgebra $\mathcal{U}_K = \mathcal{U}_\mathbf{Z} \otimes K$, Carter and Lusztig proved that $\bar{V}_\lambda$ has many important properties as a Γ-module.

Now we observe that the basis $\{\xi_{i,\ell}e_\ell \mathcal{V}^T\}$ of Green's $V_{\lambda,K}$, derived in his book [1980, 5.3], is precisely that of (3.5). What we shall do in 3.4.3(i) is to sketch Green's identification of $V_{\lambda,K}$ with the original definition of Carter and Lusztig. This process requires the Carter-Lusztig Lemma, 2.4.4(v) and Carter and Lusztig [1974, p. 211].

Theorem 3.4.3 (Properties of Weyl Modules) *Let* $\lambda \in \Lambda^+(n,r)$ *be given.*

   (i) *The Weyl module* $V(\lambda)$ *is isomorphic to Green's Weyl module* $V_{\lambda,K}$; *it is also isomorphic to the Carter-Lusztig Weyl module* $\bar{V}_\lambda$ *and to the dual induced module* $H^0(-\pi_0\lambda)^*$ *of Jantzen. In particular,* $V(\lambda)$ *is a cyclic* S_r-*module generated by* $\bar{\xi}_\lambda$ *(or by* $f_\ell = e_\ell \mathcal{V}^T$ *for the Carter-Lusztig module).*

  (ii) $V(\lambda)$ *has a unique maximal submodule* $\mathrm{Rad}\, V(\lambda)$ *and hence has an irreducible quotient* $F(\lambda)$.

 (iii) $V(\lambda)$ *has highest weight* λ.

 (iv) *In characteristic zero* $V(\lambda) = F(\lambda)$ *is an ordinary irreducible* S_r-*module, and is isomorphic to* $M(\lambda)$.

  (v) $\mathrm{ch}(V(\lambda))(\mathbf{X}) = s_\lambda(\mathbf{X})$.

Proof For (i) let $N = \mathrm{Ker}\varphi$ be as in the preamble. Recalling the Garnir relations, 2.4.4(v), define the following subsets of tensor space:

$$
\begin{aligned}
R_1 &= \{e_i : i \in I \text{ and } i_{a(p,q)} = i_{a(s,q)} \text{ for } p \neq s \text{ and some } q\}; \\
R_2 &= \{e_i - (-1)^\sigma e_{i\sigma} : i \in I, \sigma \in C(T)\}; \\
R_3 &= \{ \sum_{\sigma \in G_{X,Y}} (-1)^\sigma e_{i\sigma} : i \in I, \emptyset \neq X \subseteq C_{t+1}(T) \\
&\qquad \text{for some } 1 \leq t < r\}; \\
R &= R_1 \cup R_2 \cup R_3.
\end{aligned}
$$

The first claim is that N is the K-span R' of R. Of course $R \subseteq N$ by 2.4.4, hence $R' \le N$.

Consider the mapping $\psi : F = E^{\otimes r}/R' \to M(\lambda)$ given by $\psi(e + R') = \varphi(e)$ for all $e \in E^{\otimes r}$. It is clear that we have a well-defined epimorphism. Actually ψ is injective. For, the map $f : I \to F$ such that $f(i) = e_i + R'$ satisfies the hypotheses of 2.4.4(v) and so its image F is $\langle e_i + R' : T_i$ standard$\rangle$. Now ψ maps this set onto the set (3.1), so $\mathrm{Ker}\,\varphi = 0$. Then $N \le R'$.

We conclude that $V_{\lambda,K}$ as defined by Green is the set of all tensors having inner product zero with elements of R_1, R_2 and R_3. Since the inner product is non-singular and G-invariant, i.e. $\langle x\pi, y \rangle = \langle x, y\pi^{-1} \rangle$ for all $x, y \in E^{\otimes r}$ and $\pi \in G$, we have that $V_{\lambda,K}$ is the set of all $x \in E^{\otimes r}$ such that

$$\langle x, e_i \rangle = 0 \quad \forall i \in I \text{ and } T_i \text{ such that } i_{a(p,q)} = i_{(s,q)}$$
$$\text{for } p \neq s \text{ and some } q; \tag{3.6}$$
$$x\sigma = (-1)^\sigma x \quad \forall \sigma \in C(T); \tag{3.7}$$
$$\sum_{\sigma \in G_{X,Y}} (-1)^\sigma x\sigma^{-1} = 0 \quad \forall \emptyset \neq X \subseteq C_{t+1}(T) \ (t < r). \tag{3.8}$$

If (3.6)–(3.8) are compared with equations (28) and (29) in Carter and Lusztig [1974] it will be noticed that we have derived the Carter-Lusztig conditions defining their $\bar{V}_\lambda$.

One can identify the Carter-Lusztig generator $\bar\Phi^\lambda$ with $f_\ell = e_\ell \mathcal{V}^T$ by checking that $\langle f_\ell, r \rangle = 0$ for all $r \in R$: this is done in Green [1980, 5.3a]. Finally, the map $\bar\xi_{i,\ell} \mapsto \xi_{i,\ell} e_\ell \mathcal{V}^T$ induces an S_r-isomorphism $f : V(\lambda) \to V_{\lambda,K}$. The identification with Jantzen's construction was done just after 3.2.6.

For (ii), we give a standard argument. If W is any proper submodule of $V(\lambda)$ then it does not contain $\bar\xi_\lambda$, since this is the generator. Thus $W^\lambda = W \cap V(\lambda)^\lambda = W \cap K\bar\xi_\lambda = \{0\}$. If X denotes the sum of all proper S_r-submodules of $V(\lambda)$, then

$$X^\lambda = \xi_\lambda \sum_{W < V(\lambda)} W = \sum_W \xi_\lambda W = \sum_W W^\lambda = \{0\}.$$

Hence X is a proper submodule, and is clearly unique and maximal.

(iii) is immediate from the fact that $V(\lambda) \cong (^\lambda M)^*$ and 3.4.1.

For (iv), the map $E^{\otimes r} \to M(\lambda)_{\mathbf{Q}}$ induces a non-zero map $V(\lambda)_{\mathbf{Q}} \to M(\lambda)_{\mathbf{Q}}$, which is easily seen to be an isomorphism; we then use 3.4.1.

(v) follows since dual modules have the same character. $\qquad\square$

Remarks.

(1) The basis of $V(\lambda)$ given by (3.5) (or that given by Carter and Lusztig) and the basis of $M(\lambda)$ given in the last chapter are not dual bases in the usual sense; rather they are connected by the unimodular matrix of Désarménien (by construction).

(2) Thus we obtain another complete set $\{F(\lambda) : \lambda \in \Lambda^+(n,r)\}$ of pairwise non-isomorphic irreducible S_r-modules. In 3.4.9 we shall see that $F(\lambda) \cong L(\lambda)$.

(3) In finite characteristic the Schur and Weyl modules are not necessarily isomorphic. An example will be given in 3.5.

(4) We also have that $V(\lambda) = \Delta(\lambda)$ is standard and $M(\lambda) = \nabla(\lambda)$ is costandard in the highest weight categories $_{S_r}\mathbf{mod} = \mathcal{P}_K(n,r)$.

Contravariant duality

By construction the left Schur modules are dual to the right Weyl modules. A procedure is desirable whereby the 'dual' of a left Schur module produces some other *left* module: we give such a general procedure and show in particular that left Schur modules become 'duals' of left Weyl modules under this new process.

We begin with a few general definitions. Let S be any finite-dimensional K-algebra possessing an involutory anti-automorphism $\circ : S \to S$. This means that $\circ$ is K-linear, $^\circ 1 = 1$, $^\circ(^\circ x) = x$ and $^\circ(xy) = {}^\circ y\, {}^\circ x$ for all $x, y \in S$. Given a subalgebra R, write R° for $^\circ(R)$. Let $V \in {}_R\mathbf{mod}$, and define the dual space $V^* = \mathrm{Hom}(V, K)$. This is a right R-module in the usual way. Instead, we consider the left R°-module with action

$$(g \cdot f)(v) = f(^\circ g v) \quad (f \in V^*,\ g \in {}^\circ R,\ v \in V).$$

Definition 3.4.4 For each $V \in {}_R\mathbf{mod}$, the left R°-module V^* with the above action is called the *contravariant dual* of V (modulo $\circ$). We denote by V° the space V^* with this left R°-action. We say that V is *self-dual* if $R = R^\circ$ and $V \cong V^\circ$. Given $V \in {}_R\mathbf{mod}$, and $W \in {}_{R^\circ}\mathbf{mod}$, a K-bilinear form $(,) : W \times V \to K$ is called *contravariant* if

$$(gw, v) = (w, {}^\circ g v) \quad (\forall g \in R^\circ, w \in W, v \in N).$$

A standard theorem in this area is

Proposition 3.4.5 *There is a one-to-one correspondence between contravariant forms* $(\ ,\ ) : W \times V \to K$ *and morphisms* $\gamma : W \to V^\circ$ *in* $_{R^\circ}\mathbf{mod}$ *given by*

$$\gamma(w)(v) = (w, v) \quad (v \in V,\ w \in W).$$

The form is non-singular if and only if γ *is an isomorphism.* γ *is an* R°*-map if and only if* $(\ ,\ )$ *is contravariant.*

Contravariant duality for S_r

We take $R = S = S_r(\Gamma)$, and define $J : S \to S$ by sending $\xi_{i,j}$ to $\xi_{j,i}$ for $(i, j) \in \Omega$. It is immediate that J is an involutory anti-automorphism of S such that $J(T_{n,r}(g)) = T_{n,r}(g^{\mathrm{tr}})$ for all $g \in \Gamma$. If V is regarded as a left $K\Gamma$-module via (2.6), then V° is a left Γ-module with action:

$$(g \cdot f)(v) = f(g^{\mathrm{tr}}v) \quad (\forall g \in \Gamma,\ f \in V^\circ,\ v \in V).$$

Remark. If $V \in \mathcal{P}_K$, and one tries to turn V^* into a left Γ-module by the standard trick: $(g \cdot f)(v) = f(g^{-1}v)$, then it is not necessarily true that $V^* \in \mathcal{P}_K$.

Basic linear algebra yields:

Lemma 3.4.6 (Contravariant duality) *Take* $V, W \in {}_S\mathbf{mod}$.

  (i) $V \cong W^\circ$ *if and only if there exists a contravariant pairing* $(\ ,\ ) :$ $V \times W \to K$.

  (ii) *In this case there is an inclusion-reversing bijective correspondence between the sets* $\mathrm{Sub}(V)$ *and* $\mathrm{Sub}(W)$ *of all S-submodules of V and W as follows: given* $V_1 \in \mathrm{Sub}(V)$, *send it to its 'perp space'* $V_1^\perp = \{w \in W : (v_1, w) = 0\ \forall v_1 \in V_1\}$. *Conversely, given* $W_1 \in \mathrm{Sub}(W)$, *let it correspond to* $W_1^\perp = \{v \in V : (v, w_1)\ \forall w_1 \in W\}$.

  (iii) W *irreducible* $\Rightarrow W^\circ$ *irreducible.*

  (iv) *If* $\lambda \in \Lambda$ *then the λ-weight spaces of V and V° are isomorphic as K-spaces (see Green [1980, p. 40]).*

Of course there are comparable statements about contravariant objects in $\mathbf{mod}_S$.

Exercises.
(1) Show that $E^{\otimes r}$ is self-dual with respect to the form $\langle \, , \, \rangle : E^{\otimes r} \times E^{\otimes r} \to K$ defined by $\langle e_i, e_j \rangle = \delta_{i,j}$ for all $i, j \in I$.
(2) Show that $(S^r(E))^\circ$ is the space of symmetric tensors in $E^{\otimes r}$.
(3) By considering the map $e_{[i]} \mapsto (e_{i_1} \otimes \cdots \otimes e_{i_r})(\sum (-1)^\pi \pi))$ show that $(\Lambda^r(E))^\circ$ is the space of antisymmetric tensors.

The main result is a further observation about the pairing between A_r and S_r defined in 2.2.4.

Theorem 3.4.7 *The form $\langle \, , \, \rangle$, 2.2.4, is contravariant on the left and on the right with respect to J. Hence the map*

$$A_r \to S_r^\circ$$

defined by this form (using 3.4.5) is an (S, S)-bimodule isomorphism.

Proof Check that

$$\langle \xi' \circ c, \xi \rangle = \langle c, J(\xi')\xi \rangle = \langle c \circ J(\xi), J(\xi') \rangle.$$

for all $\xi, \xi' \in S_r$ and $c \in A_r$ (by bilinearity, we may assume $c = c_{i,j}$ and $\xi = \xi_{p,q}$). $\qquad\square$

Theorem 3.4.8 *Let $\lambda \in \Lambda^+(n, r)$. Then*

$$V(\lambda) = (M(\lambda))^\circ.$$

Proof If $\lambda \in \Lambda^+(n, r)$ then $(^-K_\lambda)^\circ \cong {}^+K_\lambda$, as S^+-modules. To see this, take $f \in (^-K_\lambda)^\circ$ and $\beta \in {}^-K_\lambda$. For $i \leq j$ we have $(\xi_{i,j}f)(\beta) = f(\xi_{j,i}\beta) = f(\beta)$ if $\xi_{i,j} = \xi_\lambda$, otherwise it is zero. We have proved that $(^-K_\lambda)^\circ$ affords χ'_λ and so $(^-K_\lambda)^\circ \cong {}^+K_\lambda$, as S^+-modules. Hence

$$V(\lambda) = S_r(\Gamma) \otimes_{S^+} {}^+K_\lambda \cong (\text{Hom}_{S^-}(S_r, {}^-K_\lambda))^\circ = M(\lambda)^\circ$$

by the adjoint isomorphism theorem (cf. 3.2.5). $\qquad\square$

Theorem 3.4.9 *If $L \in \mathcal{P}_K$ is irreducible, then L is isomorphic to a submodule of the left S_r-module A_r; moreover L is self-dual in the sense of 3.4.4.*

Proof Considering L as an S-module, L° is also irreducible by Lemma 3.4.6(iii). Take a non-zero $m^\circ \in L^\circ$ and define a map $\psi : S \to L^\circ$ by

$\xi \mapsto \xi m^\circ$ for all $\xi \in S$. Since this is an S-epimorphism, the dual map $\psi^\circ : (L^\circ)^\circ \to S^\circ \cong A$ is an S-monomorphism. But clearly $(L^\circ)^\circ \cong L$.

For the second part, suppose $L = L(\lambda)$. Again L° is irreducible so $L^\circ \cong L(\mu)$ for some $\mu \in \Lambda^+(n, r)$. Since $L(\lambda)$ and $L(\lambda)^\circ$ have the same formal character, by 3.4.6(iv), ch $L(\lambda)$=ch $L(\mu)$, whence $\lambda = \mu$. $\square$

Jantzen's contravariant forms

The discussion of Weyl modules illustrated the problem of identifying maximal submodules as in 3.4.3(ii). It is worth briefly explaining an idea of Wong [1971], [1972] and Jantzen [1973], [1980] that allows one to identify maximal submodules of Weyl modules quite easily. The details are skimped; see Green [1980, 5.5] for more.

The approach depends on which of the Weyl modules in 3.4.3(i) you start with! For definiteness, we shall start with the S_r-generator $f_\ell \in E^{\otimes r}$ of the Carter-Lusztig Weyl module (remember that $f_\ell = e_\ell \mathcal{V}^T$). One can define a contravariant form on $V = V_{\lambda, K}$ by

$$\langle\langle \xi f_\ell, \eta f_\ell \rangle\rangle = \langle \xi e_\ell, \eta f_\ell \rangle$$

for all $\xi, \eta \in S_r$. Here $\langle \, , \, \rangle$ is the contravariant form on $E^{\otimes r}$ which you considered in Exercise (1) after 3.4.6. Check that this endows $V_{\lambda, K}$ with a symmetric contravariant non-degenerate form. Choosing $\xi = \xi_{i, \ell}$ and $\eta = \xi_{j, \ell}$ yields

$$\langle\langle \xi_{i, \ell} f_\ell, \xi_{j, \ell} f_\ell \rangle\rangle = \langle \xi_{i, \ell} e_\ell, \xi_{j, \ell} f_\ell \rangle = \langle \xi_{\ell, j} \xi_{i, \ell} e_\ell, f_\ell \rangle \qquad (3.9)$$

Now we recall the action of S_r on tensor space, studied just after 2.2.6. If $\xi \in S_r$ and $i, j \in I$, we may write

$$\langle \xi e_j, e_i \mathcal{V}^T \rangle = \sum_{\pi \in C(T)} (-1)^\pi \langle \xi e_j, e_{i\pi} \rangle = \sum_{\pi \in C(T)} (-1)^\pi \langle c_{i\pi, j}, \xi \rangle.$$

Thus $\langle \xi e_j, e_i \mathcal{V}^T \rangle = \langle T^\lambda(i : j), \xi \rangle$ for $\lambda \in \Lambda^+(n, r)$. We put this into (3.9) to obtain

$$\langle\langle \xi_{i, \ell} f_\ell, \xi_{j, \ell} f_\ell \rangle\rangle = \langle T^\lambda(\ell : \ell), \xi_{\ell, j} \xi_{i, \ell} \rangle.$$

This allows one to shift from a form on the Carter-Lusztig module to our Weyl module $V(\lambda) = S\bar{\xi}_\lambda$: under the isomorphism f appearing at the end of the proof of 3.4.3(i), we obtain a contravariant form on $V(\lambda)$ by

$$(\bar{\xi}_{i,\ell}, \bar{\xi}_{j,\ell}) = \langle T^\lambda(\ell : \ell), \xi_{\ell,j}\xi_{i,\ell}\rangle$$

for all $i, j \in I$. The main result may now be stated.

Proposition 3.4.10 *The radical* $J = \{v \in V(\lambda) : \langle\langle v, V(\lambda)\rangle\rangle = 0\}$ *of* $\langle\langle\,,\,\rangle\rangle$ *coincides with the unique maximal submodule of* $V(\lambda)$.

Proof J is a proper submodule, since it does not contain f_ℓ, and clearly it has the contravariant property. Thus $J \subseteq V(\lambda)^{\max}$. But $V(\lambda)^{\max} \subseteq \bigoplus_{\alpha \neq \lambda} V(\lambda)^\alpha = V'$. Since V' is orthogonal to $V(\lambda)^\lambda = Kf_\ell$, we have

$$\langle\langle V(\lambda)^{\max}, S_r f_\ell\rangle\rangle \subseteq \langle\langle S_r V(\lambda)^{\max}, f_\ell\rangle\rangle \subseteq \langle\langle V', f_\ell\rangle\rangle = 0.$$

Thus $V(\lambda)^{\max} \subseteq J$. $\qquad\qquad\qquad\qquad\qquad\qquad\qquad\qquad\square$

Let us illustrate our results by revealing a telling example.

Examples.
(1) The reader was invited, in Exercise (2) after 3.4.6, to show that $V(r) = M(r)^\circ$ was $(\sum_{\pi \in G} \pi)E^{\otimes r}$, the space of symmetric tensors. Restriction of the canonical form on $E^{\otimes r}$ to this space gives us the form $\langle\langle\,,\,\rangle\rangle$, since $C(T^{(r)}) = \{1\}$. So the contracted form on $V(r)$ is the restriction of the canonical form on $E^{\otimes r}$. Our space has basis $\{v_\alpha\}_{\alpha \in \Lambda(n,r)}$, where $v_\alpha = \sum_{i \in \alpha} e_i$, and explicitly the form is

$$\langle v_\alpha, v_\beta\rangle = \begin{cases} 0 & \alpha \neq \beta \\ (r/\alpha)1_K & \alpha = \beta \end{cases}$$

where $(r/\alpha) = r!/\alpha_1! \ldots \alpha_n!$. The radical of this form will consist of all v_α such that $p|(r/\alpha)$. By 3.4.10 the radical J coincides with the unique maximal ideal, hence the simple quotient $F(r) = V(r)/V(r)^{\max}$ has basis $\{v_\alpha + J : \alpha \in \Lambda(n,r),\ (r/\alpha) \not\equiv 0 \pmod{p}\}$. Thus the character is

$$\mathrm{ch}_{r,K}(F(r)) = \sum_{\substack{\alpha \in \Lambda(n,r) \\ (r/\alpha) \not\equiv 0 \pmod{p}}} X_1^{\alpha_1} \ldots X_n^{\alpha_n}.$$

(2) Here is another way to identify the maximal submodule. Check that the map $V(\lambda) \to M(\lambda)$ induced by $\bar{\xi}_\lambda \mapsto T^\lambda(\ell : \ell)$ is well defined. This

produces an epimorphism onto $L(\lambda)$ and hence the kernel of the map is maximal. Then 3.4.3(ii) completes the identification.

3.5 Modular representation theory for Schur algebras

It was pointed out in the introduction that a central theme of the text was to understand the problem of computing decomposition numbers for Γ. Let us prepare the way for this by reviewing from Curtis and Reiner [1981, §16] how to arrive at a version of Brauer's theory of modular reduction that is tailored to the algebraic group Γ. After some general theory, we discuss the important result of Carter and Lusztig that Schur modules have **Z**-forms, as do their contravariant duals, the Weyl modules.

General theory of R-forms

Let R be a complete rank one discrete valuation ring (d.v.r.) with maximal ideal $\mathcal{M}$ and field of fractions Q of characteristic zero. By Curtis and Reiner [1981, 6.17] the assumptions on R guarantee that the Krull-Schmidt theorem holds for all algebras under discussion. Let π be some ring homomorphism $\pi : R \to K$.

Let S_Q be a Q-algebra with finite basis $\{\xi_i^Q : i \in \mathbf{b}\}$ (we have an additional index to indicate the dependence on Q). Consider the set S_R that is R-generated by all ξ_i^Q. If this set is multiplicatively closed we call S_R an R-order in S_Q. Let $S_K = S_R \otimes_R K$, where one thinks of K as an R-module under π (i.e. $rk \cdot = \pi(r)k$ for $r \in R$, $k \in K$). Then S_K is the K-algebra with K-basis $\{\xi_i^Q \otimes_R 1_K : i \in \mathbf{b}\}$.

At the level of objects in $_{S_Q}\mathbf{mod}$ suppose that $V_Q \in {}_{S_Q}\mathbf{mod}$ with basis $\{e_b^Q : b \in B\}$. Then one can define an R-form or admissible lattice V_R for V_Q as follows: take the R-span of some Q-basis of V_Q and assume that $S_R V_R \subseteq V_R$. It is a fact that every $V_Q \in {}_{S_Q}\mathbf{mod}$ determines some R-form, provided R is a principal ideal domain (e.g. see Green [1976, p. 159]). If we have a family $\{V_K\}_{K \in \mathcal{K}}$ of modules in S_K, where $\mathcal{K}$ is a collection of infinite fields, we say that it is defined over R or R-defined if there is a R-form V_R for V_Q and, for each K, an isomorphism $\delta_K : V_R \otimes_R K \to V_K$ (in $_{S_K}\mathbf{mod}$).

One can regard $V_K = V_R \otimes_R K$ as a left $S_K = S_R \otimes K$-module as follows: start with the matrix of coefficients $T_Q = [\alpha_{ab}^Q]$ with respect to the basis $\{e_b^Q\}$ of V_Q. Then the matrix T_K relative to the basis $\{e_b^Q \otimes 1\}$ of V_Q is just $T_K = [\alpha_{ab}^Q \otimes 1]$.

Examples.

(1) In the special case where $Q = \mathbf{Q}$, $R = \mathbf{Z}$ and $K = \mathbf{Z}_p$ and where $\pi : \mathbf{Z} \to K$ is defined by $\pi(n) = n1_K$ $(n \in \mathbf{Z})$, this process amounts to reducing the entries of $T_{\mathbf{Q}}$ modulo p.

Terminology. The procedure by which V_Q is converted via V_R into the S_K-module V_K is called *modular reduction* (it is well known that this procedure can be formalised by defining the decomposition map $d : G_0(S_Q) \to G_0(S_K)$ between respective Grothendieck groups). A given object $V_Q \in {}_{S_Q}\mathbf{mod}$ may have many non-isomorphic R-forms. It is, however, most important to remember that for any simple module $L \in {}_{S_K}\mathbf{mod}$, the composition multiplicity $[V_K : L]$ of L as a factor of V_K depends only on V_K and not on the particular R-form chosen.

Let $L = L(\mu) \in {}_{S_K}\mathbf{mod}$ be simple. If $M_{\mathbf{Q}}(\lambda)$ is an ordinary irreducible module, let $\overline{M(\lambda)}$ be its modular reduction. Then $[\overline{M(\lambda)} : L(\mu)] = d_{\lambda\mu}$ is called the *decomposition number* for the modular reduction ${}_{S_Q}\mathbf{mod} \to {}_{S_K}\mathbf{mod}$.

Now we specialise: let $S_Q = S_{\mathbf{Q}}(n, r)$ and define $S_{\mathbf{Z}}$ to be the $\mathbf{Z}$-submodule generated by all $\xi_{i,j}^{\mathbf{Q}}$. By the Product Rule this set is multiplicatively closed and $S_{\mathbf{Z}}$ is a $\mathbf{Z}$-order. Clearly, if $K \in \mathcal{K}$ there is an isomorphism of K-algebras $S_{\mathbf{Z}} \otimes K \cong S_K$ given by $\xi_{i,j}^{\mathbf{Q}} \otimes 1_K \mapsto \xi_{i,j}^{K}$. We have therefore defined a modular reduction $\mathcal{P}_{\mathbf{Q}}(n, r) \to \mathcal{P}_K(n, r)$. Our aim in what follows is to show that the modular reduction of $M_{\mathbf{Q}}(\lambda)$ is itself a Schur module defined over K.

(2) Take $V_K = E^{\otimes r}$; a suitable $\mathbf{Z}$-form of V_Q is $V_{\mathbf{Z}} = \sum_{i \in I} \mathbf{Z}e_i^{\mathbf{Q}}$. If

$$\begin{aligned} \delta_K : V_{\mathbf{Z}} \otimes K &\to V_K \\ e_i^{\mathbf{Q}} \otimes 1_K &\mapsto e_i^K \end{aligned}$$

for all $i \in I$ then it is clear that we have an isomorphism in $\mathcal{P}_K$. Hence, $\{E^{\otimes r}\}_K$ is defined over $\mathbf{Z}$.

(3) Suppose that $\{V_K\}$ is a family of modules in $\mathcal{P}_K$, that is $\mathbf{Z}$-defined by integral forms $V_{\mathbf{Z}}$ and morphisms $\{\delta_K\}$. Let $\{e_b^{\mathbf{Q}}\}$ be a (finite) basis of V_Q that generates the integral form $V_{\mathbf{Z}}$; then by assumption the $e_b^K = \delta_K(e_b^{\mathbf{Q}} \otimes 1_K)$ form a basis of V_K.

Let $\{f_b^K\}$ be the basis of $(V^{\circ})_K$ dual to $\{e_b^K\}$. A suitable $\mathbf{Z}$-form for $(V^{\circ})_{\mathbf{Q}}$ is $(V^{\circ})_{\mathbf{Z}} = \{f \in V_{\mathbf{Q}} : f(V_{\mathbf{Z}}) \subseteq \mathbf{Z}\}$, having basis $\{f_b^{\mathbf{Q}}\}$. Then

$(V^\circ)_K$ is **Z**-defined by these integral forms and by the maps

$$\delta'_K : (V^0)_{\mathbf{Z}} \otimes K \;\;\to\;\; (V^0)_K$$
$$e_b^{\mathbf{Q}} \otimes 1_K \;\;\mapsto\;\; e_b^K.$$

Integral forms for Schur modules and Weyl modules

We wish to apply the above general theory to the case of Schur and Weyl modules. So we fix $\lambda \in \Lambda^+(n, r)$. The ordinary irreducible Schur module $M(\lambda)_{\mathbf{Q}}$ has basis (3.1) over the rationals and, moreover, $M(\lambda)_{\mathbf{Q}}$ affords a matrix representation of $\Gamma_{\mathbf{Q}}$ of the form

$$T_\lambda : \Gamma_{\mathbf{Q}} \;\;\to\;\; \mathrm{Mat}_m(\mathbf{Q})$$
$$g \;\;\mapsto\;\; T_{\mathbf{Q}} = [\alpha_{j,i}(g)]$$

where the coefficients $\alpha_{j,i}$ are given by

$$g \circ T^\lambda(\ell : i) = \sum_{j \in I^\lambda} \alpha_{j,i}(g) T(\ell : j) \qquad (3.10)$$

for $g \in \Gamma_{\mathbf{Q}}$ and $i \in I^\lambda$. Somehow, we have to 'invert' this expression in order to obtain the αs: we employ the Straightening Formula to do this. From (2.16) we have a left action

$$g \circ T^\lambda(\ell : i) = \sum_{t \in I} g_{t,i} T^\lambda(\ell : t). \qquad (3.11)$$

By Corollary 2.5.8(ii) we can straighten each bideterminant on the right-hand side of (3.11) to obtain

$$T^\lambda(\ell : t) = \sum_{j \in I^\lambda} w'_{j,t} T^\lambda(\ell : j) + \gamma \qquad (3.12)$$

where the w' are integers and $\gamma \in \sum_{\mu < \lambda} \sum_{i,j \in I} \mathbf{Z} T^\mu(i : j)$. By (2.19) it is enough to take this latter sum over all $i \in I^\lambda$ and $j \in I^\mu$. But we recall that the right weight space of $T^\mu(i : j)$ equals the weight of i (Section 2.4, Example (4)). If $\mu < \lambda$, the Kostka number $K_{\mu\lambda}$ is zero (proof of 3.4.1(iii), observing that $\lambda \trianglelefteq \mu$), and so no *standard* bideterminant, $T^\mu(i : j)$, can have right weight λ. However, all terms in $T^\lambda(\ell : t)$ and $T^\lambda(\ell : j)$, (3.12), have right weight λ, forcing $\gamma = 0$.

By combining (3.11), (3.12) and comparing with (3.10) we are led to

Lemma 3.5.1 $\alpha_{j,i}(g) = \sum_{t \in I} w'_{j,t} g_{t,i}$ *for all* $i, j \in I^\lambda$ *and* $g \in \Gamma_{\mathbf{Q}}$; *hence*

$$\alpha_{j,i} = \sum_{t \in I} w'_{j,t} c_{t,i},$$

for all $i, j \in I^\lambda$.

The main point is that T_λ is a **Z**-form since the $w'_{j,t}$ are integers that are *independent of K and j*, and hence we may interpret T_λ as a representation of Γ_K after reducing the matrix coefficients modulo p.

Theorem 3.5.2 *The family* $\{M(\lambda)_K\}$ *is defined over* **Z**. *By Example (3), so too are the Weyl modules* $\{V(\lambda)_K\}$.

Proof   The **Z**-form of $M_{\mathbf{Q}}(\lambda)$ is obtained by taking the **Z**-span of the **Q**-basis elements $T^\lambda(\ell : i)$ $(i \in I)$. The maps $\delta_K : M_{\mathbf{Z}}(\lambda) \otimes K \rightarrow M_K(\lambda)$ that take $T^\lambda(\ell : i)_{\mathbf{Q}} \otimes 1_K \mapsto T^\lambda(\ell : i)_K$ are evidently $S_K(n,r)$-isomorphisms.   □

Remark. G. Higman [1967] was the first to observe that the bidetermi-nants generate a **Z**-form for the Schur module $M_{\mathbf{Q}}(\lambda)$. Explicit **Z**-forms for $V(\lambda)_{\mathbf{Q}}$ (hence also for $M(\lambda)_{\mathbf{Q}}$) are described in Green [1980, 5.6].

$S_K(2,2)$, *revisited*

We return to the examples considered in 2.5 and 1.2. In the example in 2.5 there occurred the module $M(2,0)$ with basis $B = \{c_{11}^2, c_{11}c_{12}, c_{12}^2\}$. This afforded the matrix representation $T_{(2,0)}$ in the example in 1.2. Observe that the representation $T'_{(2,0)}$ in 1.2 is defined as $T'_{(2,0)}(A) = T_{(2,0)}(A^{\mathrm{tr}})^{\mathrm{tr}}$. This is afforded by the Weyl module $V(2,0) = M(2,0)^\circ$, using the basis dual to that used to produce $T_{(2,0)}$.

If char $K \neq 2$, then $S_K(2,2)$ is semisimple by 2.2.8, so $M(2,0) = L(2,0)$ is irreducible and $V(2,0) = M(2,0)^\circ = L(2,0)^\circ \cong L(2,0)$ in this case.

If char $K = 2$ we need to compute $L(2,0) = S_r \omega_\lambda$. But $\ell = (1,1) \in I(2,2)$ and $\omega_\lambda = c_{11}^2$. From the basis B above, $L(2,0)$ has basis $\{c_{11}^2, c_{12}^2\}$. Relative to B, $M(2,0)$ affords the reduced matrix representation

$$A \mapsto \begin{bmatrix} a_{11}^2 & a_{12}^2 & a_{11}a_{12} \\ a_{21}^2 & a_{22}^2 & a_{21}a_{22} \\ 0 & 0 & a_{11}a_{22} + a_{21}a_{12} \end{bmatrix}.$$

Inspecting this matrix, one sees $L(2,0)$ in the top-left 2×2 submatrix.

Also note at the bottom right the determinant module $M_{(2,1)} = L_{(2,1)}$:
$A \mapsto a_{11}a_{22} + a_{21}a_{12} = a_{11}a_{22} - a_{21}a_{12} = \det A$.

Exercise.   Check, by taking transposes, that $M(2,0) \ncong V(2,0)$ if char $K = 2$.

4

Schur functors and the symmetric group

The equivalence of categories established in 2.2.7 provides a strong link between the representation theory of the Schur algebra and the polynomial representations of GL_n. However, there is a third player in the game, namely the symmetric group, whose rôle in this chapter is fundamental.

In Theorem 4.1.1 the symmetric group $G = \Sigma_r$ enters our considerations as follows. If $r \leq n$ there is a functor from $\mathcal{P}_K(n, r)$ to $_{KG}\mathbf{mod}$ induced by sending a polynomial module to the weight space consisting of vectors on which the element $d(t) \in T$ of (1.8), acts as left multiplication by $t_1 \ldots t_r$. This weight space may be viewed as a module for G using 1.5.2(ii). In fact this procedure is a manifestation of a quite general principle; section 4.1 contains a summary of it. To give a flavour of this, suppose we have a K-algebra S and an idempotent $e \in S$; we therefore induce a functor $\mathrm{Hom}_S(Se, -) : {}_S\mathbf{mod} \to {}_{eSe}\mathbf{mod}$ by sending V to eV. In the special case where $S = S_r$ and $r \leq n$ there exists a particular e inducing an isomorphism $eSe \cong KG$. In this incarnation the Hom functor is known as the *Schur functor*, and it has a myriad uses. For example Green [1980, §6] showed how to construct the representations of G from a knowledge of the representations of $S_r(\Gamma)$. This reverses a procedure invented by Schur [1901, III, IV], for whom, working in characteristic zero, this functor was actually a category equivalence $R_0 \leftrightarrow R_1$ (see 1.2); he was using Frobenius' work on $\mathbf{C}G$ to obtain results about modules for the complex Schur algebra.

In the remaining sections we study, for suitable r, n, certain permutation modules for G, now called Young modules. These are the indecomposable direct summands of $E^{\otimes r}$ and hence the indecomposable direct summands of the permutation modules M^λ for $\lambda \in \Lambda^+(n, r)$. Using the fact that over a field or complete valuation ring every finitely-generated projective S_r-module is in $\mathcal{F}(\Delta)$ (a result guaranteed by the quasi-hereditariness of S_r), and applying the Schur functor, we prove the existence of a filtration of Young modules whose quotients are Specht modules with well-defined multiplicities. In fact, we shall see that knowledge of decomposition numbers for the Schur algebra S_r is the same as knowledge of Weyl module decomposition numbers for $\mathrm{GL}_n(K)$; this in turn is equivalent to knowing the Σ_n-Specht module multiplicities of the Young modules in the above filtration.

4.1 The Schur functor

As mentioned above, this whole chapter is centred on the use of a certain functor, now called the Schur functor because it generalises a functor originally used by Schur to prove the equivalence of the categories $\mathcal{P}_\mathbf{C}(n,r)$ and $_{\mathbf{C}\Sigma_r}\mathbf{mod}$. Let us describe the main idea. As usual, fix integers n and r, and an infinite field K. Assume initially that $r \leq n$ (we sometimes say that n is chosen *large enough*). Let $\omega = (1,\ldots,1,0,\ldots,0) \in \Lambda^+(n,r)$ be the weight with r 1s and length n. We shall call this the *Schur weight*. Put another way, ω is the rth fundamental dominant weight $\varepsilon_1 + \cdots + \varepsilon_r$. Then $u = (1,2,\ldots,r) \in I$ has weight ω. The index u has the property that it is not fixed by any non-trivial permutation in G.

Let $V \in \mathcal{P}_K(n,r)$, and consider V as an object of $_S\mathbf{mod}$, where S is the Schur algebra $S_r(\Gamma)$. Then by 2.2.10 the ω-weight space V^ω, on which $d(t_1,\ldots,t_n)$ acts by the homothety $t_1 \ldots t_r$, equals $\xi_\omega V$, where we recall that $\xi_\omega = \xi_{u,u}$ is the element of S_r satisfying $\xi_\omega(\xi_{p,q}) = 0$ or 1, the latter provided $p = q \sim u$. Hence V^ω can be considered as a left $S(\omega)$-module, where $S(\omega) = \xi_\omega S \xi_\omega$. Now $S(\omega)$ is the span of the elements $\xi_\omega \xi_{i,j} \xi_\omega$ for $i,j \in I$, and by (2.8) $\xi_\omega \xi_{i,j} \xi_\omega = \xi_{i,j}$ provided $i,j \in \omega$; otherwise it is 0. In the case $i,j \in \omega$, there exist $\pi_1,\pi_2 \in G$ such that $i = u\pi_1$ and $j = u\pi_2$; hence $\xi_{i,j} = \xi_{u\pi,u}$ with $\pi = \pi_1\pi_2^{-1}$. But then $\xi_{u\pi,u} = \xi_{u\pi',u}$ if and only if $\pi = \pi'$; also $\xi_{u\pi,u}\xi_{u\pi',u} = \xi_{u\pi\pi',u}$ for all $\pi,\pi' \in G$. This proves the following.

Theorem 4.1.1 *The map*

$$K\Sigma_r \quad \rightarrow \quad S(\omega)$$

$$\sum_{\pi \in G} \alpha_\pi \pi \quad \mapsto \quad \sum \alpha_\pi \xi_{u\pi,u}$$

gives an isomorphism of K-algebras.

So, writing $e = \xi_\omega$ we have a K-algebra isomorphism

$$KG \cong eSe \qquad (4.1)$$

that takes π to $\xi_{u\pi,u}$ for all $\pi \in G$. We can, and will, view V^ω as a left KG-module under this isomorphism.

Definition 4.1.2 Let $f = f_e : \mathcal{P}_K \to {}_{KG}\mathbf{mod}$ be the functor sending V to $eV = V^\omega$ (viewed as a left KG-module), and sending $\theta : V \to W$ to its restriction $\theta^\omega : V^\omega \to W^\omega$. Call f the *Schur functor*.

Remarks.
(1) The choice of weight ω is important for later applications; however, starting with an arbitrary weight $\lambda \in \Lambda(n, r)$ for arbitrary n and r, we may similarly obtain an algebra $S(\lambda) = \xi_\lambda S \xi_\lambda$, which is the span of $\xi_{u'\pi,u'}$, where $u' = (1, \ldots, 1, 2 \ldots, 2, \ldots, n, \ldots, n)$ is composed of λ_1 1s, λ_2 2s etc. and π runs over a set of double coset representatives of Σ_λ. For this see Green [1980, p. 82]. We then obtain a functor $f_\lambda : \mathcal{P}_K \to {}_{S(\lambda)}\mathbf{mod}$ that sends a module to its λ-weight space. In fact, by Iwahori [1964, p. 218], $S(\lambda)$ is isomorphic to the *Hecke algebra* $\mathcal{H}_K(G, \Sigma_\lambda)$. This is defined by

$$\mathcal{H} = \mathcal{H}(G, \Sigma_\lambda) = \mathrm{End}\left(\sum_{\sigma \in \Sigma_\lambda} \sigma \, \mathbf{Z}G\right),$$

viewing the subset $(\sum_{\sigma \in \Sigma_\lambda} \sigma)\mathbf{Z}G$ of $\mathbf{Z}G$ as a right $\mathbf{Z}G$-module (Curtis and Reiner [1981, §11D]). The Hecke algebra is free, having free $\mathbf{Z}$-basis consisting of the maps

$$\chi_A : \sum_{\pi \in \Sigma_\lambda} \pi \mapsto \sum_{\pi \in A} \pi,$$

where A runs over the set $\Sigma_\lambda \backslash G / \Sigma_\lambda$ of all double-cosets of Σ_λ in G. We obtain $\mathcal{H}_K(G, \Sigma_\lambda)$ by tensoring $\mathcal{H}$ over $\mathbf{Z}$ with K. It may be checked that the map $\xi_{u'\pi,u'} \mapsto \chi_{\Sigma_\lambda \pi \Sigma_\lambda}$ $(\pi \in G)$ induces an isomorphism of K-algebras $S(\lambda) \cong \mathcal{H}_K(G, \Sigma_\lambda)$.

(2) These ideas may be extended when we work over the complete, local principal ideal domain R. If Γ_R is the general linear group scheme over R (see A.8), and V is a Γ_R-module that is finitely-generated and R-torsion free, then $f_R(V) = V^\omega$ is defined in a completely analogous way, and

gives a functor between $\Gamma_R\mathbf{mod}$ and $_{RG}\mathbf{mod}$. The construction given in the first remark yields a Hecke algebra description of $\mathrm{End}_{RG}(M^\lambda)$, where $M^\lambda = \mathrm{Ind}K_{\Sigma_\lambda}\!\uparrow^G$, and will prove vital in the quantum context discussed later.

(3) Schur [1901, p. 61] dealt with the case $r > n$ by finding a functor $\mathcal{P}_K(r,r) \to \mathcal{P}_K(n,r)$, that takes irreducibles $F(\lambda)$ to irreducibles or 0; moreover the non-trivial case occurs if and only if $\lambda \in \Lambda^+(n,r)$. Thus he was able to classify the irreducibles in all cases. We deal with this in 4.3.

General theory of Hom functors

It seems easier to state results about Schur functors in the context of a general finite-dimensional K-algebra S rather than for the particular case of Schur algebras. We outline this theory, and refer the reader to Green [1980], where more detailed proofs may be found.

Let $e \in S$ be a non-identity, non-zero idempotent. Define a covariant Hom functor $f : _S\mathbf{mod} \to _{eSe}\mathbf{mod}$ by

$$f : M \mapsto eM \cong \mathrm{Hom}_S(Se, M)$$

where $M \in _S\mathbf{mod}$, and if $\alpha : M \to M'$ is a morphism, let $f(\alpha) : eM \to eM'$ be the restriction of α to eM. We sometimes write S^e for eSe. Actually, functors of the form $\mathrm{Hom}_S(P,-)$ where P is a projective S-module, are not particularly new; they appear explicitly in work of Auslander [1974] from the early seventies. Refinements of this (to $\mathrm{SL}_n(K)$ in particular) are considered by Jantzen and Seitz [1992] and by Dipper [1990] in the quantum case.

Proposition 4.1.3

(i) f *is an exact functor.*

(ii) *If $L \in _S\mathbf{mod}$ is irreducible then fL is either an irreducible S^e-module, or 0. In the case where $f(L) \neq 0$ is irreducible, L is a homomorphic image of Se.*

(iii) *If $L \in _S\mathbf{mod}$ is irreducible with projective cover P and injective hull I and if $fL \neq 0$ then fP is the projective cover of fL and fI is the injective hull (as S^e-modules).*

Proof   (i) follows from the fact that eM is a direct summand of M. For (ii), choose some non-zero S^e-submodule N of eL. Then $N = eN$, and $SN = SeN$ is a non-zero S-submodule of L, hence $SeN = L$. So $eL = e(SeN) = (eSe)N \subseteq N$, and so $N = eL$; the last clause follows because $\text{Hom}_S(Se, L) \cong eL$. (iii) is similar.   □

An obvious move now would be to construct an inverse to f; what we shall obtain is actually only a right inverse. Define

$$h : eSe \;\rightarrow\; S$$
$$M \;\mapsto\; Se \otimes_{eSe} M.$$

Se is a left S-module and a right S^e-module, so $h(M)$ is well defined and is a left S-module, hence the definition does make sense. For a morphism $\psi : M \rightarrow M'$, let $h(\psi) = 1_{Se} \otimes \psi : h(M) \rightarrow h(M')$; thus h is a functor $_{S^e}\mathbf{mod} \rightarrow {_S}\mathbf{mod}$.

Proposition 4.1.4 *If $M \in {_{eSe}}\mathbf{mod}$, then h is a right inverse to f in the sense that $f(hM) = e\,hM \cong M$.*

Proof It is easy enough to check that

$$M \;\rightarrow\; e\,h(M)$$
$$m \;\mapsto\; e \otimes m$$

$(m \in M)$ gives the required isomorphism.   □

Now it can happen that M is simple while $h(M)$ is not. To salvage something, let $M \in {_S}\mathbf{mod}$, and let M_0 be the sum of all S-submodules U of M such that $eU = 0$. Let $a(M) = M/M_0$, and if $\theta : M \rightarrow M'$ is in $_S\mathbf{mod}$, then we have $\theta : M_0 \rightarrow M_0'$, hence θ induces a map $a(\theta) : a(M) \rightarrow a(M')$; thus we have a functor $a : {_S}\mathbf{mod} \rightarrow {_S}\mathbf{mod}$.

Theorem 4.1.5

   (i) *If $M \in {_S}\mathbf{mod}$, the canonical map $\pi : M \rightarrow a(M) = M/M_0$ induces an isomorphism*

$$f(\pi) : f(M) \rightarrow f(a(M)).$$

   (ii) *L irreducible in $eSe \Rightarrow a(h(L))$ irreducible.*

Proof The restriction $f(\pi) = \pi|_{fM}$ is onto $fa(M) = e\,a(M)$. But $\text{Ker} f(\pi) = eM \cap M_0 = \{0\}$ since M_0 is the largest S-submodule of M

contained in $(1 - e)M$. For (ii) one shows that $h(M)_0$ is the unique maximal submodule of $h(M)$. □

In particular if S is semisimple, $h(M)_0 = 0$.

Definition 4.1.6 Let $h^* = a \circ h : {}_{S^e}\mathbf{mod} \to {}_S\mathbf{mod}$.

Again, if S is semisimple $h = h^*$. The next result tells us that h^* is another right inverse to f and it also preserves irreducibles:

Theorem 4.1.7

  (i) *If $L \in {}_S\mathbf{mod}$ is irreducible and $eL \neq 0$ then $h^*(f(L)) \cong L$.*
  (ii) *Given a classification of irreducibles in ${}_S\mathbf{mod}$, say $\{L(\lambda) : \lambda \in \Lambda\}$, then $\{fL(\lambda) : \forall \lambda \text{ such that } fL(\lambda) \neq 0\}$ is a classification of irreducibles in ${}_{S^e}\mathbf{mod}$. Moreover, if $fL(\lambda) \neq 0$, $L(\lambda) \cong h^*(eL(\lambda))$.*
     Hence if S is semisimple $h = h^*$.

Proof For (i), the S-map $h(eL) \to L$ given by $a \otimes em \mapsto aem$ for all $a \in Se$, $m \in L$ is a surjection with kernel $h(eL)_0$. (ii) is a direct consequence of (i) and the results of this section. □

Proposition 4.1.8 *Let $M \in {}_S\mathbf{mod}$ and $L \in {}_S\mathbf{mod}$ be irreducible. Suppose $eL \neq 0$. Then*

$$[M : L] = [eM : eL].$$

Proof Taking a composition series $M = M_0 > M_1 > \cdots > M_s = 0$, we obtain a series $eM \supseteq eM_1 \supseteq \cdots \supseteq eM_s = 0$ of S^e-modules. By 4.1.3, $eM_{i-1}/eM_i \cong e(M_{i-1}/M_i)$ $(i \in \mathbf{s})$ as S^e-modules. Deleting those eM_{i-1} for which $eM_i = eM_{i-1}$ we obtain a composition series for eM. □

4.2 Applying the Schur functor

We return to the particular case with $S = S(n, r)$ and $r \leq n$. We could work over R or K, so for definiteness, we will choose to work here over K. Consider the Schur functor defined by the idempotent $e = \xi_\omega = \xi_{u,u}$. Identify S^e with $K\Sigma_r$ via the isomorphism of (4.1). Following the method of Green [1980], [1981], we use the classification of simple objects in $\mathcal{P}_K(n, r)$, together with general properties of Schur and Weyl modules, to classify the ordinary and modular representations of G. Other less well-known applications of Schur functors are discussed informally at the end of the section.

Ordinary representations of G

We note that since n is large enough, we can identify $\Lambda^+(n, r)$ with the set of all partitions $\Lambda^+(r)$. Fix some $\lambda \in \Lambda^+(r)$. By definition, f takes the Schur module $M(\lambda)$ ($\lambda \in \Lambda^+(r)$) to its weight space $M(\lambda)^\omega$; the latter module has K-basis consisting of all bideterminants $T^\lambda(\ell : i)$ such that $i \in I^\omega$. The aim of this section will be to show that $M(\lambda)^\omega$ is precisely S^λ, the *Specht module* for λ.

Let us recall from James and Kerber [1981, §7] the construction of the Specht module: for any tableau T the *row sum* and the *signed column sum* are the elements

$$\mathcal{H}^T = \sum_{\rho \in R(T)} \rho, \qquad \mathcal{V}^T = \sum_{\pi \in C(T)} (-1)^\pi \pi$$

of KG. We start with the permutation module $M^\lambda = KG\mathcal{H}^T$ and construct S^λ as the submodule $KG\mathcal{V}^T\mathcal{H}^T$ of M^λ, for any basic λ-tableau T^λ.

Remark. It is convenient to work with isomorphic copies of Specht modules at certain times. The most common manifestations are as follows.

(1) In order to relate the given definition to the usual one as the span of the λ-polytabloids, we note that the map sending $\mathcal{H}^T$ to the λ-tabloid $\{T\}$ is a well-defined isomorphism, which restricts to an isomorphism between the sub-ideal $KG\mathcal{V}^T\mathcal{H}^T$ and the Specht module of James [1978a, §4]. Of course a λ-polytabloid is of the form $\mathcal{V}^T\{T\}$.

(2) M^λ can be embedded inside $E^{\otimes r}$ for $\lambda \in \Lambda^+(n, r)$ via $\{T\} \mapsto e_i$ where i_j is the row index of j in $\{T\}$, indeed M^λ is a summand. Thus there is an isomorphic copy of S^λ inside $E^{\otimes r}$: it is precisely $KG\mathcal{V}^T e_\ell$ where $\ell = \ell(\lambda)$ is as in 2.4.3.

We now recall the basis (3.1) of $M(\lambda)$. All the basis elements lie in the right λ-weight space $A_r\xi_\lambda = {}^\lambda A_r$ of A_r. Hence $fM(\lambda)$ lies in the left ω-weight space ${}^\lambda A_r{}^\omega = \xi_\omega A_r\xi_\lambda$ of ${}^\lambda A_r$. It is easily seen from (3.11)—(3.14) that this has basis $\{c_{\ell, u\pi} : \pi \in G\}$. We make the observation that $c_{\ell, u\pi} = c_{\ell, u\pi'}$ if and only if $\pi\mathcal{H}^T = \pi'\mathcal{H}^T$ to infer that we have an isomorphism between the spaces ${}^\lambda A_r{}^\omega$ and M^λ given by

$$c_{\ell,u\pi} \;\mapsto\; \pi \mathcal{H}^T \qquad\qquad (4.2)$$

for all $\pi \in G$. The right-hand side is a left KG-module (being a left ideal of KG). Further, the left-hand side is a KG-module under the isomorphism (4.1); (explicitly from (2.11), $\sigma c_{\ell,u\pi} = \xi_{u\sigma,u} \circ c_{\ell,u\pi} = c_{\ell,u\sigma\pi}$ for $\sigma \in G$). Thus (4.2) is actually a left KG-isomorphism.

We know that $M(\lambda)^\omega$ has basis $\{T^\lambda(\ell:i) : i \in I^\omega\}$. Let $i \in \omega$. Then there exists a unique $\pi \in G$ such that $i = u\pi$. Now $T^\lambda(\ell:i) \in {}^\lambda A_r{}^\omega$ so we may apply the isomorphism (4.2) to map

$$T^\lambda(\ell:u\pi) \;\mapsto\; \sum_{\sigma \in C(T)} (-1)^\sigma \pi\sigma\mathcal{H}^T = \pi\mathcal{V}^T\mathcal{H}^T.$$

Hence $fM(\lambda)$ is the left KG-submodule $KG\mathcal{V}^T\mathcal{H}^T = S_K^T$, the *Specht module* over K corresponding to $T = T^\lambda$. We shall use the notation of James and Kerber [1981, 7.1.4] and write this as S_K^λ, where λ has bijective λ-tableau T^λ. We may obtain the well-known result that the standard polytabloids give a basis of the Specht module over any field (James [1978a, 7.2]):

Theorem 4.2.1 S_K^λ *has K-basis*

$$\{\pi\mathcal{V}^T\mathcal{H}^T : T_{u\pi}^\lambda \text{ is standard}\}.$$

In characteristic zero, $S_\mathbf{Q}^\lambda$ is irreducible, and, choosing a bijective λ-tableau T^λ for each λ, the various Specht modules $S_\mathbf{Q}^\lambda$ ($\lambda \in \Lambda^+(r)$) give a complete set of irreducible $\mathbf{Q}G$-modules up to isomorphism.

Proof We apply the isomorphism (4.2) to the basis of $M(\lambda)^\omega$. Over $\mathbf{Q}$, the Schur modules form a complete set of irreducible objects of $\mathcal{P}_\mathbf{Q}(n,r)$, by 3.4.2. Since the images $fM(\lambda) = S^\lambda$ are non-zero for each $\lambda \in \Lambda^+(r)$, then, by 4.1.7(ii), the proof is complete. $\qquad\square$

We may follow a similar path to describe $fV(\lambda)$. In 3.4 we learned that the Weyl module is a subspace of $E^{\otimes r}$, hence $fV(\lambda)$ is a subspace of $(E^{\otimes r})^\omega = \xi_\omega E^{\otimes r} = \sum_{i \in \omega} Ke_i = \sum_{\pi \in G} Ke_{u\pi}$, on which the G-action is given by $\sigma e_{u\pi} = \xi_{u\sigma,u}e_{u\pi} = e_{u\sigma\pi} = e_{(u\sigma)\pi}$ for all $\sigma, \pi \in G$ by the example after 2.2.6. Thus there is a left KG-isomorphism

$$(E^{\otimes r})^\omega \quad \to \quad KG = M^\omega$$
$$e_{u\pi} \quad \mapsto \quad \pi \tag{4.3}$$

for all $\pi \in G$. Recall that 3.4.3 gave a K-basis for $V(\lambda)$: from this it is clear that $V(\lambda)^\omega$ has basis $\{v_i = \xi_{i,\ell} f_\ell : i \in I^\omega\}$. Putting $i = u\pi$ yields $v_{u\pi} = e_u \pi \mathcal{H}^T \mathcal{V}^T$ and under (4.3) this basis element $v_{u\pi} i = e_u \pi \mathcal{H}^T \mathcal{V}^T$ of $V(\lambda)^\omega$ is sent to $\pi \mathcal{H}^T \mathcal{V}^T \in KG$. Hence $fV(\lambda) = KG\mathcal{H}^T \mathcal{V}^T = S_\lambda$, the *dual Specht module* for λ, choosing a bijective λ-tableau $T = T^\lambda$. We have derived a theorem analogous to 4.2.1.

Theorem 4.2.2 $S_{\lambda,K}$ *has K-basis*

$$\{\pi \mathcal{H}^T \mathcal{V}^T : T^\lambda_{u\pi} \text{ is standard}\}.$$

In characteristic zero $S_{\lambda,\mathbf{Q}}$ is irreducible and, choosing a bijective λ-tableau T^λ for each λ, the various dual Specht modules $S_{\lambda,\mathbf{Q}}$ ($\lambda \in \Lambda^+(r)$) form a complete set of irreducible $\mathbf{Q}G$-modules up to isomorphism.

Let us explicitly record the fact, implicit in the proof of 4.2.1, that $S_r e \cong E^{\otimes r}$; for, by (4.3),

$$S_r e \cong \mathrm{Hom}_G(eE^{\otimes r}{}_{KG}, E^{\otimes r}{}_{KG}) \cong \mathrm{Hom}_G(KG, E^{\otimes r}) \cong E^{\otimes r}.$$

Exercises.
(1) Refer to Example (1) in 3.4. Use (4.2) to show that the S_r-map $1 \otimes_{\Sigma_\lambda} 1 \mapsto \bar{e}_u$ induces an isomorphism between M^λ and $fS^\lambda(E)$.
(2) Prove that S^λ_K and $S_{\lambda,K}$ are dual modules in the usual sense. (At least two different proofs are available in James and Kerber [1981, p. 318] and Green [1980, p. 92]).

Remarks.
(1) All the above remains true if we replace K by $\mathbf{Z}$ (or more generally by our complete d.v.r. R). The isomorphisms (4.2) and (4.3) send $M_{\mathbf{Z}}(\lambda)^\omega$ to $\mathbf{Z}G\mathcal{V}^T\mathcal{H}^T$ and $V_{\mathbf{Z}}(\lambda)^\omega$ to $\mathbf{Z}G\mathcal{H}^T\mathcal{V}^T$, i.e. to $\mathbf{Z}$-forms of the $\mathbf{Q}G$-modules $S^\lambda_{\mathbf{Q}}$ and $S_{\lambda,\mathbf{Q}}$.
(2) Assume char$K = p$. When λ is p-restricted, $S_{\lambda,K}$ has a unique maximal submodule, with quotient isomorphic to $KG\mathcal{V}^T\mathcal{H}^T\mathcal{V}^T$. Denote this module by D_λ. Similarly, when λ is p-regular, denote the irreducible quotient, $KG\mathcal{H}^T\mathcal{V}^T\mathcal{H}^T$ of S^λ by D^λ.

Modular representations of G

Assume throughout that K is a field of non-zero characteristic p and $r \leq n$. We already have a complete set of irreducibles $\{F(\lambda) : \lambda \in \Lambda^+(r)\}$ in $\mathcal{P}_K(n, r)$, namely, the irreducible heads of the Weyl modules (or the irreducible socles of the Schur modules). Let Λ' be the subset of $\Lambda^+(r)$ consisting of those $\lambda \in \Lambda^+(r)$ satisfying $F(\lambda)^\omega \neq 0$. Then by 4.1.7 the set $\{F(\lambda)^\omega : \lambda \in \Lambda'\}$ is a complete set of p-modular irreducible KG-modules. How may we relate these to the modules D^λ and to the D_λ defined in the last remark? First we state the theorem of Clausen [1979] and James [1980] that tells us that Λ' is precisely the set $\Lambda^{(p)}(r)$ of all p-restricted partitions of r.

Theorem 4.2.3 (Clausen, James) $\Lambda' = \Lambda^p(r)$.

Proof ($\Rightarrow$). By 3.4.1(iii), $L(\lambda) = S_r(\Gamma)\omega_\lambda$ is K-generated by all elements $\xi_{i,j} \circ \omega_\lambda$ $(i, j \in I)$. By (2.18) these elements are zero, unless $j \sim \ell$. Now

$$\xi_{i,\ell} \circ \omega_\lambda = \sum_{h \in I} \langle c_{h,\ell}, \xi_{i,\ell} \rangle T^\lambda(\ell : h) = \sum_{h \in iR(T)} T^\lambda(\ell : h)$$

for all i, and this is an element of $L(\lambda)^\mu$, where μ is the weight of i. So $L(\lambda)^\omega$ is K-spanned by all $\xi_{i,\ell} \circ \omega_\lambda$, with $i \in \omega$. Notice also that the elements $i\sigma$ $(\sigma \in R(T))$ are all distinct if i has weight ω, and so

$$\xi_{i,\ell} \circ \omega_\lambda = \sum_{\sigma \in R(T)} T^\lambda(\ell : i\sigma). \tag{4.4}$$

Let C be the subgroup of $R(T)$ consisting of all permutations that take each column of the basic tableau T to a (possibly different) column of T. Let $\mu = \lambda'$. Then $\theta \in C$ is specified by a sequence $\theta_1, \theta_2, \ldots$ such that for each $s \geq 1$, θ_s permutes $W_s = \{t \geq 1 : \mu_t = s\}$. So under θ, the (α, β)th entry of T goes to the $(\alpha, \theta_s(\beta))$th entry $(\alpha \geq 1, \beta \in W_s)$. Since $|W_s| = (\lambda_s - \lambda_{s+1})$, $|C| = (\lambda_1 - \lambda_2)!(\lambda_2 - \lambda_3)! \ldots$. We have $T^\lambda(\ell : i) = T^\lambda(\ell : i\theta)$ for $i \in I$, $\theta \in C$. Breaking (4.4) into C-orbits, it is divisible by $|C|$. But if λ is not p-restricted, p divides $|C|$, hence every term in (4.4) is zero, which means $F(\lambda)^\omega = \{0\}$.

($\Leftarrow$). Suppose now $\lambda \in \Lambda^{(p)}(r)$. By (4.4), we need to prove that

$$\xi_{u,\ell} \circ \omega_\lambda = \sum_{\sigma \in C(T)} \sum_{\tau \in R(T)} (-1)^\sigma c_{\ell\sigma, u\tau} \tag{4.5}$$

is non-zero.

Let $\pi \in C(T)$ reverse the order of the entries in each column of T_ℓ, i.e. $\pi : a(p,q) \mapsto a(s+1-p, q)$ $(p \geq 1, q \in W_s)$. Then we claim that the coefficient of $c_{\ell\pi,u}$ in (4.5) is non-zero.

For suppose there is some $\sigma \in C(T)$ and $\tau \in R(T)$ such that $c_{\ell\pi,u} = c_{\ell\sigma,u\tau}$. Then there exists $\gamma \in G$ such that $\ell\pi\gamma = \ell\sigma$, $u\gamma = u\tau$. So $\gamma = \tau$ and $\ell\pi\tau = \ell\sigma$. Let M be the maximum entry in T_ℓ. In T_ℓ and hence in $T_{\ell\sigma}$ all entries equal to M are in columns $t \in W_M$. Also, in $T_{\ell\pi}$ and hence in $T_{\ell\pi\tau}$, all entries equal to M are in the first row. But $T_{\ell\pi\tau} = T_{\ell\sigma}$, so all entries equal to M in $T_{\ell\sigma}$ are in the same places as the entries M in $T_{\ell\pi}$.

We focus successively on the places occupied by $M-1, M-2, \ldots$, and notice that $T_{\ell\pi} = T_{\ell\sigma}$ so $\pi = \sigma$. Further $|\{\tau \in R(T) : \ell\pi\tau = \ell\pi\}| = w = \prod_{s \geq 1}((\lambda_s - \lambda_{s+1})!)^s$ and p and w are coprime since $\lambda \in \Lambda^{(p)}(r)$. Therefore the coefficient of $c_{\ell\pi,u}$ in (4.5) must be $(-1)^\pi w 1_K$, and this is non-zero, as required. $\qquad\square$

Corollary 4.2.4 *The set* $\{fL(\lambda) : \lambda \in \Lambda^{(p)}(r)\}$ *is a complete set of inequivalent irreducible KG-modules.*

Recall that we have made the universal assumption that $r \leq n$. Consider the left ideal $S_r \xi_\omega$ of S_r, which one can also view as a right KG-module via (4.1). Clearly $S_r \xi_\omega$ has K-basis $\{\xi_{i,u} : i \in I\}$.

Corollary 4.2.5 *Let* $V \in \mathcal{P}_K(n,r)$ *be irreducible. Then*

$$fV \neq 0 \Leftrightarrow V \cong \text{ a submodule of } E^{\otimes r}.$$

Proof Observe first that the K-isomorphism

$$S_r \xi_\omega \quad \to \quad E^{\otimes r}$$
$$\xi_{i,u} \quad \mapsto \quad e_i$$

$(i \in I)$ is both a left S_r-map and a right KG-map. Now $V^\omega \neq 0$ if and only if V is a homomorphic image of $S_r \xi_\omega$, by 4.1.3(ii), and hence of $E^{\otimes r}$. By 3.4.9, V, and by the first exercise in 3.4, $E^{\otimes r}$, are self-dual. $\qquad\square$

Corollary 4.2.6 $E^{\otimes r}$ *is both projective and injective.*

Proof Recall from the proof of 4.2.5 that $E^{\otimes r}$ is isomorphic to $S_r \xi_\omega$, which is a direct left ideal summand of S_r. Hence it is projective. Being self-dual, it must also be injective. □

Combining 4.2.5 with 4.2.3 yields:

Corollary 4.2.7 $F(\lambda)$ *is isomorphic to a submodule of* $E^{\otimes r}$ *if and only if* $\lambda \in \Lambda^{(p)}(r)$.

This is a result of James [1980, 3.2].

Two conjectures on modular representations

We round off this section with two important conjectures concerning modular representations of G and Γ respectively. Both involve Schur functors, at least indirectly.

The Mullineux Conjecture. The first conjecture suggests a possible relationship involving the labellings of the various $fF(\lambda)$, $fL(\lambda)$, D^λ and D_λ. Recall that at the end of 3.4 Jantzen's contravariant form $\langle\langle \, , \, \rangle\rangle$ on $E^{\otimes r} \mathcal{V}^T$ was defined. Restricting this form to the ω-weight space $(E^{\otimes r})^\omega \mathcal{V}^T$, we may then use (4.3) to map into $KG\mathcal{V}^T$. This will define for us a symmetric invariant form on $KG\mathcal{V}^T$ which we denote by $(\, , \,)$. Explicitly, for $\pi, \sigma \in G$,

$$(\pi \mathcal{V}^T \sigma \mathcal{V}^T) = \begin{cases} (-1)^{\pi^{-1}\sigma} & \text{if } \pi^{-1}\sigma \in C(T^\lambda) \\ 0, & \text{otherwise} \end{cases} . \qquad (4.6)$$

By 3.4.10 the radical of the contracted form on $V(\lambda)$ is identified as the unique maximal submodule of $V(\lambda)$, so applying f yields

Theorem 4.2.8 *The restriction of the form (4.6) to* $S_{\lambda,K}$ *is non-zero if and only if* λ *is p-restricted. In the case where* λ *is p-restricted,* $\mathrm{Rad}(\, , \,) = S_{\lambda,K}^{\max}$ *and*

$$D_\lambda = S_{\lambda,K}/S_{\lambda,K}^{\max} \cong fF(\lambda).$$

Now let $s : KG \to KG$ be the K-algebra automorphism defined by $s(\pi) = (-1)^\pi \pi$ $(\pi \in G)$. Define the 1-dimensional KG-module K_a with basis 1_a via the action $\pi k = (-1)^\pi k$ for $k \in K$ and $\pi \in G$: this is the *sign module* for G. If M is any left ideal of KG then we have an isomorphism of left KG-modules $M \otimes_K K_a \cong s(M)$ sending $m \otimes 1_a \mapsto s(m)$; observe that $s(\mathcal{V}^{T^\lambda}) = \mathcal{H}^{T^{\lambda'}}$ and $s(\mathcal{H}^{T^\lambda}) = \mathcal{V}^{T^{\lambda'}}$. The form (4.6) on $KG\mathcal{V}^T$ translates under s to a symmetric bilinear form $(\ ,\ )$ on $KG\mathcal{H}^{T'}$, where T' is the associated λ'-tableau. Explicitly, if $\pi, \sigma \in G$ then

$$(\pi\mathcal{H}^{T'}\sigma\mathcal{H}^{T'}) = \begin{cases} 1 & \text{if } \pi^{-1}\sigma \in R(T^{\lambda'}) \\ 0 & \text{otherwise} \end{cases}. \qquad (4.7)$$

Since $s(S_{\lambda,K}) = S_K^{\lambda'}$ we have $S_K^{\lambda'} \cong S_{\lambda,K} \otimes K_a$ and we now obtain the Specht module analogue of 4.2.8. Choose any $\mu \in \Lambda^+(r)$.

Theorem 4.2.9 *The restriction of the form (4.7) to S_K^μ is non-zero on S_K^μ if and only if μ is p-regular. In the case where μ is p-regular, $\mathrm{Rad}(\ ,\ ) = (S_K^\mu)^{\max}$ and*

$$D^\mu = S_K^\mu/(S_K^\mu)^{\max} \cong fF(\mu') \otimes_K K_a.$$

Thus the labellings are related as follows:

$$D^{\mu'} \cong D_\mu \otimes_K K_a \qquad \forall \mu \in \Lambda^{(p)}(r).$$

If λ is p-regular then $D^\lambda \otimes K_a \cong D^{\lambda^*}$ for some p-regular λ^*. The question of an explicit expression for the involutory map $\lambda \mapsto \lambda^*$ now arises. Mullineux [1979] defined a bijection between $\Lambda_{(p)}(r)$ and a set of *Mullineux symbols* (see Bessenrodt and Olsson [1991/92]). For convenience we recall the main points of this procedure. The *p-rim* of $[\lambda]$ is a part of the rim of $[\lambda]$ composed of *p-segments* and each p-segment, except possibly the last, comprises p nodes. The first p nodes of the rim (starting at the top right) comprise the first p-segment (so if the rim has at most p nodes we have the entire rim). The next p-segment is obtained by starting in the row immediately below where the last p-segment ended. Continue until the last row is reached. Suppose there are a_1 nodes in the p-rim of $\lambda = \lambda^1$, a partition with r_1 rows, say. Removing the p-rim of λ^1 exhibits a partition $\lambda^2 \in \Lambda_{(p)}(r - a_1)$. Define a_2 and r_2 for λ^2 and continue in this manner: eventually we obtain a sequence $\lambda = \lambda^1, \lambda^2, \ldots, \lambda^t$ of partitions with $\lambda^t \neq 0$ and $\lambda^{t+1} = 0$. The Mullineux symbol is then

$$M_p(\lambda) = \begin{pmatrix} a_1 & a_2 & \cdots & a_t \\ r_1 & r_2 & \cdots & r_t \end{pmatrix}.$$

One reconstitutes λ from $M_p(\lambda)$ by starting with λ^t (given by a_t and r_t) and working backwards from the bottom row upwards. Finally,

Mullineux defined a 'conjugation' map on his symbols: let $\lambda^M \in \Lambda_{(p)}(r)$ be the partition with Mullineux symbol

$$M_p(\lambda^M) = \left(\begin{array}{cccc} a_1 & a_2 & \cdots & a_t \\ s_1 & s_2 & \cdots & s_t \end{array} \right)$$

where $s_i = a_i - r_i + \varepsilon_i$. Here $\varepsilon_i = 1$ if $p \nmid a_i$, otherwise $\varepsilon_i = 0$.

Example. Take $p = 5$ and $\lambda = (8, 6, 5, 5)$. Number the nodes of successive 5-rims by 1, 2, 3 and 4:

```
4  4  3  2  2  1  1  1
4  3  3  2  1  1
3  3  2  2  1
2  1  1  1  1
```

$$M_5(\lambda) = \left(\begin{array}{cccc} 10 & 6 & 5 & 3 \\ 4 & 4 & 3 & 2 \end{array} \right)$$

```
4  4  3  3  3  2  2  1  1  1
4  3  3  2  2  2  1  1
2  1
1  1
1
1
```

$$M_5(\lambda^M) = \left(\begin{array}{cccc} 10 & 6 & 5 & 3 \\ 6 & 3 & 2 & 2 \end{array} \right)$$

Hence $(8, 6, 5, 5)^M = (10, 8, 2, 2, 1, 1)$.

Miraculously (or unfortunately) it appears that $\lambda^M = \lambda^*$ in all cases checked. Several authors have tried, with varying amounts of success, to reconcile the tensoring operation with the Mullineux map, e.g. see Donkin [1987], Martin [1989] and Bessenrodt and Olsson [1991/92]. Young modules allow an interesting reformulation, which I shall mention later. In Chapter 7 there even appears a 'quantum' Mullineux Conjecture!

(3) Let $\lambda \in \Lambda^+(n, r)$. By considering the S_r-map $1 \otimes_{\Sigma_\lambda} 1_a \mapsto \bar{e}_u \in \Lambda^\lambda(E)$, show that $f\Lambda^\lambda(E) \cong (K_a) \uparrow_{\Sigma_\lambda}^G$. Compare this with Exercise (1) above.

Doty's conjecture. Assume in this section that K is algebraically closed. We return once again to the problem of computing the irreducible p-modular characters of Γ. Here it is profitable to consider polynomial representations of Γ as rational representations of the affine algebraic monoid over K, $M = \mathrm{Mat}_n(K)$. Clearly M acts on E by linear substitution, and the action of $\Gamma = \mathrm{GL}_n(K)$ on E given in Example (1) after

1.5.6 is just the restriction of the M-action. The space $A_K(n)$ generated by the coordinate functions is identified as the coordinate ring, $K[M]$, of M. We recall the definition of the symmetric algebra $S = K[e_1, \ldots, e_n]$ of E, and its rth homogeneous graded piece $S^r(E)$; let $\alpha \in \mathbf{N}_0^n$, and write $S^\alpha(E)$ for the tensor product of symmetric powers. What follows is based on Doty [1991], and Doty and Walker [1992b].

Now, M acts on $K[M]$ by left and right translation exactly as in 1.3.1. We can identify the K-algebras $K[M] = A_K(n)$ and $S^{\otimes n}$ via the correspondence

$$\psi : K[M] \rightarrow S^{\otimes n}$$
$$c_{ij} \mapsto 1 \otimes \cdots \otimes e_j \otimes \cdots \otimes 1$$

with e_j in the ith slot $(i, j \in \mathbf{n})$. We let M act diagonally on $S^{\otimes n}$: for $g \in M$,

$$g \circ (1 \otimes \cdots \otimes e_j \otimes \cdots \otimes 1) = \sum_{k=1}^{n} g_{kj}(1 \otimes \cdots \otimes e_k \otimes \cdots \otimes 1),$$

which corresponds under ψ to the left action of right translation on $K[M]$. Thus ψ is an isomorphism of left M-modules. Under this correspondence it is also clear that we have a decomposition of the Schur coalgebra $A_K(n, r) = \bigoplus_{\mu \in \Lambda(n,r)} S^\mu(E)$.

(4) Use this identification to give another proof that the Schur module $M(\lambda)$ $(\lambda \in \Lambda^+(n, r))$ embeds into $S^\lambda(E)$. It is tempting to think that there is a surjection $S^\lambda(E) \rightarrow V(\lambda)$. But take $\lambda = (r)$; if this statement were true, there would have to be an isomorphism $S^r E = M(r, 0, \ldots, 0) \cong V(r, 0, \ldots, 0)$. But the exercise at the very end of Chapter 3 already demonstrates that $M(2, 0)$ and $V(2, 0)$ are not isomorphic as $GL_2(K)$-modules if char $K = 2$.

Note that $\mathbf{N}_0^n$ is identified with the character monoid, $X(D)$, of the diagonal matrices, D, in M. Indeed, restriction of characters to T induces an embedding of monoids $X(D) \rightarrow X(T)$, therefore $X(D)$ is regarded as the set of polynomial weights (or compositions). We write $X^+(D) = X(D) \cap X^+(T)$ for the dominant polynomial weights of D and identify this as the set of partitions. Regarding $K[M]$ as a right M-module under left translation, we can decompose $K[M]$ into its weight spaces $K[M]^\alpha$, $\alpha \in X(D)$. The weight spaces are left M-modules under right translation (since left and right translation commute). Left translation by the diagonal element $d(t) \in D$ of the form (1.8) sends the generator c_{ij} to $t_i c_{ij}$, hence in $K[M]$ the α-weight space is K-spanned by all monomials $\prod_{i,j} c_{ij}^{\alpha_{ij}}$ such that $\sum_{j=1}^n \alpha_{ij} = \alpha_i$ for each $i \in \mathbf{n}$. Under ψ this corresponds to $S^\alpha(E)$. Actually, we may assume that α is dominant since the symmetric power is independent of permutation of the factors.

The conjecture we want to discuss is motivated by the situation in characteristic zero, so we begin by assuming $K = \mathbf{C}$. The reader is asked to believe that the usual theory of induction for algebraic groups as in Jantzen [1987] extends to the scenario of algebraic monoids and to check that (just as for Schur modules) if α is a weight of length at most n (viewed as a 1-dimensional D-module), then $K[M]^\alpha \cong \operatorname{Ind}\uparrow_D^M (\alpha)$. Let $\lambda, \mu \in X^+(D)$ and let $L(\mu)$ be the irreducible M-module of highest weight μ. The proof of the next result will be redeployed in 4.5.7: it is due ultimately to Donkin [1987, (2.4)].

Lemma 4.2.10 (Kostka Duality) *With the above notation*

$$[S^\lambda(E) : L(\mu)]_M = \dim L(\mu)^\lambda.$$

Proof The left-hand side equals

$$
\begin{aligned}
\dim \operatorname{Hom}_M(L(\mu), S^\lambda(E)) &= \dim \operatorname{Hom}_M(L(\mu), K[M]^\lambda) \\
&= \dim \operatorname{Hom}_M(L(\mu), \operatorname{Ind}\uparrow_D^M (\lambda)).
\end{aligned}
$$

By Frobenius Reciprocity for monoids this is just $\dim \operatorname{Hom}_D(L(\mu), \lambda)$, which equals the right-hand side. Using Exercise (1) and Young's Rule, this number equals $K_{\mu\lambda}$, the (μ, λ)th Kostka number. □

Consider the matrix, having rows and columns indexed by the dominant polynomial weights $X^+(D)$ (in some fixed order), whose (λ, μ)th entry is the composition multiplicity of $L(\mu)$ in $S^\lambda(E)$. This matrix then is the transpose of the usual Kostka matrix, and has inverse whose rows are given by the usual Jacobi-Trudi identity (Macdonald [1979, I§6]).

What is the analogue of this matrix over algebraically closed fields K of characteristic p? From the work of Krop [1986, 1988] we consider so-called 'truncated' symmetric powers. Let J be the ideal of S generated by all pth powers of the e_j and denote by $T^r(E)$ the image of $S^r(E)$ in the quotient S/J. Now J is closed to the M-action and so $T^r(E)$ is an M-module; moreover Krop had already observed that it is irreducible. Taking $\alpha = (\alpha_1, \ldots, \alpha_n) \in X(D)$, we set $T^\alpha(E) = T^{\alpha_1}(E) \otimes \cdots \otimes T^{\alpha_n}(E)$.

Following Doty and Walker [1992b, 2.2] it is easy to characterise the $T^\alpha(E)$ as the weight space of some affine algebra: taking $M_0 = F^{-1}(0)$ as the additive kernel of the Frobenius morphism on M, so that $K[M_0]$ is just the truncated algebra $K[M]/\langle c_{ij}^p \rangle \cong (S/J)^{\otimes n}$, a similar argument to that for $S^\alpha(E)$ gives an M-module isomorphism between $K[M_0]^\alpha$ and $T^\alpha(E)$.

We need to assume that α is $n(p-1)$-bounded, that is, no part exceeds $n(p-1)$. This condition ensures only that $T^r(E) = 0$ if $r > n(p-1)$ and hence that $T^\alpha(E) = 0$ unless α is $n(p-1)$-bounded. Consider then the matrix with rows and columns indexed by the $n(p-1)$-bounded weights of $X^+(D)$, whose (λ, μ)th entry is

$$K'_{\mu\lambda} = [T^\lambda(E) : L(\mu)]_M.$$

Purely by analogy with the classical case we call these *p-modular Kostka numbers*.

Conjecture 4.2.11 (Doty) *The modular Kostka matrix is non-singular for all n and p. Equivalently, the irreducible characters are determined by the truncated coordinate ring of* $\mathrm{Mat}_n(K)$.

This has been checked and found to be true for several infinite families of (n, p). Now, in the ring S_n, for any λ which is $n(p-1)$-bounded dominant polynomial, there are equations

$$\mathrm{ch}\, T^\lambda(E) = \sum_\mu K'_{\mu\lambda} \mathrm{ch}\, L(\mu)$$

over all $n(p-1)$-bounded dominant weights μ. If 4.2.11 is true one could in principle invert these equations to express the $\mathrm{ch} L(\lambda)$ as a $\mathbf{Q}$-linear combination of the $\mathrm{ch} T^\mu(E)$ for $n(p-1)$-bounded dominant polynomial weights λ, μ of length at most n.

Conjecture 4.2.11 provides a way to compute the modular characters for Γ and hence could lead to a proof of the Lusztig Conjecture for Γ. For, recall that if F is the Frobenius map induced by raising to the pth power in K, then the fixed points under F in M, or respectively Γ, are $M^F = \mathrm{Mat}_n(p)$, or respectively $\Gamma^F = \mathrm{GL}_n(p)$. By the result of Steinberg (see Curtis [1970]) we get simple modules for Γ^F and M^F by restricting simple Γ- and M-modules respectively; the simple $K\Gamma^F$-modules are indexed by all partitions that are p-restricted and with last part $0 \leq \lambda_n \leq p-2$. This strange condition on λ_n is necessary because the $(p-1)$th power of the determinant representation is the same as the trivial representation for Γ^F. Such weights form a subset of the indexing set for the irreducibles for M^F: we simply relax the condition on the last part, since the $(p-1)$th power of the determinant and the trivial representation are inequivalent for $\text{Mat}_n(p)$ but are equivalent for $GL_n(p)$; see Harris and Kuhn [1988]. This latter set is in turn a subset of the $n(p-1)$-bounded polynomial dominant weights of D. The upshot is that the modular Kostka numbers determine the formal characters of the simples of both M^F and Γ^F, and hence of the simples for Γ (by the Steinberg Tensor Product theorem), and hence of M.

(5) There is an analogue of the result of Exercise (4). With λ p-restricted and $\lambda_n \leq p-2$, consider the Γ-homomorphism $L(\lambda) \to T^\lambda(E)$ given by composing the embedding $L(\lambda) \to H^0(\lambda)$ with the truncation $H^0(\lambda) \to T^\lambda(E)$. This latter map is the composite of the map of Exercise (4) with the truncation map $S^\lambda(E) \to T^\lambda(E)$. Show that $L(\lambda)$ embeds into $T^\lambda(E)$ by showing that any maximal weight vector in $L(\lambda)$ has non-zero image in $T^\lambda(E)$ under the truncation map. So $H^0(\lambda)$ embeds into $T^\lambda(E)$ and by taking duals, $T^\lambda(E)$ surjects onto $V(\lambda)$.

Remarks.

(1) The characters of the truncated symmetric powers are known: Doty and Walker [1992b, 3.2] define $h'_{n,r}(\mathbf{X}) = \sum m_\lambda(\mathbf{X})$, where the sum is over all $\lambda \in \Lambda^{(p)}(n,r)$ (a type of 'p-modular analogue' of the complete symmetric function). Taking λ to be a partition with at most n parts, it is clear that if $h'_{n,\lambda} = h'_{n,\lambda_1} \ldots h'_{n,\lambda_n}$, then ch $T^\lambda(E) = h'_{n,\lambda}(\neq 0)$ if λ is $n(p-1)$-bounded. It is now a short step to show that knowing the p-modular Kostka numbers is effectively the same as knowing the decomposition numbers (defined properly in 4.4) of the Weyl modules. Using the Schur functor, we will show in 4.4.3 that this will be enough to compute the decomposition numbers of the symmetric groups.

(2) Conjecture 4.2.11 is equivalent to the linear independence, in $\mathcal{S}_n$, of the $h'_{n,\lambda}$. Suppose, instead, we work with infinitely many variables: let $\mathcal{S} = \bigoplus \mathcal{S}^r$ be the ring of all symmetric functions in $X_1, X_2, \ldots$, where $\mathcal{S}^r = \varprojlim \mathcal{S}_{n,r}$. Write $\mathcal{S}_{\mathbf{Q}}$ for the algebra $\mathcal{S} \otimes_{\mathbf{Z}} \mathbf{Q}$ and denote by $h'_1, h'_2, \ldots$, the resulting symmetric functions in $\mathcal{S}$. Explicit calculations reveal that the $h'_1, h'_2, \ldots$, are algebraically independent and that $\mathcal{S}_{\mathbf{Q}} = \mathbf{Q}[h'_1, h'_2, \ldots]$. (In fact this latter result is sufficient to prove that the conjecture is true when n is large enough.) In the case $p=2$, $h'_r = e_r$ so one can think of this result as a modular version of the 'fundamental theorem of symmetric functions'. As a consequence, if one considers

tensors of arbitrarily many truncated symmetric powers, then it is a fact
that the irreducibles are then determined by the corresponding decompositions; 4.2.11 then amounts to the assertion that no more than n
tensor factors are required.

4.3 Hom functors for quasi-hereditary algebras

There still remains the problem of dealing with the case where n is not
large enough. Thankfully there is a procedure, pioneered by Schur [1901,
p. 61] that relates the irreducibles of the two categories $\mathcal{P}_K(N, r)$ and
$\mathcal{P}_K(n, r)$ for $N \geq n$. The procedure, as explained by Green [1980, 6.5], is
to manufacture a functor $\mathcal{P}_K(N, r) \to \mathcal{P}_K(n, r)$ using the general results
of 4.1. Rather than repeat this we prefer to generalise the argument
to make it valid for quasi-hereditary algebras. We read off the required
results for S_r at the end.

General set-up

We start with any finite-dimensional quasi-hereditary algebra S. We
choose a complete set $\{L(\lambda) : \lambda \in \Lambda\}$ of simple S-modules and a set
of orthogonal primitive idempotents $\{e_\lambda : \lambda \in \Lambda\}$ in S. Now, $e_\lambda L(\lambda) =$
$L(\lambda)$ and $e_\lambda L(\mu) = 0$ for all $\lambda \neq \mu \in \Lambda$ (because a primitive idempotent
is represented by zero in all types of simple representations of S except
for one). For $\Lambda' \subseteq \Lambda$, define $e = \sum_{\lambda \in \Lambda'} e_\lambda$ and put $S^e = eSe$ as usual.
We need to suppose that $\Lambda \backslash \Lambda'$ is a decreasing saturation; equivalently
suppose that Λ' is itself an increasing saturation. Then

Proposition 4.3.1 (S^e, Λ') *is quasi-hereditary with* $\{e\Delta(\beta) : \beta \in \Lambda'\}$
as standard modules and $\{e\nabla(\beta) : \beta \in \Lambda'\}$ *as costandard modules.*
Moreover, if $\beta \in \Lambda \backslash \Lambda'$ *then*

$$e\Delta(\beta) = e\nabla(\beta) = eP(\beta) = eI(\beta) = 0.$$

Proof Theorem 4.1.7 provides a complete list of simple modules $\{eL(\beta) : \beta \in \Lambda'\}$.

It is convenient to write β^* if $\beta \in \Lambda'$ labels some S^e-module (for
instance $L(\beta^*) = eL(\beta)$). Suppose initially that $\alpha \in \Lambda \backslash \Lambda'$, so that by
saturation all $\alpha' \leq \alpha$ do not lie in Λ'. Thus $e\Delta(\alpha) = e\nabla(\alpha) = 0$, and
also $eP(\alpha) = eI(\alpha) = 0$.

Now take $\beta \in \Lambda'$. If we can prove the next two statements we will have shown that $_{S^e}\mathbf{mod}$ is a highest weight category:

(1) The indecomposable projectives are $P(\beta^*) = eP(\beta)$. This has a filtration by the $e\Delta(\gamma)$ with $\gamma \in \Lambda'$ and $\gamma \geq \beta$ in which $e\Delta(\beta)$ appears only at the top.

(2) For $\beta \in \Lambda'$, $e\Delta(\beta)$ has a filtration by $L(\gamma^*)$ $(\gamma \leq \beta)$ in which $L(\beta^*)$ appears only at the top.

Now, all composition factors of $e\Delta(\beta)$ are isomorphic to $L(\gamma^*)$ where $\gamma \in \Lambda'$ satisfies $\gamma \leq \beta$. Thus $e\Delta(\beta)$ is a factor of $\Delta(\beta^*)$. By the definition of standardness in 3.3.5 and 3.3.6, $\Delta(\beta) \cong Se_\beta/U(\beta)$ where $U(\beta)$ is the good radical. Form the short exact sequence

$$0 \to U(\beta) \to P(\beta) \to \Delta(\beta) \to 0$$

in $_S\mathbf{mod}$ with the property that $U(\beta)$ has a Δ-good filtration by $\Delta(\gamma)$ with $\gamma > \beta$. Passing to $_{S^e}\mathbf{mod}$ produces the short exact sequence

$$0 \to eU(\beta) \to P(\beta^*) \to e\Delta(\beta) \to 0,$$

where $eU(\beta)$ has an $e\Delta$-good filtration by $e\Delta(\gamma)$ with $\gamma > \beta$ and $\gamma \in \Lambda'$. Thus $e\Delta(\beta) \cong \Delta(\beta^*)$ is standard and $P(\beta^*)$ has the required filtration. $\qquad\square$

We fix K, and also fix $r \geq 0$. Let $N \geq n$. We identify $I(n,r)$ with a subset of $I(N,r)$ via the inclusion $\mathbf{n} \subseteq \mathbf{N}$. By means of this identification we can relate the Schur algebras $S(n,r)$ and $S(N,r)$.

Lemma 4.3.2 *There exists an idempotent e such that $eS(N,r)e \cong S(n,r)$.*

Proof View $S(n,r)$ is a subobject of $S(N,r)$ as follows: a basis for $S(N,r)$ is $B = \{\xi_{i,j} : i,j \in I(N,r)\}$, moreover $S(n,r)$ is that K-subspace spanned by all $\xi_{i,j}$ with $i,j \in I(n,r)$. By (2.8) for $\xi_{i,j}, \xi_{p,q} \in B$, if $i,j,p,q \in I(n,r)$ then the coefficient of $\xi_{\alpha,\beta}$ in the product is zero unless both $\alpha, \beta \in I(n,r)$, while if $\alpha, \beta \in I(n,r)$ the coefficient is the same as if $\xi_{i,j}\xi_{p,q}$ were computed inside $S(n,r)$.

There is an injection $* : \Lambda(n,r) \to \Lambda(N,r)$ given by sending $\lambda = (\lambda_1, \ldots, \lambda_n)$ to $\lambda^* = (\lambda_1, \ldots, \lambda_n, 0 \ldots, 0)$. If i has weight $\beta \in \Lambda(n,r)^*$ then $i \in I(n,r)$, so we see that the image $\Lambda(n,r)^*$ is just the set of G-orbits of $I(N,r)$ lying in $I(n,r)$. Let

$$e = \sum_{\lambda \in \Lambda(n,r)^*} \xi_\lambda,$$

so that $e \in S(N, r)$ is idempotent. To see this, note that e is induced by the projection of an N-dimensional K-space onto E. Now $e\xi_{i,j} = \xi_{i,j}$ or 0, according to whether $i \in I(n, r)$ or not; and $\xi_{i,j}e = \xi_{i,j}$ or 0, according to whether $j \in I(n, r)$ or not. It follows that $eS(N, r)e = S(n, r)$. $\square$

The labours of section 4.1, with $S = S(N, r)$, guarantee the existence of a functor:

Definition 4.3.3 The *deflating Schur functor*

$$d_{N,n} : \mathcal{P}_K(N, r) \to \mathcal{P}_K(n, r)$$

is the functor $\mathrm{Hom}_{S(N,r)}(S(N, r)e, -) : V \mapsto eV$ where e is as above.

To apply 4.3.1, take e as just defined, $S = S(N, r)$, $S^e = S(n, r)$ and $B = \Lambda^+(n, r)^* \subseteq A = \Lambda^+(N, r)$. Identify $B = \Lambda^+(n, r)^*$ with $\Lambda^+(n, r)$. Then it is clear that $A \backslash B$ is a decreasing saturation.

Corollary 4.3.4 *Let $V \in \mathcal{P}_K(N, r)$. Then we have*

(i) $eV = \displaystyle\sum_{\alpha \in \Lambda(N,r) \cap \Lambda(n,r)^*} V^\alpha$.

(ii) $\mathrm{ch}(eV)(X_1, \ldots, X_n) = \mathrm{ch}(V)(X_1, \ldots, X_n, 0, \ldots, 0)$.

(iii) *Let $\lambda \in \Lambda^+(N, r)$. Let $X(\lambda)$ be any one of $M(\lambda)$, $V(\lambda)$, $F(\lambda)$ or $L(\lambda)$. Then*

$$d_{N,n}(X(\lambda)) \neq 0 \Leftrightarrow \lambda \in \Lambda(n, r)^*,$$

i.e. $d_{N,n}(X(\lambda)) = 0$ if and only if λ has more than n parts.

Proof Part (i) is clear since if $\alpha \in \Lambda(N, r)$ then $eV^\alpha = e\xi_\alpha V$, and this is V^α or 0 according to whether $\alpha \in \Lambda(n, r)^*$ or not. We have (ii) from (i) and general properties of characters as in 1.5; (iii) is immediate from 4.3.1. $\square$

We now see that the the simple $S(N, r)^e$-modules are labelled by the same indexing set as the set of simple $S(n, r)$-modules. This was proved for $K = \mathbf{C}$ by Schur.

Theorem 4.3.5

(i) $\{d_{N,n}F(\mu^*) : \mu \in \Lambda^+(n, r)\}$ *is a complete set of irreducible objects in $\mathcal{P}_K(n, r)$.*

(ii) $d_{N,n}F(\mu^*) \cong F(\mu)$ *as $\mathcal{P}_K(n, r)$-objects.*

(iii) *For all* $\mu \in \Lambda^+(n,r)$ *the following character equation is true in any characteristic:* $ch_{\mu,p}(\mathbf{X}) = ch_{\mu^*,p}(\mathbf{X},0,\ldots,0)$.

Proof For (i) note that for $\lambda \in \Lambda^+(N,r)$, $\lambda \in \Lambda(n,r)^* \Leftrightarrow \lambda = \mu^*$ for some $\mu \in \Lambda^+(n,r)$. Then we use 4.3.1. The other two parts are clear since both involve irreducible modules having a character with top term $\mathbf{X}^\mu$. □

Remark. For $N \geq n \geq r$, $\lambda \mapsto \lambda^*$ is a bijection between $\Lambda^+(n,r)$ and $\Lambda^+(N,r)$. So, given $\lambda \in \Lambda^+(N,r)$ there is a unique $\mu \in \Lambda^+(n,r)$ such that $\mu^* = \lambda$. Now, 4.3.5 shows that $d_{N,n} : \mathcal{P}_K(N,r) \to \mathcal{P}_K(n,r)$ induces a bijection between irreducible objects in both categories. Actually, this is a Morita equivalence:

Theorem 4.3.6 *Let* $N \geq n \geq r$. *The deflating Schur functor d defines an equivalence of categories between* $\mathcal{P}_K(N,r)$ *and* $\mathcal{P}_K(n,r)$. *Equivalently, the K-algebras* $S_K(n,r)$ *and* $S_K(N,r)$ *are Morita equivalent.*

Proof Take the partial inverse h to the general Schur functor f_e. That $hd \sim 1$ and $dh \sim 1$ is easily checked. □

Exercise

As an exercise in wielding Hom functors, the reader is invited to begin proving a useful theorem of James [1981], the end result of which leads to a *principle of row and column removal* given formally in 4.4.6 and 4.4.7.

Let $I_1(n,r)$ be the subset of $I(n,r)$ consisting of all indices for which $i_{r-m+1} = i_{r-m+2} = \cdots = i_r = 1$ and all other $i_\rho \neq 1$. Given $i \in I_1(n,r)$ let $\hat{\imath} \in I(n-1,r-m)$ be obtained after deleting the last m parts equal to 1 and changing i_ρ to $i_\rho - 1$ for $1 \leq \rho \leq r - m$. Thus $i \to \hat{\imath}$ is a bijection between $I_1(n,r)$ and $I(n-1,r-m)$ and we have a basis $\{\xi_{\hat{\imath},\hat{\jmath}} : i,j \in I_1(n,r)\}$ for $S_K(n-1,r-m)$. Put

$$e = \sum_{I_1(n,r)} \xi_{i,i}.$$

(1) Show that e is idempotent, and find a basis for eSe.
(2) Check that η annihilates any $V(\alpha)$ or $F(\alpha)$ when $\alpha_1 < m$, but does not annihilate $F(\alpha)$ when $\alpha_1 = m$. If μ has $\mu_1 = m$, and $\alpha_1 > m$, compute $[V(\mu) : F(\alpha)]$.

(3) Use the results about Hom functors studied above to classify all the possible $\eta S_K(n,r)\eta$ composition factors of $\eta V(\mu)$. Compute the multiplicities.

4.4 Decomposition numbers for G and Γ

The relationship between decomposition numbers for Σ_n and GL_n will be investigated using Schur functors. The highlight is James' theorem which states that the decomposition matrix for Σ_r is a submatrix of that for Γ.

Modular reduction

We return again to the general set-up of 3.5, where we considered the process of modular reduction. Thus R is a principal ideal domain with field of fractions Q and there is a ring homomorphism $\pi : R \to K$. Let S_Q be some finite-dimensional Q-algebra containing an R-order S_R.

Choose an idempotent $e = e_R \in S_Q$ lying in S_R. Then, if we have a full set of irreducibles $\{X(\lambda)_Q : \lambda \in \Lambda\}$ in $_{S_Q}\mathbf{mod}$, and if $\Lambda' = \{\lambda \in \Lambda : eX(\lambda)_Q \neq 0\}$, then $\{eX(\lambda)_Q : \lambda \in \Lambda'\}$ is a full set of irreducibles in $_{eS_Qe}\mathbf{mod}$. Similarly, $e_K = e_R \otimes 1_K$ is an idempotent in the K-algebra $S_K = S_R \otimes K$. Hence, if we have a full set of irreducibles $\{U(\delta)_K : \delta \in \Delta\}$ of $_{S_K}\mathbf{mod}$, and if $\Delta' = \{\delta \in \Delta : eU(\delta)_K \neq 0\}$, then $\{e_K U(\delta)_K : \delta \in \Delta'\}$ is a full set of irreducibles in $_{e_K S_K e_K}\mathbf{mod}$.

Now eS_Re is an R-order in the Q-algebra eS_Qe (it is an R-module direct summand of S_R); identifying $eS_Re \otimes K$ and $e_K S_K e_K$, we have the modular reduction

$$eS_Qe\mathbf{mod} \longrightarrow e_K S_K e_K \mathbf{mod}.$$

Thus it makes sense to talk of decomposition numbers $d_{\lambda\delta}(eSe)$ for $\lambda \in \Lambda', \delta \in \Delta'$. In fact Martins [1982] proved that there is a simple connection between these numbers and the decomposition numbers $d_{\lambda\delta}(S) = [\overline{X(\lambda)_Q} : U(\delta)_K]$ ($\lambda \in \Lambda, \delta \in \Delta$) defined in 3.5. Her result is

Theorem 4.4.1 *If* $\lambda \in \Lambda'$ *and* $\delta \in \Delta'$,

$$d_{\lambda\delta}(S) = d_{\lambda\delta}(eSe).$$

Proof We choose an R-form X_R of $X_Q = X(\lambda)_Q$. Since $X_R = eX_R \oplus$

$(1-e)X_R$, eX_R is an R-form of eX_Q. As usual, we identify the $e_K S_K e_K$-module $eX_R \otimes K$ with $e_K X_K$ where $X_K = X_R \otimes K$. By 4.1.8,

$$[e_K X_K : e_K U(\delta)_K] = [X_K : U(\delta)_K].$$

$\square$

Now we specialise: take $Q = \mathbf{Q}$, $R = \mathbf{Z}$ and K to be an infinite field of characteristic p. Let $\pi : \mathbf{Z} \to K$ be such that $\pi(n) = n1_K$. For fixed n, r, let $S_\mathbf{Q} = S(\Gamma_\mathbf{Q})$, $S_\mathbf{Z} = S(\Gamma_\mathbf{Z})$ and $S_K = S(\Gamma_K) \cong S_\mathbf{Z} \otimes K$. Corresponding to the sets above take $\{V(\lambda)_\mathbf{Q} : \lambda \in \Lambda^+(n,r)\}$ and $\{F(\lambda)_K : \lambda \in \Lambda^+(n,r)\}$, so here $\Lambda = \Delta = \Lambda^+(n,r)$. The relevant decomposition numbers are those integers $d_{\lambda\mu} = d_{\lambda\mu}(\Gamma)$ appearing in the character decomposition (see the end of 1.6):

$$\mathrm{ch}_{\lambda,0}(\mathbf{X}) = \sum_{\mu \in \Lambda^+(n,r)} d_{\lambda\mu} \mathrm{ch}_{\mu,p}(\mathbf{X}).$$

As intimated above, the proof of James' theorem requires the use of Schur functors, so we bring these into action now.

Proposition 4.4.2 *Let* $N \geq n$ *and* $* : \Lambda(n,r) \to \Lambda(N,r)$ *the injection appearing just after the proof of 4.3.2. Then*

(i) $d_{\lambda\mu}(\mathrm{GL}_n) = d_{\lambda^* \mu^*}(\mathrm{GL}_N)$ *for all* $\lambda, \mu \in \Lambda^+(n,r)$.

(ii) $d_{\lambda\mu}(\mathrm{GL}_N) = 0$ *for all* $\lambda \in \Lambda^+(N,r) \setminus \Lambda^+(n,r)$ *and* $\mu \in \Lambda(n,r)^*$.

Proof For (i) apply 4.4.1 with $\Lambda' = \Delta' = \Lambda^+(N,r) \cap \Lambda(n,r)^*$ and $e = e_\mathbf{Z} \in S_\mathbf{Q}$ as above. Note in addition that $e_K = e \otimes 1_K \in S_K$ is also as in 4.3.3. For (ii) note by 4.3.4(iii) that $e_K V(\lambda)_K = 0$. By 4.1.8 $[V(\lambda)_K : F(\mu)_K] = [e_K V(\lambda)_K : eF(\mu)_K]$, i.e. $d_{\lambda\mu}(\mathrm{GL}_N) = 0$, so (ii) is proved. $\square$

This proposition implies that the integers $d_{\lambda\mu}(\mathrm{GL}_n)$ (all n) are entries in the matrix

$$[d_{\lambda\mu}(\mathrm{GL}_r)]_{\lambda,\mu \in \Lambda^+(r)} \qquad (4.8)$$

For, with $r = n$ then $N \geq r$ and $* : \Lambda^+(r) \to \Lambda^+(N,r)$ is a bijection so that the decomposition matrix for GL_N is identical modulo $*$ to (4.8) by 4.4.2(i). With $r = N$, $n \leq r$, the map $*$ sends $\Lambda^+(n,r)$ bijectively onto itself, so that the decomposition matrix for GL_n is identical (modulo this bijection) to the submatrix of (4.8) obtained after deleting rows and columns indexed by those partitions having more than n parts.

Theorem 4.4.3 (James) *Let $r \leq n$. Given partitions λ, μ of r, where μ is p-restricted,*

$$d_{\lambda\mu}(\mathrm{GL}_n) = [\,\overline{S_{\lambda,\mathbf{Q}}} : D_{\mu,K}].$$

Thus, the p-modular decomposition matrix for Σ_r is a submatrix of (4.8) obtained by deleting those columns of (4.8) that are not labelled by p-restricted partitions.

Proof The irreducible $\mathbf{Q}\Sigma_r$ and $K\Sigma_r$ modules are respectively $\{S_{\lambda,\mathbf{Q}} : \lambda \in \Lambda^+(r)\}$ and $\{D_{\mu,K} : \mu \in \Lambda^{(p)}(r)\}$. Apply 4.4.1 with $S = S_{\mathbf{Q}}$, $e = \xi_\omega \in S_{\mathbf{Q}}$, $e_K = \xi_\omega \in S_K$. We know by 4.2.2 and the paragraph following 4.2.9 that $S_{\lambda,\mathbf{Q}} = eV(\lambda)_{\mathbf{Q}}$ and $D_{\mu,K} = e_K F(\mu)_K$ and so the theorem is proved. $\quad\square$

Remarks.
(1) Lists of matrices (4.8) for various characteristics are reproduced in James [1990b]. Decomposition matrices for G up to $r = 13$ for the primes 2 and 3 occur in James and Kerber [1981].
(2) Generalisations of James' theorem for the q-Schur algebras are detailed in Chapters 6 and 7.

Applications

We shall conclude this section with various important applications of the above theory. First of all we obtain, as a corollary of 4.4.3, a criterion for the Weyl module to be irreducible:

Corollary 4.4.4 *$V(\lambda)$ is irreducible if and only if λ is restricted and S_λ is irreducible.*

Proof One way is clear. Assume that $\lambda \in \Lambda^{(p)}(r)$ and that S_λ is irreducible. Hence the only factor of $V(\lambda)$ isomorphic to some $F(\mu)$ with $\mu \in \Lambda^{(p)}(r)$ is $F(\lambda)$ itself, by James' theorem. Since $F(\lambda)$ is a top factor of $V(\lambda)$, there can be no other factors of the form $F(\mu)$ with μ not restricted, since by 4.2.7 no Γ-submodule of $E^{\otimes r}$ can be isomorphic to $F(\mu)$. Thus $V(\lambda) = F(\lambda)$. $\quad\square$

An arithmetic criterion for the Specht module (and/or its dual) to be irreducible was conjectured by Carter and proved by James (see James [1978b] and James and Murphy [1979]). As we need this in Chapter 5, we give the statement.

Theorem 4.4.5 (Carter's Criterion) *Let $\nu_p : \mathbf{Z} \to \mathbf{N}_0$ be the usual p-adic valuation and let $\lambda \in \Lambda^+(n,r)$, with $n \geq r$. Then λ is p-regular and $S^\lambda = fM(\lambda)$ is simple if and only if, for all i,j,k with $(i,k),(j,k) \in [\lambda]$,*

$$\nu_p(h_{ik}^\lambda) = \nu_p(h_{jk}^\lambda).$$

In particular, λ is a p-core if and only if every h_{ij}^λ is coprime to p.

Remark. This result is actually a special case of Schaper [1981]. A classification of all irreducible Weyl modules for the reductive group Γ appears in Jantzen [1973].

Exercise (continued)

We now complete the exercise started at the end of 4.3. The reader should convince herself of the following steps before continuing:

(4) Let S_1 be the subspace of $\eta S \eta$ spanned by $\{\xi_{i,j} : i,j \in I_1(n,r)\}$. By careful application of Rule (2.8) show that $S_1 \cong S_K(n-1, r-m)$ (as algebras) via a map θ_1 sending $\xi_{i,j}$ to $\xi_{\bar{i},\bar{j}}$.

(5) Consider the subspace $E_1 = \langle e_1, \ldots, e_{n-1} \rangle$ of E. There is a linear map $\theta : E^{\otimes r} \to E_1^{\otimes(r-m)}$ that sends e_i to e_i if $i \in I_1$, otherwise to 0. Check that $\theta(\xi v) = \theta_1(\xi)\theta(v)$ for all $\xi \in S_1$ and $v \in E^{\otimes r}$.

(6) By analysing the action of θ on a basis of the Weyl module, find a basis of $\eta V(\lambda)$ and a basis of $V(\bar{\lambda})$ (where $\bar{\lambda}$ denotes λ with $\lambda_1 = m$ removed). Hence show that $\bar{\theta}$, the restriction of θ to $\eta V(\lambda)$, is an isomorphism into $V(\bar{\lambda})$.

(7) Use 3.4.10 to compute $\bar{\theta}(\eta V(\lambda)^{\max})$. Deduce that

$$\eta V(\lambda)/(\eta V(\lambda)^{\max}) = \eta F(\lambda)$$

is irreducible both as an S_1-module and as an S^η-module.

The next result proves enormously useful when doing examples, and will also be applied frequently in Chapter 5.

Theorem 4.4.6 (Principle of Row Removal) *We recall our assumption that $\lambda = (m, \lambda_2, \ldots, \lambda_n)$ and $\mu = (m, \mu_1, \ldots, \mu_n)$. Then $[\mu : \lambda] = [\bar{\mu} : \bar{\lambda}]$.*

Proof By step (3) of the exercise $\eta V(\mu)$ has an S^η-composition series with factors isomorphic to $\eta F(\hat{\mu})$ where $\hat{\mu}$ has first part equal to m. By step (6), this is an S_1-series. Then

$$[\mu : \lambda]_S = [\eta V(\mu) : \eta V(\lambda)]_{S^\eta} = [\eta V(\mu) : \eta F(\lambda)]_{S_1} = [\bar{\mu} : \bar{\lambda}]_{S(n-1,r-m)}.$$

The last integer equals $[\bar{\mu} : \bar{\lambda}]$, using 4.4.2. $\square$

Theorem 4.4.7 (Principle of Column Removal) *If* $\lambda, \mu \in \Lambda^+(n, r)$ *both have m non-zero parts, let* $\hat{\lambda} = (\lambda_1 - 1, \lambda_2 - 1, \ldots, \lambda_m - 1, \lambda_{m+1}, \ldots, \lambda_n)$, *and similarly for* μ. *Then* $[\mu : \lambda] = [\hat{\mu} : \hat{\lambda}]$.

Proof Take $n = m$ first of all, and tensor a series for $V(\hat{\mu})_K$ with the determinant module. For the general case, apply 4.4.2. $\square$

Remarks.
(1) By applying 4.4.3, one can obtain analogues of these results for Specht modules (James [1981, §4]).
(2) A generalisation of these results valid for algebraic groups appears in a note of Donkin [1983].

On Ext *groups*

We have already mentioned the difficulty of computing projective resolutions of Weyl modules, and have touched on Woodcock's version of the Kempf Vanishing Theorem (Woodcock [1992, Corollary 5.2]). Let us expand on this with a few thoughts about Ext groups for Γ and G. It should be mentioned that when we write $\mathrm{Ext}^*_{\Gamma_K}$ we are referring to the Hochschild cohomology $\mathrm{Ext}^*_{K[\Gamma]}$.

A result of Donkin [1986, (2.2)d], proved originally in the context of 'generalised Schur algebras', but which is actually clear from 3.3.7 and 2.2.7, shows that for $M, N \in \mathcal{P}_K$ there is a natural isomorphism

$$\mathrm{Ext}^*_{S_r}(M, N) \cong \mathrm{Ext}^*_{\Gamma_K}(M, N) \qquad (4.9)$$

where we pass between the two categories by means of 2.2.7 and inflation. This isomorphism can be 'understood' in terms of an embedding of certain derived categories (Parshall [1987, (3.3)] or CPS [1988a, §5]), but we shall not attempt a discussion here.

While we are investigating Schur functors, let us measure the effect of f on extension classes. Suppose one has an extension class Ξ defined by

$$0 \to M \to X \to N \to 0$$

in $\mathrm{Ext}^1_{S_r(\Gamma)}(N, M)$, and let Ξ' be the exact sequence

$$0 \to eM \to eX \to eN \to 0$$

where $e = \xi_\omega$. We now have a well-defined map $f\Xi = \Xi'$. Note that if M and N are simple and eM and eN are non-zero, then

Proposition 4.4.8 f *is injective on extension classes.*

This proves (Fettes [1985], Martin [1991]) the following:

Proposition 4.4.9

$$\mathrm{Ext}^1_{S_r}(F(\lambda), F(\mu)) \neq 0 \Rightarrow \mathrm{Ext}^1_G(D_\lambda, D_\mu) \neq 0$$

for restricted weights λ, μ *such that* $\lambda \triangleright \mu$ *and for* n *large enough.*

Nothing appears to be known about if or when the implication in 4.4.9 can be reversed: are there conditions under which the Ext^1 groups are isomorphic? The extent of our ignorance is vast: for symmetric groups in odd characteristic it is not even known whether or not $\mathrm{Ext}^1_G(D_\lambda, D_\lambda)$ is always zero, for λ restricted. It is true, however, that $\mathrm{Ext}^1_\Gamma(F(\lambda), F(\lambda)) = 0$ if $\lambda \in \Lambda^+(n, r)$, e.g. by 3.3.9.

4.5 Δ- and ∇-good filtrations

Bearing 4.4.3 in mind, it should appear natural to seek an interpretation of *all* the columns in the decomposition matrix (4.8) in terms of G. We know, of course, that columns of the decomposition matrix for G correspond to the components (i.e. the indecomposable direct summands) of the module $M^{(1^r)}$. The idea is to prove that every column of (4.8) gives a component of some M^λ. These components are trivial source modules, now known to the world as *Young modules*.

We do not follow the direct approach of James [1983, 3.1], which in any case only gives a weak version of 4.6.4 below. Rather we adapt ideas of Donkin on ∇-good filtrations to the representation theory of the quasi-hereditary algebra $S_K(n, r)$. As in 3.5 we work over the ring R, a local, complete, principal ideal domain with residue field k and with $K \supseteq k$.

Some filtrations

We recall the general framework for studying the quasi-hereditary algebra S with some given ordering of simple modules $(\Lambda, \leq)$; see 3.3 and 4.3.1. In particular let us formally restate one aspect implicit in the remark following 3.3.5.

Definition 4.5.1 We say that $M \in {}_S\mathbf{mod}$ has a *∇-good filtration* if $M \in \mathcal{F}(\nabla)$. Dually, M has a *Δ-good filtration* if $M \in \mathcal{F}(\Delta)$.

Remark. Originally Donkin defined 'good filtrations' of rational modules for connected reductive algebraic groups as increasing, exhaustive filtrations with quotients either zero or isomorphic to $H^0(\lambda)$ for some $\lambda \in X^+$. Of course, in characteristic zero $L(\lambda) = H^0(\lambda)$ and so every rational module of finite dimension has a good filtration (its composition series).

One can show that a direct sum of modules in $\mathcal{F}(\nabla)$ belongs to $\mathcal{F}(\nabla)$ and also that a direct summand of a module in $\mathcal{F}(\nabla)$ belongs to $\mathcal{F}(\nabla)$, e.g. see Donkin [1985].

Example. If $S = S_r$ then $\Lambda^a(E) \in \mathcal{F}(\nabla) \cap \mathcal{F}(\Delta)$ for all $a \in \mathbf{N}_0$. To see this, note that $\Lambda^0(E) = K$, $\Lambda^a(E) \cong \nabla(1^a) \cong \Delta(1^a)$ for all $a \in \mathbf{n}$ and $\Lambda^a(E) = 0$ for $a > n$.

Remark. Thorough investigations into $\mathcal{F}(\Delta)$ and $\mathcal{F}(\nabla)$ were conducted by Ringel [1991]. Either category has Auslander-Reiten sequences, and $\mathcal{F}(\Delta) \cap \mathcal{F}(\nabla)$ is the category add T of all direct sums of direct summands of a module T. This module T is a *(full) tilting module* and from it the quasi-hereditary structure of S is recoverable. In fact $T = \bigoplus_{\lambda \in \Lambda} T(\lambda)$ where $T(\lambda)$ is an indecomposable relative injective module in $\mathcal{F}(\Delta)$, with $\Delta(\lambda)$ embedding into it and with all factors of $T(\lambda)/\Delta(\lambda)$ of the form $L(\mu)$, $\mu < \lambda$. Donkin [1993] analysed the special case of the Schur algebra: one obtains one indecomposable summand $T(\lambda)$ of T for each element $\lambda \in \Lambda = \Lambda^+(n,r)$, as indicated in Remark (4) at the end of Chapter 1. The set of $T(\lambda)$ form a complete set of inequivalent indecomposable modules having the property that they are self-dual and lie in $\mathcal{F}(\Delta)$. They are called the *partial tilting modules* for S_r and are actually useful in that the multiplicity (see below) of $\nabla(\mu)$ in a ∇-good filtration of the partial tilting module of type λ is the decomposition number $d_{\mu'\lambda'}$, (Donkin [1993, p. 56]).

We return to our main theme. It is a fact (CPS [1988b, 3.8]) that in (S, Λ), for arbitrary $i \geq 0$ and $\lambda, \mu \in \Lambda$,

$$\text{Ext}_S^i(\Delta(\lambda), \nabla(\mu)) \cong \begin{cases} K & \text{if } i = 0 \text{ and } \lambda = \mu \\ 0 & \text{otherwise} \end{cases}. \qquad (4.10)$$

Thus $\text{Hom}_S(\Delta(\alpha), \nabla(\lambda)) = \text{End}(L(\lambda))$ if $\lambda = \mu$, otherwise it is zero. We conclude that if $M \in \mathcal{F}(\Delta)$ or $M \in \mathcal{F}(\nabla)$ then the *multiplicity* of $\Delta(\alpha)$ or $\nabla(\alpha)$ in the filtration (e.g. the number $|\{i > 0 : M_i/M_{i-1} \cong \nabla(\alpha)\}|$, in the case of ∇-good filtrations), is independent of the filtration chosen. Writing the multiplicity of $\Delta(\alpha)$ in a Δ-good filtration of $M \in \mathcal{F}(\Delta)$ as $(M : \Delta(\alpha))$ (and similarly for multiplicities in ∇-good filtrations), we have that $(M : \Delta(\alpha)) = \dim \text{Hom}_S(M, \nabla(\alpha))$, and $(M : \nabla(\alpha)) = \dim \text{Hom}(\Delta(\alpha), M)$. We now have the relation

$$(I(\lambda) : \nabla(\alpha)) = [\Delta(\alpha) : L(\lambda)].$$

The Authorities are divided on what to call this fact: some call it the 'Brauer-Humphreys reciprocity', others the 'Bernstein-Gelfand-Gelfand reciprocity' (depending on one's upbringing—see CPS [1988b p. 99]). It amounts to the assertion that the Cartan matrix for S has determinant 1. One might also view $(M : \nabla(\alpha))$ as the coefficient of the Weyl character $\text{ch } V(\alpha) = \text{ch } H^0(\alpha)$ in the expression for the formal character of M as a linear combination of Weyl characters.

It usually makes no difference whether one works over K or over some complete valuation ring. As stated in A.8, the general linear group scheme over R, Γ_R, is obtained from that over $\mathbf{Z}$ by base change; thus left modules M for Γ_R possess a structure map $M \to M \otimes_R R[\Gamma_\mathbf{Z}]$. One can define $S_r(\Gamma_R)$ as the dual to the coalgebra $A_R(n, r) = A_\mathbf{Z}(n, r) \otimes_\mathbf{Z} R$, where $A_\mathbf{Z}(n, r)$ is the rth homogeneous piece of $A_\mathbf{Z}(n) = \mathbf{Z}[c_{ij}]$.

We are particularly interested in what form these filtrations take in the case where $S = S_r(\Gamma_K)$. An example of this phenomenon the next theorem. The proof of this theorem makes use of a result of Jantzen that states that if L is algebraically closed and M is a finite-dimensional $S_r(\Gamma_L)$-module with a Δ-good filtration and v is a maximal weight vector of highest weight λ, then $S_r(\Gamma_L)v \cong V_L(\lambda)$ and $M/S_r(\Gamma_L)v$ has a Δ-good filtration. For the second assertion see Donkin [1985, (11.5.1)].

Theorem 4.5.2 *Let M be an object of $_{S(\Gamma_R)}\text{mod}$ free of finite rank. Then M has a Δ-good filtration as an $S_r(\Gamma_R)$-module if and only if $K \otimes_R M$ has a Δ-good filtration as a $S_r(\Gamma_K)$-module for every homomorphism*

$R \to K$, *where K is some algebraically closed field containing the residue class field $k = R/\mathcal{M}$.*

Proof The proof is by induction on the rank of M. Assume that Q (the quotient field of R) is algebraically closed. Suppose that $M \neq 0$, and assume that the result is true for all modules of smaller rank. Let $\bar{v}$ be a maximal weight vector of weight λ of $\bar{M} = M/\mathcal{M}M$, and lift it to a maximal weight vector v of M. Then v generates some $S_r(\Gamma_R)$-module, say V, and $\bar{v}$ generates some Weyl module in $\bar{M}$. Now V_Q and $S_r(\Gamma_Q)(v \otimes_R 1)$ can be identified (as submodules of M_Q) since tensoring with Q over R is exact. Hence $V_Q \cong V_Q(\lambda)$, a Weyl module over Q, by Jantzen's result referred to in the preamble. But, then, since V is R-free, we may identify it with the submodule $S_r(\Gamma)v \otimes 1 = S_r(\Gamma)(v \otimes 1)$ of V_Q. Hence $V \cong V_R(\lambda)$.

Consider the mapping $V_K \to M_K$ induced by the map $V \to M$. The image $S_r(\Gamma_K)(v \otimes 1)$ of this does not vanish, hence it must be isomorphic to $V_K(\lambda)$ (according to Jantzen again). Finally, since this latter module is isomorphic to V_K, there exists an exact sequence

$$0 \to V_K \to M_K \to (M/V)_K \to 0.$$

Thus M/V is torsion-free (over K), having a Δ-good filtration by induction since $(M/V)_K \cong M_K/S_r(\Gamma_K)(v \otimes 1)$ has a Δ-good filtration. The result follows. ☐

Let $\lambda \in \Lambda^+(n,r)$. Shifting our focus to the language of comodules, let $I_k(\lambda)$ be the injective hull of $L(\lambda)$ as an $A_k(n,r)$-comodule. In what follows, A_k refers to $A_k(n,r)$ and A_R refers to $A_R(n,r)$. We regard $I_k(\lambda)$ as a rational Γ-module, i.e. as a $k[\Gamma]$-comodule, 'by inflation'. Namely let i be the inclusion of A_k in $k[\Gamma]$. If $I_k(\lambda)$ has the structure map ω_I as an A_k-comodule, then $(1 \otimes i)\omega_I$ gives $I_k(\lambda)$ the structure of a $k[\Gamma]$-comodule.

By Green [1976, 2.4c] the injective hull is liftable to some summand $I_R(\lambda)$ of A_R; we think of this as a Γ_R-module by inflation. Given a homomorphism $R \to K$, as in 4.5.2, the Γ_K-module $I_R(\lambda)_K$ is isomorphic to a direct summand of A_K. Hence $I_R(\lambda)_K$ has a ∇-good filtration (A_K has a ∇-good filtration being a direct sum of modules with ∇-good filtrations, and direct summands of objects with ∇-good filtrations have ∇-good filtrations; see Donkin [1985, 3.2.5]). Hence $I_R(\lambda)$ has a ∇-good filtration by 4.5.2. Now $L(\lambda)$ is an absolutely irreducible Γ_k-module, and so $I_R(\lambda)$ has a ∇-good filtration with $(I_R(\lambda) : M_R(\mu)) = [M_K(\mu) : L(\lambda)]$, from the above reciprocity rules. Recalling the ordering (3.3) on the dominant weights $\Lambda^+(n,r)$, we deduce using 3.3.9 (or rather a version valid over R) an important result of Donkin [1987, (2.1)]:

Theorem 4.5.3 $I = I_R(\lambda)$ *has a ∇-good filtration* $0 = I_0, I_1, \ldots, I_t = I$ *where I_i/I_{i-1} is a direct sum of $[M_K(\lambda_i) : L(\lambda)]$ copies of $M_R(\lambda_i)$, for $1 \leq i \leq t = \Lambda^+(n,r)$.*

Now we quickly arrive at the dual situation: let $P_R(\lambda)$ be the projective indecomposable $S_r(\Gamma_R)$-module, with simple top $L(\lambda)$. Since $I_R(\lambda)$ is a component of A_R, the contravariant dual $I_R(\lambda)^\circ$ is a summand of $A_R^\circ = S_r(\Gamma_R)$, and hence is a projective indecomposable $S_r(\Gamma_R)$-module. But by 4.5.3 $M_R(\lambda)$ embeds in $I_R(\lambda)$, so we must have

$$I_R(\lambda)^\circ \cong P_R(\lambda).$$

The dual statement to 4.5.3 now reads as follows.

Theorem 4.5.4 $P = P_R(\lambda)$ *has a Δ-good filtration* $P = P_1, P_2, \ldots,$ $P_{t+1} = 0$ *where P_i/P_{i+1} is a direct sum of $[V_K(\lambda_i) : L(\lambda)]$ copies of $V_R(\lambda_i)$, for $1 \leq i \leq t$.*

Hence, over a field or a complete valuation ring every finitely-generated projective S_r-module lies in $\mathcal{F}(\Delta)$.

Hom functors and Hom spaces

Returning to the general case we now consider a generalisation of Donkin [1987, 2.3].

Theorem 4.5.5 *Suppose S is quasi-hereditary and contains an idempotent e with the property that Se is a faithful S-module. Assume that for each $\beta \in \Lambda$, $\nabla(\beta)$ is a homomorphic image of of the dual of S_S. If $M, N \in \mathcal{F}(\nabla)$ then the natural map*

$$\mathrm{Hom}_S(M, N) \to \mathrm{Hom}_{eSe}(eM, eN)$$

is injective.

Proof By the left exactness of $\mathrm{Hom}_S(M, -)$ and induction on $\dim N$, we may take $N = \nabla(\beta)$ for some $\beta \in \Lambda$. By the right exactness of $\mathrm{Hom}_S(-, N)$ and induction on $\dim M$ we take $M = \nabla(\alpha)$ for some $\alpha \in \Lambda$. Now S^* is an injective cogenerator and so there is an embedding

$\nabla(\beta) \to S^*$. By assumption there is a surjection $S^* \to \nabla(\alpha)$. Thus we merely need to show that the natural map

$$\operatorname{End}_S(S^*) \to \operatorname{End}_{eSe}(eS^*) \qquad (4.11)$$

is injective.

To see this note that the left-hand side of (4.11) is isomorphic as a K-space to $_sS$ under the identification of $s \in S$ with the map φ_s, where $\varphi_s(f) = fs$ $(f \in S^*)$. So, if φ_s is sent to zero under (4.11), then φ_s is identically zero on eS^*. We conclude that for all $f \in S^*$, $\varphi_s(ef) = (ef)s = 0$; so for all $\xi \in S$, $((ef)s)(\xi) = f(s\xi e) = 0$, whence $sSe = 0$. Since Se is faithful, $s = 0$. $\qquad\square$

Consequences for Schur algebras

We can now recast 4.5.5 in a form appropriate for Schur algebras. We assume henceforth that $r \leq n$, so that there is a Schur functor $f = f_e$ defined by $e = \xi_\omega$. Now $S_re \cong E^{\otimes r}$ and hence is faithful. Also $\nabla(\lambda) = M(\lambda)$ is a homomorphic image of $(S_rS_r)^* = A_r$, since there is an epimorphism from the injective module $E^{\otimes r}$ onto $\nabla(\lambda)$. So the assumptions of 4.5.5 hold and we have

Corollary 4.5.6 *Let* $M, N \in \mathcal{P}_K(n, r)$ *be finite-dimensional, with* ∇-*good filtrations. Then the natural map*

$$\operatorname{Hom}_{\Gamma_K}(M, N) \to \operatorname{Hom}_G(fM, fN)$$

induced by $f = f_K$ *is an injection.*

Proposition 4.5.7 *For* $\lambda, \mu \in \Lambda^+(n, r)$, *the natural map*

$$\operatorname{Hom}_\Gamma({}^\lambda A_K, {}^\mu A_K) \to \operatorname{Hom}_G(M_K^\lambda, M_K^\mu)$$

is an isomorphism.

Proof (Basically, this is Kostka Duality again, cf. 4.2.10.) Both right λ- and μ-weight spaces for A_K have ∇-good filtrations, being summands of objects having ∇-good filtrations; hence we certainly have an injection, by 4.5.6. We need to compare dimensions of the domain and codomain.

If ${}^\mu K[\Gamma]$ is the right μ-weight space of the coordinate ring of Γ then we have ${}^\mu A_K = A_K \cap {}^\mu K[\Gamma]$ and the inclusion $\operatorname{Hom}_\Gamma({}^\lambda A_K, {}^\mu A_K) \to \operatorname{Hom}_\Gamma({}^\lambda A_K, {}^\mu K[\Gamma])$ is actually an isomorphism. But ${}^\mu K[\Gamma]$ is isomorphic to the induced module $K_\mu \uparrow_T^\Gamma$. Therefore Frobenius Reciprocity implies the existence of an isomorphism

$$\text{Hom}_\Gamma(^\lambda A_K, {}^\mu A_K) \cong \text{Hom}_T(^\lambda A_K, K_\mu).$$

Hence the dimension equals $\dim {}^\lambda A_K{}^\mu$, the dimension of the left μ-weight space of $^\lambda A_K$. This in turn equals $K_{\lambda\mu}$, the (λ, μ)th Kostka number, which is also $\dim \text{Hom}_G(M_K^\lambda, M_K^\mu)$, by 'Young's Rule' (James [1978a, 13.19]), completing the proof. □

4.6 Young modules

We start with $E^{\otimes r}$ and decompose it as a direct sum of transitive permutation modules, one for each G-orbit on I; the set $\Lambda(n, r)$ parametrises the G-orbits. That is, $\lambda \in \Lambda(n, r)$ consists of all $i \in I$ of weight λ. Writing $M^\lambda = \langle e_i : i \in \lambda \rangle$ for the orbit, we have a decomposition $E^{\otimes r} = \bigoplus_{\lambda \in \Lambda(n,r)} M^\lambda$. So one can think of ξ_λ as the projection of $E^{\otimes r}$ onto M^λ with kernel $\bigoplus_{\alpha \neq \lambda} M^\alpha$. Clearly, $M^\lambda \cong M^\mu$ if and only if λ and μ lie in the same $\alpha \in \Lambda^+(n, r)$. Collecting together the multiplicities we obtain

$$E^{\otimes r} = \bigoplus_{\lambda \in \Lambda^+(n,r)} m_\lambda M^\lambda \tag{4.12}$$

where $m_\lambda \in \mathbf{N}$ are determined (for those who want to know) in Klyachko [1983/84, p. 51]. Since S_r is the endomorphism ring of (4.12) the importance of collecting detailed information on the indecomposable G-module summands of $E^{\otimes r}$ and/or the components of M^λ, together with the multiplicities, is suggested.

Definition 4.6.1 Let $M_R^\lambda = \bigoplus_{i=1}^s Y_i$ be a decomposition into indecomposable summands. Then there is a unique i such that $S^\lambda \subseteq Y_i$ and this Y_i is unique up to isomorphism. Y_i is called the *Young module* for λ, and we denote it by Y^λ.

For uniqueness, we recall the non-degenerate bilinear form, Φ, on M_R^λ such that the standard basis is orthonormal (James and Kerber [1981, 7.1.7]). Now for any submodule X, either $S^\lambda \subseteq X$ or $X \subseteq (S^\lambda)^\perp$ (the 'Submodule Theorem'). Since Φ is non-singular there must exist some index i such that $Y_i \not\subseteq (S^\lambda)^\perp$, hence $S^\lambda \subseteq Y_i$. Further, Y_i is unique since the sum in 4.6.1 is direct.

These modules were investigated first by Klyachko [1983/84]; the theory was developed by Grabmeier [1985], James [1983], Green [1985/86], Donkin [1987] and Erdmann [1993a]. We list some of the basic properties and then investigate in detail the effects of the Schur functor on Young modules. When the Schur functor is applied we need to assume that $r \leq n$.

Lemma 4.6.2

(i) *Young modules are liftable, hence they have an ordinary character. This character is well defined.*

(ii) *Young modules are self-dual.*

(iii) *$fI_K(\lambda)$ is a Young module; conversely every Young module is isomorphic to $fI_K(\lambda)$ for some $\lambda \in \Lambda^+(n,r)$.*

(iv) *If I is an indecomposable summand of A_R then fI is a Young module; conversely every Young module over R has the form fI for some indecomposable summand I of A_R.*

Proof For (i), since M_K^λ is liftable, so is every summand of it, hence these modules have ordinary characters associated to them. So every Young module over K has the form Y_K for some R-Young module Y and, conversely, Y_K is a K-Young module for every Young module Y over R. Also, given summands X of M^λ and Y of M^μ, $\dim_{QG}(X,Y)$ is the character inner product $(\operatorname{ch} X, \operatorname{ch} Y)$, and so the character is well defined.

For (ii), if Y is the Young module associated to M^λ then Y^* is a summand of $(M^\lambda)^*$. But M^λ is self-dual and $\operatorname{ch} Y = \operatorname{ch} Y^*$, since ordinary characters are self-dual. But the uniqueness of Y implies that $Y \cong Y^*$.

For (iii), let Y be a K-Young module and hence an indecomposable summand of M_K^λ for some $\lambda \in \Lambda^+(n,r)$. Taking $\lambda = \mu$ in 4.5.7 we deduce that primitive orthogonal decompositions of the identities of $\operatorname{End}_\Gamma(^\lambda A_K)$ and of $\operatorname{End}_G(M_K^\lambda)$ have the same length. The existence of a ∇-good K-filtration in 4.5.3 ensures that $fI \neq 0$ for each non-zero direct summand I of $^\lambda A_K$. Thus fI is a non-zero direct summand of M_K^λ, and every component appears as some fI. Since $I_K(\lambda)$ is an indecomposable summand of A_K, and since all components of A_K are injective indecomposable A_K-comodules, we have $Y \cong fI_K(\lambda)$ for some $\lambda \in \Lambda^+(n,r)$. Clearly, now, $fI_K(\lambda)$ is a Young module for any $\lambda \in \Lambda^+(n,r)$.

For (iv), we know that I_K is an indecomposable summand of A_K by the familiar idempotent lifting in $\mathrm{End}_{\Gamma_R}(A_R)$, hence $I_K = I_K(\lambda)$ for some $\lambda \in \Lambda^+(n,r)$; fI_K is indecomposable, therefore so too is fI. Now A_R is a sum of its weight spaces, hence fI is an indecomposable summand of $f\,^\lambda A_R \cong M_R^\lambda$ and hence is a Young module. The second clause follows similarly. $\qquad\square$

Remark. Thus we have yet another manifestation of the Schur algebra, namely as

$$S_R(n,r) = \mathrm{End}_{RG}\left(\bigoplus_{\lambda \in \Lambda^+(n,r)} n_\lambda Y_R^\lambda \right) \tag{4.13}$$

where the n_λ are multiplicities. For reasons that will now be explained, knowledge of these modules and the integers n_λ sheds light on the calculation of decomposition numbers for Γ and G.

Digression: Green Correspondence

The next result can be considered as a version of the well-known Green Correspondence for finite groups. The procedure is as follows. Let $\mathcal{Y}$ be the collection of all Young subgroups of G. Grabmeier shows that every Young module over K has a 'Young vertex', i.e. a minimal element $\Sigma_\rho \in \mathcal{Y}$ such that Y^λ is a component of M^ρ.

First, notice that any $\lambda \in \Lambda^+(n,r)$ has a unique 'p-adic expansion' of the form

$$\lambda = \sum_{i=1}^{s} \lambda_{(i)} p^{\alpha_i}$$

where $\lambda_{(i)} \in \Lambda^{(p)}(n,r)$ and $\alpha_1 > \ldots > \alpha_s = 0$. Here is an example to illustrate the procedure:

Example. Let $\lambda = (21, 8^2, 6, 5^3, 1^2) \in \Lambda(60)$, and take $p = 3$. Stripping off horizontal 3-hooks successively we have

```
x x x x x x  x-x-x  x-x-x  x-x-x  x-x-x  x-x-x
x x x x x x  x-x-x                           x x  x-x-x
x x x x x x  x-x-x                           x
x x x  x-x-x                                 x
x x  x-x-x                                   x
x x  x-x-x                                   x
x x  x-x-x                                   x
x                                            x
x
```

which gives a 3-adic expansion $\lambda = (6, 5^2, 3, 2^3, 1^2) + 3(2, 1^6) + 3^2(1)$.

The Green Correspondence (Grabmeier [1985, 7.8]) between $N_G(\Sigma_\rho)$ and G may now be stated.

Theorem 4.6.3 (Klyachko, Grabmeier) *Given λ and its p-adic expansion as above, define*

$$\rho = (p^{\alpha_1}, \ldots, p^{\alpha_1}, p^{\alpha_2}, \ldots, p^{\alpha_2}, \ldots, p^{\alpha_s}, \ldots, p^{\alpha_s})$$

with weight $(\beta_1, \ldots, \beta_s)$, where $\sum_j \lambda_{(i)j} = \beta_i$ for each i. Observe that the normaliser is $N = N_G(\Sigma_\rho) = \Sigma_\rho . \Sigma_\beta$.

(i) **(Green Correspondence for Young modules)** *The Green Correspondent $f(Y^\lambda)$ in N is*

$$f(Y^\lambda) = \begin{cases} trivial & as\ \Sigma_\rho\text{-module} \\ Y^{\lambda_1} \otimes \cdots \otimes Y^{\lambda_s} & as\ \Sigma_\beta\text{-module} \end{cases} .$$

Here Y^{λ_i} is the Young module of Σ_{β_i} corresponding to $\lambda_{(i)}$. Further, given a Young vertex Σ_ρ where $\Sigma_\rho \leq \Sigma_\lambda$ for $\Sigma_\lambda \in \mathcal{Y}$, the summands of M^λ with Young vertex Σ_ρ are precisely the Correspondents of the summands of $(K_{N_1}) \uparrow^{\bar{N}}$ where $\bar{N} = N/\Sigma_\rho$ and $N_1 = N_{\Sigma_\lambda}(\Sigma_\rho)/\Sigma_\rho$.

(ii) **(Klyachko Multiplicity Formula)** *Given $\lambda, \mu \in \Lambda^+(r)$, where μ has at most t non-zero parts, and writing $(Y^\lambda | M^\mu)$ for the multiplicity of Y^λ in a direct sum decomposition of M^μ, we have*

$$(Y^\lambda | M^\mu) = \sum_{\mu_{ij}} \prod_{j=1}^{s} (Y^{\lambda_{(j)}} | M^{(\mu_{1j}, \ldots, \mu_{tj})})$$

summed over all μ_{ij} satisfying $\sum_i \mu_{ij} = \beta_j$ and $\sum_j \mu_{ij} p^j = \mu_i$.

Proof For (i), the proof in Grabmeier [1985, 7.8] is straightforward. For (ii) see Klyachko [1983/84, 9.2], but another method is available using the Kostka duality 4.2.10, 4.5.7. For, one proves that $(I(\lambda)|S^\mu(E)) = \dim L(\lambda)^\mu$, and hence applying f and using Exercise (1) in 4.2, $\dim L(\lambda)^\mu = (Y^\lambda|M^\mu)$. Now, we apply the Steinberg Tensor Product Theorem; see Jantzen [1987, II.3.17]. □

The result (i) of 4.6.3 is used only in one of the examples in 5.6, but these combinatorial results perhaps have intrinsic interest for aesthetic reasons.

In the example above, by 4.6.3, Y^λ has a trivial source and Young vertex $\Sigma_{(9,3^8,1^{29})}$; also $fY^\lambda = Y^{(1)} \otimes Y^{(2,1^6)} \otimes Y^{(6,5^2,3,2^3,1^2)}$.

Young modules have Specht filtrations

Now we are in a position to prove the James-Donkin theorem. We know that I is an indecomposable direct summand of A_R if and only if I° is an indecomposable direct summand of $A_R{}^\circ = S_R$, that is, if and only if I° is a projective indecomposable $A_R^\circ = S_R$-module. But, using 4.6.2(ii) the dual of a Young module is a Young module, so we now have that $fP_R(\lambda)$ is a Young module and that every Young module is isomorphic to some $fP_R(\lambda)$. Let $Z_R^\lambda = fP_R(\lambda)$.

Theorem 4.6.4 (James, Donkin) Z_R^λ *has a (dual) Specht filtration, in the sense that there is a filtration*

$$Z_R^\lambda = Y_1 > Y_2 > \cdots > Y_{t+1} = \{0\}$$

where each section is a direct sum of $[M(\lambda_i) : L(\lambda)]$ *copies of* $S_{\lambda_i, R}$. *Thus any principal indecomposable module for* RG *has a (dual) Specht filtration.*

Proof Apply f to 4.5.4. The last clause of 4.6.4 holds since for λ restricted Z_R^λ is the projective cover of $fL(\lambda)$ and we have all indecomposable projectives this way. □

The genesis of this remarkable result is interesting: James [1983] proved that the character of the Young module is precisely the one predicted by 4.6.4. Then Donkin [1987] exhibited a filtration to realise the numerical result. Another comment worth making is to the effect that in general the filtration multiplicity of a KG-module with a Specht series is not well defined. Take $r = 2$ and $p = 2$ and let $M = Y^{(1^2)} = K\Sigma_2$. Then there are three short exact sequences

$$
\begin{array}{ccccccccc}
0 & \to & S^{(2)} & \to & M & \to & S^{(2)} & \to & 0 \\
0 & \to & S^{(2)} & \to & M & \to & S^{(1^2)} & \to & 0 \\
0 & \to & S^{(1^2)} & \to & M & \to & S^{(1^2)} & \to & 0
\end{array}
$$

So $S^{(1^2)}$ occurs with multiplicities 0,1 and 2 in M, respectively.

Example. Take $n = r = 7$ and suppose K is a field of characteristic 3. The decomposition matrix was defined as in (4.8) to be the matrix whose (λ, μ)th entry is $[V(\lambda) : F(\mu)]$. In our case it can be computed by hand or gleaned from published tables (James [1980]).

	1	2	3	4	5	6	7	8	9
(1^7)	**1**								
$(2,1^5)$		1							
$(2^2,1^3)$	**1**	1							
$(2^3,1)$		1	1						
$(3,1^4)$			1						
$(3,2,1^2)$	**1**	1	1	1					
$(3,2^2)$			1	1					
$(3^2,1)$	**1**				1				
$(4,1^3)$			1		1				
$(4,2,1)$	**1**	1	1		1	1			
$(4,3)$	**1**					1	1		
$(5,1^2)$				1			1		
$(5,2)$		1			1	1	1	1	
$(6,1)$			1					1	
(7)	**1**							1	1

The columns in bold type (which are indexed by 3-restricted partitions) form the submatrix comprising the 3-modular decomposition matrix of Σ_7, i.e. the entry in row λ and column μ is $[S_\lambda : D_\mu]$.

Now, for example, the Young module $Y^{(1^7)}$ has a dual Specht filtration with filtered quotients $S_{(1^7)}, S_{(2^2,1^3)}, S_{(3,2,1^2)}, S_{(4,2,1)}, S_{(4,3)}$ (reading down the first column), and similarly we can read off the filtered quotients of any of the others. Here are a few more:

$$
\begin{array}{ccl}
Y^{(3,2^2)} & \sim & S_{(3,2^2)} + S_{(6,1)} \\
Y^{(3^2,1)} & \sim & S_{(3^2,1)} + S_{(5,1^2)} \\
Y^{(4,1^3)} & \sim & S_{(4,1^3)} + S_{(4,2,1)} + S_{(5,2)}
\end{array}
$$

$$Y^{(5,2)} \quad \sim \quad S_{(5,2)} + S_{(7)}$$
$$Y^{(6,1)} \quad \sim \quad S_{(6,1)}$$

At this point, one could use Young's Rule to obtain the Specht factors of any M^λ, and then decompose M^λ into its Young components using the above list.

Highest weights for PIMs

Here is a reformulation of the Mullineux Conjecture. Recalling 4.2.9 *et seq.* one can define the involution i on $\Lambda^{(p)}(r)$ by putting $D_\lambda \otimes K_a \cong D_{i(\lambda)}$ where λ is p-restricted. Hence we have a bijection $\lambda \mapsto i(\lambda)'$ between the set of p-restricted partitions and the set of p-regular partitions.

We fix some restricted λ and deduce from 4.4.3 that, for $\alpha \in \Lambda^+(n, r)$,

$$d_{\alpha\lambda} = [S_\alpha : D_\lambda] = [S_\alpha \otimes K_a : D_\lambda \otimes K_a] = [S_{\alpha'} : D_{i(\lambda)}] = d_{\alpha',i(\lambda)}$$

(recalling that the sign character transposes labels of ordinary Specht modules and then dualises, and using self-duality of the simple KG-modules, cf. James [1978a, 6.7, 11.5]). Thus if $d_{\alpha\lambda} \neq 0$ we have $\alpha' \geq i(\lambda)$, proving $\alpha \leq i(\lambda)'$. But $d_{i(\lambda)',\lambda} \neq 0$, and so from 4.6.4 and 4.6.2, we have the following connection between weights of PIMs for Schur algebras, in terms of the involution i:

Proposition 4.6.5 *If* $\lambda \in \Lambda_{(p)}(r)$ *then* $i(\lambda)'$ *is the unique highest weight of* $P_K(\lambda)$. *Dually, if* $\lambda \in \Lambda^{(p)}(r)$ *then* $i(\lambda)$ *is the unique highest weight of* $I_K(\lambda)$.

We complete this chapter with some facts about $E^{\otimes r}$ and its summands that will prove useful in our analysis of blocks in Chapter 5. In this part, we work over K.

Proposition 4.6.6 *Assume* $n \geq r$.

(i) $E^{\otimes r} = \bigoplus I(\lambda)^{d_\lambda}$ *where* $d_\lambda = \dim fL(\lambda)$ *and the sum is over all p-restricted weights.*

(ii) *If* $\lambda \in \Lambda^{(p)}(r)$, $I(\lambda)$ *is self-dual; if* $\lambda \in \Lambda_{(p)}(r)$, $P(\lambda)$ *is self-dual.*

(iii) *The filtration multiplicity* $(E^{\otimes r} : M(\mu))$ *equals* $\dim fM(\mu)$.

Proof For (i), there is certainly a decomposition

$$E^{\otimes r} = \bigoplus_{\Lambda^+(n,r)} I(\lambda)^{d_\lambda}$$

for certain d_λ since $E^{\otimes r}$ is both projective and injective by 4.2.6. For $M \in \mathcal{P}_K(n,r)$,

$$\begin{aligned}
\mathrm{Hom}_{S_r}(M, E^{\otimes r}) &\cong \mathrm{Hom}_{S_r}(E^{\otimes r}, M^\circ) \cong \mathrm{Hom}_{S_r}(S_r e, M^\circ) \\
&\cong (M^\circ)^\omega \cong M^\omega = fM
\end{aligned}$$

where $e = \xi_\omega$. In particular, choosing $M = L(\lambda)$ with $\lambda \in \Lambda^+(n,r)$, and applying 'Schur's Lemma' yields $d_\lambda = \dim fL(\lambda)$. Thus $d_\lambda = 0$ if $\lambda \notin \Lambda^{(p)}(n,r)$.

For (ii), using 4.2.5, we embed $L(\lambda)$ into $E^{\otimes r}$. Using the fact that tensor space is self-dual, and using 3.4.3, $L(\lambda)$ is also an epimorphic image of $E^{\otimes r}$. Since $E^{\otimes r}$ is projective, $P(\lambda)$ is a component of $E^{\otimes r}$. As $E^{\otimes r}$ is self-dual, $P(\lambda)^\circ$ is also a component of $E^{\otimes r}$, and so $P(\lambda)^\circ$ is a component of $S_K(n,r)$. It follows that $P(\lambda)^\circ$ is a projective indecomposable of $S_K(n,r)$, so, since it is also isomorphic to a direct summand of $E^{\otimes r}$, 4.2.5 implies that $P(\lambda)^\circ \cong P(\mu)$ for some p-restricted μ. As both have the same characters by 3.4.6(iv), and $i(\lambda)' = i(\mu)'$ by 4.6.5, we get $P(\lambda)^\circ \cong P(\lambda)$. The result for $I(\lambda)$ is obtained dually.

For (iii), the multiplicity is independent of the characteristic, so without loss, we can work in characteristic zero. All S_r-modules are now completely reducible by 2.2.8, so $M(\lambda) = I(\lambda)$ for all $\lambda \in \Lambda^+(n,r)$. By (ii) $(E^{\otimes r} : M(\mu)) = \dim M(\mu)^\omega = \dim fM(\mu)$. This remains valid in finite characteristic as well. □

The last result is included for interest only. It gives a criterion for socles of Weyl modules to be simple, (see for example Donkin [1987, (2.10)]).

Corollary 4.6.7 *If λ is p-restricted then $V(i(\lambda)')$ has a simple socle. Thus if $\mu \in \Lambda^+(n,r)$ is p-regular $V(\mu)$ has a simple socle.*

Proof By construction, $V_K(i(\lambda)')$ embeds into $P_K(\lambda)$. Thus $V_K(i(\lambda)')^\circ$ is a homomorphic image of $P_K(\lambda)$, the latter module possessing a simple head. □

5

Block theory

We have seen in 3.5 and 4.4 that Brauer's theory of modular reduction for finite groups has analogues both for $\mathrm{GL}_n(K)$ and for the Schur algebras. Another important ingredient in Brauer's modular theory is the notion of a block. Since block decompositions exist for any finite-dimensional algebra, it is natural to seek an understanding of the blocks of the Schur algebras. In the light of our study of Schur functors, there is enough evidence to suspect that the answer may have something to do with the block decomposition of symmetric groups and of general linear groups.

Let us make the assumption, valid throughout Chapter 5, that K is algebraically closed. Now a block decomposition of KG induces a block decomposition of $E^{\otimes r}$, and hence a block decomposition of $S(n,r) = \mathrm{End}_{KG}(E^{\otimes r})$. The reader may recall that the blocks of G are described by the combinatorial rule stated originally by Nakayama. In his paper, Donkin [1987, (2.12)] demonstrated that the same rule applied provided $r \leq n$. It took another five years for the case $r > n$ to be completed— here a modified version of Nakayama's rule holds. The proof given below is based on Donkin [1992], which in turn is based on the determination of the blocks of a semisimple simply connected affine algebraic group (Donkin [1980]).

The proof is lengthy, though the techniques are simple enough. We concentrate on those Γ-blocks that 'look like' G-blocks and show that the Nakayama rule holds for them. Then we show that any block can be obtained from these so-called primitive blocks by tensoring with the Steinberg module and twisting by the frobenius map. This induces an equivalence of categories. At points, we need to quote, without proof, a selection of results pertaining to block decompositions of $K\Gamma$; this hinges on properties of the affine Weyl group that we shall discuss in 5.1.

The second part of this chapter deals with families of S_r-blocks. In her paper, Scopes [1991] verified the Donovan Conjecture for G by finding a sufficient combinatorial condition for Morita equivalence between two G-blocks of the same weight. We seek an analogue of this result for the Schur algebras. We close the chapter with a few explicit examples of blocks of Schur algebras (in fact we concentrate on components of Schur algebras coming from blocks of KG of finite type). This includes a survey of recent work of Erdmann [1993a], Xi [1991], [1992] and Donkin and Reiten [1993].

5.1 Summary of block theory

Let us work initially with an arbitrary K-algebra S. If we have an orthogonal central idempotent decomposition $1_S = \sum_i e_i$ of the identity, summed over some finite indexing set $\mathbf{m}$, say, then we obtain a direct sum decompositon of $S = \bigoplus_{i \in \mathbf{m}} B_i$ into two-sided ideals $B_i = e_i S$, and vice versa. Assuming at least that S is artinian, such a decomposition into indecomposable ideals is unique up to ordering of the summands: these ideals are the *blocks* of S (or more correctly the *p-blocks*, since the decomposition will depend on p=char$K$). If M is an indecomposable S-module then $M = \bigoplus_{i \in \mathbf{m}} e_i M$ and so there is some i such that $e_i M = M$ and $e_j M = 0$ for $j \neq i$: one says that M *belongs to* the block B_i. This distributes the simple modules and indecomposable projective modules into blocks. Also if M belongs to B then so do all its composition factors; see for example Curtis and Reiner [1987, §56].

It is usual in a first course on modular representations to find alternative criteria for membership of blocks. We remind the reader of two such criteria now.

Linkage for PIMs. The PIMs Q_1 and Q_2 for S are *linked* if there is a sequence of PIMs

$$Q_1 = A_1, A_2, \ldots, A_m = Q_2$$

such that for each i, A_i and A_{i+1} have a common composition factor. This is an equivalence relation on the set of indecomposable projectives of S: if $C_1, \ldots, C_n$ are the linkage classes, then one identifies the block B_i as the direct sum of all indecomposable projective left ideals belonging to C_i. Two PIMs are linked if and only if they are isomorphic to direct summands of a single block of S (Curtis and Reiner [1987, §56.12]), therefore no two distinct blocks have a common composition factor.

Simples in blocks. At least if S is a symmetric algebra, it is well known that the simple S-modules $L_1 \neq L_2$ lie in the same block if and only if there is a sequence of simple modules

$$L_1 = F_1, F_2, \ldots, F_r = L_2$$

such that for each i, $\mathrm{Ext}_S^1(F_i, F_{i+1}) \neq 0$.

In this chapter we deal with the blocks of G, Γ and S_r. Simple modules for any of these structures are in one-to-one correspondence with certain partitions of r, so we tend to regard blocks as subsets of the respective indexing sets.

Let us move slowly from the well-known to the less well-known. We recall that if $\lambda \in \Lambda^+(r)$ then the p-core, $\tilde{\lambda}$, of λ is the partition whose Young diagram is obtained from the diagram of λ by removing as many rim p-hooks (a connected part of the edge consisting of p nodes) as possible. All elements of a block have the same p-core. We conclude that the p-blocks of G (henceforth called G-blocks) are parametrised by p-cores, so that a G-block corresponds to a complete *core class* of partitions. The precise statement reads:

Theorem 5.1.1 (Nakayama Rule) $\lambda, \mu \in \Lambda^+(r)$ *are in the same p-block of* Σ_r *if and only if there exists a finite permutation* σ *such that for all* i

$$\lambda_i - i \equiv \mu_{\sigma(i)} - \sigma(i) \pmod{p}.$$

For a proof of this result, which is known to Mankind as the 'Nakayama Conjecture', but which we prefer to call the *Nakayama Rule*, and for references to all previous proofs, see Meier and Tappe [1976].

We now identify an $S(n, r)$-block with a subset of $\Lambda^+(n, r)$ as follows. Let $B_S(\lambda)$ be the $S(n, r)$-block containing λ and let $B_\Gamma(\lambda)$ be the Γ-block containing λ. By asserting that the partition μ belongs to $B_\Gamma(\lambda) \cap \Lambda^+(n, r)$, we mean that there is an integer t and a sequence of indecomposable rational Γ-modules $M_1, \ldots, M_t$ such that $L(\lambda)$ is a composition factor of M_1, $L(\mu)$ is a composition factor of M_t and for all $i < t$, M_i and M_{i+1} have some composition factor in common (i.e. are linked). That μ belongs to $B_S(\lambda)$ means that all the M_i can be chosen in $\mathcal{P}_K(n, r)$. Since $\mathcal{P}_K(n, r)$ is a full subcategory of $_{K\Gamma}\mathbf{mod}$, elements of $\Lambda^+(n, r)$ in the same $S(n, r)$-block are also in the same Γ-block. Hence it is immediate that $B_\Gamma(\lambda) \cap \Lambda^+(n, r)$ is a union of $S(n, r)$-blocks—the force of 5.1.3 below is in its assertion that $B_\Gamma(\lambda) \cap \Lambda^+(n, r)$ is *precisely* $B_S(\lambda)$.

Definition 5.1.2 Call $\lambda = (\lambda_1, \ldots, \lambda_n) \in \Lambda^+(n, r)$ $d(\lambda)$-*weighted*, if

$$d(\lambda) = \max\{d \geq 0 : \lambda_i - \lambda_{i+1} \equiv -1 \pmod{p^d} \ \forall i \in \mathbf{n}\}.$$

The block B is *primitive* if some λ belonging to B is 0-weighted. This would imply that all partitions in B are 0-weighted by 5.1.6 below.

Remark. In general I call this the *weighting* of a block, not to be confused with its *p-weight* (by which we mean the number of rim *p*-hooks removed from $\lambda \in B$ in order to reach the *p*-core $\tilde{\lambda}$). In the sequel, we perform the usual operations on Young diagrams, such as using the hook graph, stripping off hooks, etc. Such a diagram calculus is set out in James and Kerber [1981, 6.2].

We may now state the theorem of Donkin [1992]:

Theorem 5.1.3 (Donkin) $\lambda, \mu \in \Lambda^+(n, r)$ *belong to the same S-block if and only if they are both* $d = d(\lambda)$-*weighted and there is a finite permutation* σ *such that for all* $i \in \mathbf{n}$

$$\lambda_i - i \equiv \mu_{\sigma(i)} - \sigma(i) \pmod{p^{d+1}}.$$

Thus, for $\lambda \in \Lambda^+(n, r)$

$$B_S(\lambda) = B_\Gamma(\lambda) \cap \Lambda^+(n, r).$$

Blocks of Γ and the affine Weyl group

We begin by examining blocks in the context of rational representations of Γ. We use the notation given in the Appendix, but we shall need to quote a few additional results from the literature. Remember that we are regarding blocks of Γ as subsets of the dominant weight lattice X^+ via the identification $\lambda \leftrightarrow L(\lambda)$; see A.3.

Lemma 5.1.4 *If* λ *is dominant and minimal in its block then* $H^0(\lambda)$ *is irreducible.*

Proof Since λ is minimal, $H^0(\lambda)$ has no composition factors other than its socle. Moreover the socle occurs with multiplicity one. $\square$

Before classifying the Γ-blocks, let us recall the affine Weyl group and its 'dot action' on $X_{\mathbf{R}} = \mathbf{R}^n$ (see A.6). Namely, for fixed $i \in \mathbf{n}$, define the element $\chi_i \in X_{\mathbf{R}}^*$ by $\chi_i(\lambda) = \lambda_i - i$. Then W_p may be defined as the group of affine transformations w of $X_{\mathbf{R}}$ such that there exists a permutation $\pi \in G$ with the property that

$$\chi_i(w(\lambda)) = \chi_{\pi(i)}(\lambda) + \kappa_i p$$

($i \in \mathbf{n}$), where the κ_i are integers whose sum is zero. This action may be defined concisely by defining $w \cdot x = w(x + \delta) - \delta$ for $w \in W_p$, $x \in \mathbf{R}^n$ where the shift is by the weight $\delta = (n-1, n-2, \ldots, 0)$.

Now, the weight $\lambda \in X_{\mathbf{R}}$ is p-regular if and only if it fails to belong to any affine hyperplane $H_{ij} = \{\mu \in X_{\mathbf{R}} : \chi_i(\mu) - \chi_j(\mu) = \kappa p\}$ for $1 \leq i < j \leq n$ and $\kappa \in \mathbf{Z}$. The p-regular elements comprise a non-connected open set in $X_{\mathbf{R}}$, whose connected components are the alcoves. In A.6 we denoted the standard alcove by $C_0 = \{\lambda \in X_{\mathbf{R}} : \chi_1(\lambda) > \cdots > \chi_n(\lambda), \chi_1(\lambda) - \chi_n(\lambda) < p\}$, and its closure by $\overline{C_0}$.

An important result in reductive group theory is the 'Linkage Principle' for the reductive group $\mathbf{G}$ (Jantzen [1987, 6.17]). This states that if λ, μ are dominant then

$$\mathrm{Ext}^1_{\mathbf{G}}(L(\lambda), L(\mu)) \neq 0 \Rightarrow \lambda \in W_p \cdot \mu.$$

Taking $\mathbf{G} = \Gamma$ we deduce the following.

Corollary 5.1.5 *If* $\lambda \in X^+$ *then* $B_\Gamma(\lambda) \subseteq W_p \cdot \lambda \cap X^+(T)$.

We shall need the classification of the Γ-blocks given by Donkin [1980], [1992, §1(2)]. Only a sketch proof is offered.

Proposition 5.1.6 (The blocks of $K\Gamma$) *If* λ *is dominant,*

$$B_\Gamma(\lambda) = (\Sigma_n.\lambda + p^{d(\lambda)+1}\mathbf{Z}\Phi) \cap X^+.$$

Proof   Assume that this is true for the blocks of SL_n; this is justified by Donkin [1980]. We claim that the result for SL_n implies the result for Γ.

Recall the notation set out in A.7. Let $R = \mathrm{Ker}(X(T) \to X(T_1))$ so that $R \cap \mathbf{Z}\Phi = 0$ (by identifying Φ with the root system of SL_n with respect to T_1). If $\lambda \in X(T)$, let $\lambda \downarrow$ be its restriction to T_1. If $\mu \in X(T_1)^+$, let $\bar{H}^0(\mu) = K_\mu \uparrow^{\mathrm{SL}_n}_{B_1}$ have socle $\bar{L}(\mu)$. It is a fact that if $\lambda \in X(T)^+$ then $H^0(\lambda) \downarrow_{\mathrm{SL}_n} \cong \bar{H}^0(\lambda)$ (see Donkin [1985, 3.2.7]), and further $L(\lambda) \downarrow_{\mathrm{SL}_n} \cong \bar{L}(\lambda)$ since $\Gamma = \mathrm{SL}_n.Z(\Gamma)$. Thus, if $\lambda, \mu \in X(T)^+$ and $\lambda - \mu \in \mathbf{Z}\Phi$ we have the following reciprocity:

$$[H^0(\lambda) : L(\mu)] = [\bar{H}^0(\lambda) : \bar{L}(\mu)]. \tag{5.1}$$

So the result follows if we can prove that

Restriction $X(T) \to X(T_1)$ *induces a bijection that takes* $B_\Gamma(\lambda)$ *to* $B_{\mathrm{SL}}(\lambda \downarrow)$, *the* SL_n-*block containing* $\lambda \downarrow$. Of course this restriction gives a bijection modulo $\mathbf{Z}\Phi$. But the dominant weight μ is in the block of λ if and only if we can find a linkage chain of elements

$$\lambda = \sigma_1, \ldots, \sigma_m = \mu$$

in $X(T)^+$ such that $[H^0(\sigma_i) : L(\sigma_{i+1})] \neq 0$ or $[H^0(\sigma_{i+1}) : L(\sigma_i)] \neq 0$ for each $i < m$. One consequence is that $B_\Gamma(\lambda) \subseteq \lambda + \mathbf{Z}\Phi$, so $B_\Gamma(\lambda) \to X(T_1)$ is injective. But we showed initially that if $\mu \in X^+$ is in the same block as λ then $\mu \downarrow$ is in the block of $\lambda \downarrow$. Hence $B_\Gamma(\lambda) \to B_{\mathrm{SL}}(\lambda \downarrow)$ is injective. For surjectivity, pick $\xi \in B_{\mathrm{SL}}(\lambda \downarrow)$, so that there is a chain

$$\lambda \downarrow = \pi_1, \ldots, \pi_m = \xi$$

in $B_{\mathrm{SL}}(\lambda \downarrow)$ with $[\bar{H}^0(\pi_i) : L(\pi_{i+1})] \neq 0$ or $[\bar{H}^0(\pi_{i+1}) : L(\pi_i)] \neq 0$ for $i < m$. Let σ_i be the unique weight in $\lambda + \mathbf{Z}\Phi$ that restricts to π_i ($i \in \mathbf{m}$). By (5.1) $\sigma_m \downarrow = \xi$ is in the image of restriction $B_\Gamma(\lambda) \to B_{\mathrm{SL}}(\lambda \downarrow)$, as required. $\qquad\square$

We recall from A.6 that $\overline{C_0}$ is a fundamental domain for the action of W_p on $X_{\mathbf{R}}$, and that in A.6 we picked the element $w_0(\lambda) \in \overline{C_0}$ for each λ. A particular consequence of 5.1.6 (for SL_n) is

Lemma 5.1.7 *The primitive weights* $\lambda, \mu \in X_1^+$ *are in the same* SL_n-*block if and only if* $w_0(\lambda) = w_0(\mu)$.

Rewriting 5.1.6 another way, we see that λ and μ are in the same Γ-block if and only if they are both d-weighted and there is a permutation $\pi \in \Sigma_n$ such that $\lambda_i - i \equiv \mu_{\pi(i)} - \pi(i) \pmod{p^{d+1}}$ for all $i \in \mathbf{n}$. We give this a more combinatorial formulation as follows. Recall that $\omega = (1^n)$ is the Schur weight, and let δ be the weight $(n-1, n-2, \ldots, 0)$.

Corollary 5.1.8 *Let* $\lambda, \mu \in \Lambda^+(n, r)$. *Let* $a = \min\{\lambda_n, \mu_n\}$. *Then the following are equivalent:*

(i) λ *and* μ *are in the same* Γ-*block;*

(ii) λ *and* μ *are* d-*weighted and* α *and* β *have the same* p-*core where* α *and* β *are defined by*

$$\lambda = a\omega + (p^d - 1)\delta + p^d\alpha, \qquad \mu = a\omega + (p^d - 1)\delta + p^d\beta.$$

The next lemma in this section may be deduced from Humphreys and Jantzen [1978, p. 503, Corollary]. It is seen to be valid also for GL_n by re-writing their proof (which was for semisimple groups) in terms of Schur functions. Another proof appears in Donkin [1992, §1(5)].

Lemma 5.1.9 *If $\lambda \in X^+$ is primitive and $H^0(\lambda)$ is irreducible then $(\lambda, \check{\alpha}) < p$ for all $\alpha \in \Pi$, that is, λ is p-restricted (and dominant).*

Low-rank examples

We now prove a useful lemma concerned with groups of Lie rank at most two. It suggests the importance of minimal primitive weights in Γ-blocks, and indeed 5.1.10 will be the base step in an induction proof of 5.1.3.

Lemma 5.1.10 *Let $n \in 3$ and write* **G** *for any of the six groups* GL_n *and* SL_n. *Let $\lambda \in X^+$ be primitive. Then*

  (i) *The induced module $H^0(\lambda)$ is simple if and only if λ is minimal in its block.*

  (ii) *Each primitive* **G**-*block has a unique minimal member.*

Proof Since SL_1 is trivial and GL_1 is K^*, the result is immediate in the case $n = 1$. We concentrate on the SL cases, since the GL arguments are similar.

Type A_1: Consider $n = 2$. Now SL_2 has rank one and $X(T_1)$ is a free **Z**-module of rank one generated by the character $\operatorname{diag}(k, k^{-1}) \mapsto k$ where $k \in K^*$. So identify $r \in$ **Z** with $r\rho \in X$, to obtain an identification of **Z** with X, and hence of N_0 with X^+.

If $H^0(r)$ is simple, then as λ is primitive, $0 \le r < p$ by 5.1.9. Thus $B(r) = (\Sigma_2.r + 2p\mathbf{Z}) \cap N_0 = (r + 2p\mathbf{Z}) \cap N_0 \cup (-r - 2 + 2p\mathbf{Z}) \cap N_0$. This implies that r is minimal in its block.

Conversely, if λ is minimal in its block then $H^0(r)$ is simple by 5.1.4.

Take r primitive and minimal, to get $0 \le r < p$. If s is also minimal in $B(r)$, then also $0 \le s < p$. Since $w_0(r) = w_0(s)$ by 5.1.7, $r = s$.

Type A_2: Let us now consider the case of rank two. In this case the simple roots are $\Pi = \{\alpha, \beta\}$ and fundamental dominant weights are $\{\omega_\alpha, \omega_\beta\}$. Identify X and $\mathbf{Z}^2$ and X^+ with N_0^2 in the usual way. If $H^0(\lambda)$ is simple and $\lambda = (a, b)$ is primitive, then $0 \le a, b < p$ and a, b cannot both be $p - 1$. So λ is minimal.

If $\lambda \in \bar{C}_0$ then, of course, $H^0(\lambda)$ is irreducible. Otherwise $a+b+2 > p$, say $a + b + 2 = p + q$. If neither of a or b equals $p - 1$ then $[H^0(\lambda) : L(\mu)] = 1$ where $\mu = \lambda - q(\omega_\alpha, \omega_\beta)$, by direct calculation. Hence, putting $Z = \{\lambda \in X^+ : \lambda$ is primitive and $H^0(\lambda)$ is irreducible $\}$, we have

$$Z \subseteq \bar{Z} = \{(a,b) : 0 \leq a + b + 2 \leq p\}$$
$$\cup \{(p-1,b) : 0 \leq b < p\} \cup \{(a, p-1) : 0 \leq a < p - 1\} \subseteq \mathbf{N}_0^2.$$

Finally, the map $X^+ \to \overline{C_0}$ sending λ to $w_0(\lambda)$ restricts to a bijection between $\bar{Z}$ and $\overline{C_0} \cap X$, so after observing that the elements of $\bar{Z}$ are minimal in their block we have completed the proof. $\qquad\square$

5.2 Return of the Hom functors

In this section we discuss the Schur functor and its action on modules. We recall that if $\lambda \in \Lambda^+(n,r)$, the induced module $H^0(\lambda)$ and the Schur module $M(\lambda) = \nabla(\lambda)$ coincide by 3.2.6. As usual write $[\lambda : \mu]$ for the composition multiplicity $[H^0(\lambda) : L(\mu)]$. Using the ∇-good filtration result of Theorem 4.5.3, we obtain a first description of the $S(n,r)$-blocks, namely

Proposition 5.2.1 *Elements $\lambda, \mu \in \Lambda^+(n,r)$ belong to the same $S(n,r)$-block if and only if there exists a chain*

$$\lambda = \theta_1, \ldots, \theta_t = \mu$$

in $\Lambda^+(n,r)$ such that for all $1 \leq i < t$, $[\theta_i : \theta_{i+1}] \neq 0$ or $[\theta_{i+1} : \theta_i] \neq 0$.

The next result follows from 4.4.2 and 5.2.1.

Lemma 5.2.2 *If $\lambda, \mu \in \Lambda^+(n,r)$ are in the same $S(n,r)$-block then λ, μ are in the same $S(m,r)$-block for all $m \geq n$.*

In what follows, we take $r \leq n$. Let us now exhibit some injective modules for the Schur algebra that are irreducible as modules for the symmetric group (under the Schur functor).

Theorem 5.2.3 *Let $\lambda \in \Lambda^+(n,r)$ be such that $[\mu : \lambda] = 0$ for all $\mu \in \Lambda^+(n,r)$ with $\mu > \lambda$. Then*

(i) *for all $N \geq n$, the Schur module $M_N(\lambda)$ for $S(N,r)$ is injective;*
(ii) *the weight λ is p-regular;*

(iii) *for all $N \geq \max\{n,r\}$, $f M_N(\lambda)$ is an irreducible $K\Sigma_N$-module.*

Proof Choose $N \geq n$ and let $\mu \in \Lambda^+(N,r)$ be such that $\mu \geq \lambda$. Since $r \geq \sum_{i=1}^{n} \mu_i \geq \sum_{i=1}^{n} \lambda_i = r$ we see that, in fact, $\mu \in \Lambda^+(n,r)$. So the element μ of $\Lambda^+(N,r)$ possesses all the properties that λ has as an element of $\Lambda^+(n,r)$.

For (i), the injective hull $I = I_N(\lambda)$ of the simple $S(N,r)$-module $L_N(\lambda)$ has a ∇-good filtration (as an $S(n,r)$-module) $0 = I_0, I_1, \cdots, I_m = I$ with quotients $I_i/I_{i-1} \cong M_N(\tau_i)$, for $\tau_i \in \Lambda^+(n,r)$ occurring with multiplicity $[\tau_i : \lambda]$ by 4.5.3. By hypothesis $M_N(\lambda) \cong I$ is injective.

For (ii), we choose $N \geq \max\{n,r\}$ and let $f_N : \mathcal{P}_K(N,r) \to {}_{\Sigma_r}\mathbf{mod}$ be the Schur functor. We know from 4.6.6(iii) that $M_N(\lambda)$ occurs $\dim f M_N(\lambda)$ times in a ∇-good filtration of $E^{\otimes r}$. This multiplicity is non-zero by the work preceding 4.2.1. Hence $(I_N(\mu) : M_N(\lambda)) \neq 0$ by 4.6.6(i) for a certain p-restricted $\mu \in \Lambda^+(N,r)$.

To complete the proof of (ii) we demonstrate that $\lambda = i(\mu)'$, in the notation of 4.6.5. Suppose this is not the case. As usual find a ∇-good filtration $\{I_j\}$ of $I = I_N(\mu)$ with $I_j/I_{j-1} \cong M_N(\lambda_j)$ for some $\lambda_j \in \Lambda^+(N,r)$, as in 4.5.3. We may always choose this filtration subject to (3.3) and such that $M_N(\lambda_j)$ occurs $[\lambda_j : \mu]$ times. We deduce the existence of submodules $A < B < C$ of I with $B/A \cong M_N(\lambda)$ and $I/C \cong M_N(i(\mu)')$. By (i), $M_N(\lambda)$ is injective, so I has a quotient $M_N(\lambda) \oplus M_N(i(\mu)')$ which of course is false since I has a simple top, being self-dual by 4.6.6(ii). In particular $\lambda = i(\mu)'$ and $M_N(\lambda)$ occurs in a ∇-good filtration of $I_N(\mu)$ with multiplicity one; see 4.6.5. In particular, λ has to be p-regular, as required.

For (iii), the proof of (ii) implies that, given any restricted $\tau \in \Lambda^+(n,r)$, the Schur module $M_N(\lambda)$ occurs in a ∇-good filtration for $I_N(\tau)$ with multiplicity one, if $\lambda = i(\tau)'$, or zero, if not. Hence

$$(E^{\otimes r} : M_N(\lambda)) = \dim f M_N(\lambda) \ (= \dim f L_N(\mu))$$

for $\lambda = i(\mu)'$. But $[M_N(\lambda) : L_N(\mu)] \neq 0$ and $\dim f M_N(\lambda) = \dim f L_N(\mu)$ so it follows that $f M_N(\lambda) = f L_N(\mu)$. The latter module is simple by 4.2.4. $\square$

5.3 Primitive blocks

We recall that the plan was first to study primitive blocks (those containing partitions that are 0-weighted) and then pass to general blocks by applying a suitable functor.

Let us begin with a few easy but important facts concerning the Schur weight. Suppose first that $\lambda \in \Lambda^+(n,r)$ is not primitive: thus there exists some $\kappa_i \in \mathbf{N}_0$ such that $\lambda_i - \lambda_{i+1} = (p-1) + \kappa_i p$ for $i < n$. Stripping off a p-hook from λ results in a partition $\mu = \lambda - p\varepsilon_i$ for some $i < n$, and hence μ cannot be primitive either. Thus $d(\tilde{\lambda}) > 0$. We shall call a core class *primitive* if it consists of primitive elements; see 5.1.2. The following statements are simple consequences of 5.1.8 and 5.1.1.

Lemma 5.3.1

  (i) *Every core class is a union of $S(n,r)$-blocks.*
  (ii) *If $\lambda \in \Lambda^+(n,r)$ is primitive then its core class is primitive.*
  (iii) *If $\omega = (1^n)$ then $\lambda, \mu \in \Lambda^+(n,r)$ are in the same core class if and only if $\lambda + \omega, \mu + \omega \in \Lambda^+(n, r+n)$ are in the same core class.*

In fact much more is true, namely

Theorem 5.3.2 *The primitive blocks of $S(n,r)$ are precisely the primitive core classes.*

The proof of 5.3.2 is a lengthy induction on n and will take up the next few sections. We first introduce a lemma required for this proof.

Row and column removal

We now recall the Principle of Row and Column removal enshrined in 4.4.6 and 4.4.7. If $\gamma \in \Lambda^+(m,r)$ and $\delta \in \Lambda^+(n,s)$ with $\gamma_m \geq \delta_1$, we shall write $(\gamma|\delta)$ for the *concatenation* $(\gamma_1, \ldots, \gamma_m, \delta_1, \ldots, \delta_n) \in \Lambda^+(m+n, r+s)$. The next simple lemma will be used repeatedly in the proof of Theorem 5.3.2.

Lemma 5.3.3

  (i) *If $\lambda, \mu \in \Lambda^+(n,r)$ are in the same $S(n,r)$-block then $\lambda+\omega, \mu+\omega \in \Lambda^+(n, r+n)$ are in the same $S(n, r+n)$-block.*
  (ii) *For the concatenation $(\gamma|\delta)$ above, assume that δ is the unique maximal element in its $S(n,s)$-block and that $\sigma \in \Lambda^+(n,s)$ belongs to the block containing δ. Then $(\gamma|\sigma)$ lies in the $S(m+n, r+s)$-block containing $(\gamma|\delta)$.*

Proof For (i): a chain $\lambda = \theta_1, \theta_2, \ldots, \theta_t = \mu$ in $\Lambda^+(n,r)$ may be found such that $[\theta_i : \theta_{i+1}] \neq 0$ or $[\theta_{i+1} : \theta_i] \neq 0$ for all $i < t$. Hence there is a chain

$$\lambda + \omega = \theta_1 + \omega, \theta_2 + \omega, \ldots, \theta_t + \omega = \mu + \omega$$

in $\Lambda^+(n, r+n)$. Moreover for any $\zeta, \eta \in \Lambda^+(n, r)$, $[\zeta + \omega : \eta + \omega] = [\zeta : \eta]$.

For (ii): as above, there is a chain $\delta = \theta_1, \theta_2, \ldots, \theta_t = \sigma$ in $\Lambda^+(n, s)$ such that $[\theta_i : \theta_{i+1}] \neq 0$ or $[\theta_{i+1} : \theta_i] \neq 0$ for all $i < t$. But by 4.4.6 applied m times, for $\zeta, \eta \in \Lambda^+(n, s)$ with $\zeta_1 \leq \gamma_m$ and $\eta_1 \leq \gamma_m$ we have $[(\gamma|\zeta) : (\gamma|\eta)] = [\zeta : \eta]$. Since $\theta_i \leq \delta$ by maximality, $\theta_1 \leq \delta_1 \leq \gamma_m$ for each $i \leq t$. Hence there is a chain

$$(\gamma|\delta) = (\gamma|\theta_1), (\gamma|\theta_2), \ldots, (\gamma|\theta_h) = (\gamma|\sigma)$$

as required.   □

The proof of 5.3.2

This follows the inductive proof of Donkin [1992, §3 Theorem]. I have split the argument into several steps.

(1) *The case $n \leq 3$.* By 4.4.5, the set of primitive weights is a disjoint union of core classes $C_1, \ldots, C_k$, and each core class is a union of blocks by 5.3.1(i). The distinct primitive blocks $B_1, \ldots, B_l$ of S_r may be listed in such a way that $B_i \subseteq C_i$ ($i \in \mathbf{k}$). Aiming for a contradiction, let us suppose that $l > k$. Then if $\mu \in B_{k+1}$ is minimal, μ lies in some C_i ($i \leq k$). Pick $\nu \in B_i$ minimal and thereby obtain two distinct minimal members μ, ν of the same Γ-block, contrary to 5.1.10(ii).

Henceforth we assume $n > 3$. Assume inductively that the result 5.3.2 holds for all primitive blocks of $S(m, s)$ with $m \leq n$ and $m < n$ or $s < r$. Let C be a primitive core class of dominant weights with p-core γ and p-weight w. The class C contains a unique maximal weight $M = \gamma + pw\varepsilon_1$, hence consists of a unique block provided that for each $\lambda \in C$, λ and M are in the same block. Let M belong to the block B_M. We will assume that $C \neq B_M$, and derive a contradiction. We pick a maximal $\lambda \in C \backslash B_M$ with λ_1 as large as possible.

The strategy employed below is to focus on the possible forms of λ. The most frequently used weapons are 5.3.1 and 5.3.3, which will be deployed without explicit reference.

(2) *The case $\lambda_n \neq 0$.* Let $\mu = \lambda - \omega = (\lambda_1 - 1, \ldots, \lambda_n - 1)$, and suppose that $C' \subseteq \Lambda^+(n, r - n)$ is its core class. Let $\theta \geq \mu$ be maximal in C'. By induction θ is in the same $S(n, r - n)$-block as μ, and hence $\lambda = \mu + \omega$ and $\theta + \omega$ are in the same $S(n, r)$-block. Since λ is maximal

in $C \setminus B_M$, $\lambda = \theta + \omega$. Hence $\mu = \theta$ is maximal in C', so has the form $\mu = \tilde{\mu} + \mathrm{wt}(\mu)p\varepsilon_1$ (here $\mathrm{wt}(\mu)$ is the p-weight of μ).

Computing valuations on the hook graph gives, for entries not in the first row of the Young diagram $[\mu]$, $\nu_p(h_{ij}^\mu) = \nu_p(h_{ij}^{\tilde{\mu}}) = 0$ by 4.4.5. Since $\lambda = \mu + \omega$, we also have $\nu_p(h_{ij}^\lambda) = 0$ for entries of $[\lambda]$ away from the first row and column. Now, λ is maximal in its block, by definition, so applying 5.2.3, λ is regular and $fM_N(\lambda)$ is irreducible for large enough N. By Carter's Criterion, $\nu_p(h_{ij}^\lambda) = \nu_p(h_{i1}^\lambda)$ for all $(i,j) \in [\lambda]$.

(2a) $\nu_p(h_{i1}^\lambda) \neq 0$ *for* $i \in \mathbf{n}$. If we can show this, then by definition $h_{i1}^\lambda = \lambda_i + n - i \equiv 0 \pmod{p}$ for all $i \in \mathbf{n}$. Then

$$(\lambda_i + n - i) - (\lambda_{i+1} + n - (i+1)) = \lambda_i - \lambda_{i+1} + 1 \equiv 0 \pmod{p}$$

for $i < n$, contradicting the primitivity of λ.

Assume (2a) is false, so that $\nu_p(h_{i1}^\lambda) = 0$ for some $i \in \mathbf{n}$. Now if $2 \leq j \leq \lambda_2$, $\nu_p(h_{1j}^\lambda) = \nu_p(h_{2j}^\lambda) = 0$. Setting $\lambda_1 - \lambda_2 = ap + b$ for $0 \leq b < p$ we may write λ in the form $\zeta + ap\varepsilon_1$ where $\zeta \in \Lambda^+(n, r - ap)$ satisfies $\nu_p(h_{ij}^\zeta) = \nu_p(h_{ij}^\lambda) = 0$ for all $(i,j) \in [\zeta]$. Hence ζ is a core. In fact $\zeta = \gamma(= \tilde{\lambda})$, $a = w$, forcing $\lambda = M$ which is the final contradiction.

(3) *The case* $\lambda_n = 0$. This part requires the most care. The assumption allows one to view λ as an element of $\Lambda^+(n-1, r)$. Since $M \geq \lambda$, $M_n = 0$ and we take $M \in \Lambda^+(n-1, r)$ also.

(3a) λ *is primitive.* Immediately from 5.3.1(ii), M is primitive. By induction λ and M are in the same $S(n-1, r)$-block, so they are automatically in the same $S(n, r)$-block by 5.2.2.

(3b) λ *is not primitive.* This amounts to assuming that $\lambda_i - \lambda_{i+1} \equiv -1$ $\pmod{p}$ for all $i \leq n - 2$. Now if it happens that $\lambda_i - \lambda_{i+1} < p$ for all $i \leq n - 1$ then $\lambda_i - \lambda_{i+1} = p - 1$ for $i \leq n - 2$ and $\lambda_{n-1} < p$. So one cannot strip a hook from λ, which is to say that $\lambda = \tilde{\lambda}$. Consequently its core class $C = \{\lambda\}$, which is impossible (since also $M \in C$).

We can therefore assume that for some $i \leq n - 1$, $\lambda_i - \lambda_{i+1}$ is at least p. Pick the maximal such i, and observe that $i \geq 1$ (otherwise λ is maximal in C). The immediate goal is now to show that, in fact, $i = 2$.

(3bi) *The case:* $i = n - 1$. Assume first that $\lambda_{n-1} \geq 2p$. Evidently we can write $\lambda = (\mu|\sigma)$ where $\mu = (\lambda_1, \ldots, \lambda_{n-2})$ and $\sigma = (\lambda_{n-1}, 0)$. By part (1) above (with $n = 2$), σ and $\theta = (\lambda_{n-1} - p, p)$ are in the same block. Thus λ and $\xi = (\mu|\theta)$ are in the same block.

Note that ξ has n non-zero parts, so we can express it as $\psi + \omega$ where $\psi \in \Lambda^+(n, r - n)$. By induction, the block containing ψ is the core class containing it: in particular this block contains $\chi = \psi - p\varepsilon_{n-2} + p\varepsilon_1$. Now

$\chi + \omega$ and $\xi = \psi + \omega$ are in the same block, namely the one containing λ. But χ has first part equal to $\lambda_1 + p > \lambda_1$, a contradiction.

Assume now that $\lambda_{n-1} = p + a < 2p$. This time we have $\lambda = (\mu|\sigma)$ where μ is as above and $\sigma = (p + a, 0)$. Observe that σ and $\theta = (p - 1, a + 1)$ are in the same core class, hence in the same block by part (1). Then λ and $\xi = (\mu|\theta)$ are in the same block. A similar argument to the one just given constructs χ in the same block as λ with $\chi_1 > \lambda_1$.

We now conclude that $2 \le i \le n - 2$.

(3bii) *The case: $i > 2$.* We can write $\lambda = (\mu|\sigma)$ with $\mu = (\lambda_1, \ldots, \lambda_{i-1})$ and $\sigma = (\lambda_i, \ldots, \lambda_n)$. We renumber, to express $\sigma = (\sigma_1, \ldots, \sigma_{n+1-i})$. The first two parts of σ differ by $(p - 1) + \kappa p$ for some $\kappa > 0$; the next successive $(n - 1 - i)/2$ parts all differ by $p - 1$; $\sigma_{n-i} = a < p$; $\sigma_{n+1-i} = \lambda_n = 0$. Hence $\tilde{\sigma} = \sigma - \kappa p \varepsilon_1$ and σ is the unique maximal element in its core class and hence in its block.

Clearly both σ and $\theta = (\sigma_1 - p, \sigma_2, \ldots, \sigma_{n-1-i}, p - 1, a + 1)$ have the same p-core, so by induction they both lie in the same block of $S(n + 1 - i, \sum \sigma_i)$. Moreover $\lambda = (\mu|\sigma)$ and $\xi = (\mu|\theta)$ are in the same block. Put $\xi = \psi + \omega$ where $\psi = (\lambda_1 - 1, \psi_2, \ldots, \lambda_{i-1}, \lambda_i - p, \psi_{i+1}, \ldots, \psi_n) \in \Lambda^+(n, r - 1)$. Notice that ψ and $\psi + p\varepsilon_1 - p\varepsilon_{i-1}$ have the same core class, and hence by induction lie in the same block. Then $\xi = \psi + \omega$ and $\chi = \psi + p\varepsilon_1 - p\varepsilon_{i-1} + \omega$ are in the same block. We have shown that λ and χ are in the same block, whereas $\chi_1 = \lambda_1 + p > \lambda_1$, a contradiction.

(3biii) *The case $i = 2$.* If $\lambda_2 - \lambda_3$ were of the form $(p - 1) + \kappa p$ with $\kappa \ge 2$, we could imitate the argument of (3bii) to get a contradiction. Thus the parts of λ satisfy:

$$
\begin{aligned}
\lambda_1 - \lambda_2 &= (p - 1) + \kappa p \\
\lambda_2 - \lambda_3 &= 2p - 1 \\
\lambda_j - \lambda_{j+1} &= p - 1 \quad (3 \le j \le n - 1) \\
\lambda_{n-1} &= a < p - 1 \\
\lambda_n &= 0
\end{aligned}
$$

Evidently $\gamma = \tilde{\lambda}$ has parts satisfying $\gamma_1 = \lambda_1 - (\kappa + 1)p$, $\gamma_2 = \lambda_2 - p$ and $\gamma_i = \lambda_i$ for $i \ge 3$. Put another way, the parts satisfy $\gamma_n = 0$ and $\gamma_i = a + (n - 1 - i)(p - 1)$ for $i \le n - 1$.

(3c) *Any restricted weight $\phi \in \Lambda^+(n, r)$ lying in the same core class as γ has n non-zero parts.* This yields to the familiar contradiction argument. For, if $\phi_n = 0$ we obtain ϕ from γ by adding p-hooks. There must exist a sequence of partitions

$$\gamma = \theta_1, \theta_2, \ldots, \theta_t = \phi$$

where, for each $i < t$, θ_{i+1} is obtained from θ_i by adding a p-hook and $\theta_i = ((\theta_i)_1, \ldots, (\theta_i)_n)$ has $(\theta_i)_n = 0$ for $i \leq t$.

A simple calculation reveals that $(\theta_i)_j \geq (\theta_i)_{j+1} + (p-1)$ for all $j \leq n-1$, and $\theta_{i+1} = \theta_i + p\varepsilon_q$ for some $q \leq n-1$. Hence $\phi = \gamma + p\varepsilon_{q_1} + p\varepsilon_{q_2} + \cdots + p\varepsilon_{q_s}$ where $q_1 \leq \cdots \leq q_s \leq n-1$. Now $\phi_{q_s} - \phi_{q_s+1} \geq p + \gamma_{q_s} - \gamma_{q_s+1}$ which shows that ϕ is not restricted.

(4) *Completion of proof.* Let μ be minimal in the block of λ and σ minimal in the block of M. By 5.1.4 the Schur modules $M(\mu)$ and $M(\sigma)$ are simple and so μ and σ are restricted. Since both μ and σ are in the core class of γ, $\mu - \omega$ and $\sigma - \omega$ are in the same core class. By induction they are in the same $S(n, r-n)$-block. Hence μ and σ are in the same $S(n,r)$-block. Finally, λ and M are in the same $S(n,r)$-block, as required.

5.4 General blocks

Theorem 5.1.3 follows from 5.3.2 if we are able to prove that every $S(n,r)$-block is Morita equivalent to a primitive $S(n,q)$-block for some $q \leq r$. The next definition associates to any $\lambda \in \Lambda^+(n,r)$ a unique $\mu \in \Lambda^+(n,q)$ for some $q \leq r$. We write $\delta = (n-1, n-2, \ldots, 0)$ as usual.

Definition 5.4.1 Define $a(\lambda)$ by the condition $\lambda_n \equiv a(\lambda) \pmod{p^{d(\lambda)}}$. We will take $0 \leq a(\lambda) < p^{d(\lambda)}$. Then there exists a uniquely determiwined (primitive) $\mu \in \Lambda^+(n,q)$ and an expansion

$$\lambda = (p^{d(\lambda)} - 1)\delta + a(\lambda)\omega + p^{d(\lambda)}\mu \tag{5.2}$$

where q is defined by the relation $r = (p^{d(\lambda)} - 1)\sum \delta_i + na(\lambda) + p^{d(\lambda)}q$. The form (5.2) will be termed the *canonical form* of λ.

Steinberg weights

We recall that St_d, the dth Steinberg module, is the simple projective module $H^0(\sigma) = L(\sigma)$ where $\sigma = (p^d - 1)\delta$. The first result is the assertion that every d-weighted block is equivalent to a primitive block. In the proof we need to quote a straightforward result from Jantzen [1987].

Proposition 5.4.2 *Let B be a primitive block. Then $B^* = \{\sigma + p^d\mu :$*
$\mu \in B\}$ is a d-weighted block. The functor $F_1 : {}_\Gamma\text{mod} \to {}_\Gamma\text{mod}$ that
sends the rational Γ-module M to $\text{St}_d \otimes M^{F^d}$ restricts to an equivalence
of categories between ${}_B\text{mod}$ and ${}_{B^*}\text{mod}$. Conversely, any d-weighted
block is of the form B^* for some primitive block B.

Proof The first and third assertions come from 5.1.6. For the second,
F_1 is an equivalence between ${}_\Gamma\text{mod}$ and the category of all $M \in {}_\Gamma\text{mod}$
whose composition factors are of the form $\text{St}_d \otimes L(\mu)^F$ for some $\mu \in X^+$,
by a suitably modified proof of Jantzen [1987, II.10.5]. Then by 5.1.6,
$F_1 : {}_B\text{mod} \to {}_{B^*}\text{mod}$, and this induces the desired equivalence.   □

Let D be the 1-dimensional Γ-module affording the determinant representation.

Theorem 5.4.3 *Suppose that we are given a block B of $S(n,q)$ where*
$q \geq 0$. Let $d \geq 0$ and $0 \leq a < p^d$ also be given. Denote by σ the weight
$(p^d - 1)\delta + a\omega$.

(i) *The collection of canonical weights $B^* = \{\sigma + p^d\mu : \mu \in B\}$*
forms an $S(n,r)$-block, where $r = (p^d - 1)\sum \delta_i + na + p^d q$.

(ii) *The functor*

$$F_2 : {}_{S(n,q)}\text{mod} \to {}_{S(n,r)}\text{mod}$$
$$M \mapsto \text{St}_d \otimes D^{\otimes a} \otimes M^{F^d}$$
$$(\theta \mapsto 1 \otimes 1 \otimes \theta)$$

restricts to an equivalence of categories

$${}_B\text{mod} \simeq {}_{B^*}\text{mod}.$$

(iii) *In particular, given any block b of $S(n,r)$ and $\lambda \in b$ with canonical*
form (5.2) such that $\sum \mu_i = q$, then b is Morita equivalent to the
primitive block B of $S(n,q)$ containing μ.

Proof Further to 5.4.2, there is a corresponding definition of the morphism $F_2 = F_1 \otimes D^{\otimes a}$ from ${}_\Gamma\text{mod}$ to ${}_\Gamma\text{mod}$. But $\text{St}_d \otimes D^{\otimes a} \cong H^0(\sigma)$
since both are simple modules with the same highest weight. Thus F_2
induces an equivalence ${}_{B_\Gamma(\lambda)}\text{mod} \simeq {}_{B_\Gamma(\sigma + p^d\lambda)}\text{mod}$ where $\lambda \in X^+$ is
primitive by 5.4.2.

For (i), let $\lambda \in B$ and choose any $\mu \in B_S(\lambda)$. There is a sequence of indecomposable $S(n, q)$-modules $M_1, \ldots, M_t$ such that $L(\lambda)$ is a composition factor of M_1, $L(\mu)$ is a composition factor of M_t, and for all $i < t$, M_i and M_{i+1} have a common composition factor. Applying F_2 we have a sequence of indecomposable $S(n, r)$-modules $\text{St}_d \otimes M_1^{F^d}, \ldots, \text{St}_d \otimes M_t^{F^d}$ such that $L(\sigma + p^d \lambda) \cong \text{St}_d \otimes L(\lambda)^{F^d}$ is a factor of $\text{St}_d \otimes M_1^{F^d}$, $L(\sigma + p^d \mu) \cong \text{St}_d \otimes L(\mu)^{F^d}$ is a factor of $\text{St}_d \otimes M_t^{F^d}$, and for all $i < m$, $\text{St}_d \otimes M_i^{F^d}$ and $M_{i+1}^{F^d}$ have a common factor. Hence every element of B^* belongs to the block of $\sigma + p^d \lambda$.

Conversely, let $\theta \in \Lambda^+(n, r)$ be in the $S(n, r)$-block of $\sigma + p^d \lambda$. Then a fortiori $\theta \in B_\Gamma(\sigma + p^d \lambda)$ so $\theta = \sigma + p^d \mu$ for some $\mu \in B_\Gamma(\lambda)$ by 5.4.2. For $i \leq m$ we have $\theta_i = \sigma_i + p^d \mu_i$. Notice that $\theta_n = \sigma_n + p^d \mu_n = a + p^d \mu_n \geq 0$ and $0 \leq a < p^d$, so $\mu_n \geq 0$. Since we assume that θ is dominant, μ is decreasing with positive parts. Since $\theta = \sigma + p^d \mu$, $r = \sum \theta_i = \sum \sigma_i + p^d \sum \mu_i = \sum \sigma_i + p^d \sum \lambda_i$. Thus μ and λ partition the same integer and $\mu \in B_\Gamma(\lambda) \cap \Lambda^+(n, r) = B$. Hence B^* is a block.

For (ii), let F be the restriction of $F_1 \otimes D^{\otimes a}$. Again, in view of 5.4.2, one needs to show that each $Y \in {}_{B^*}\textbf{mod}$ is isomorphic to $F(M)$ for some $M \in {}_B\textbf{mod}$. Now $Y \cong \text{St}_d \otimes M^{F^d}$ for some $M \in {}_{B_\Gamma(\lambda)}\textbf{mod}$ by 5.4.2.

If $L(\mu)$ is a factor of this M (where $\mu \in X^+$), $L(\sigma + p^d \mu) \cong \text{St}_d \otimes L(\mu)^{F^d}$ is a factor of Y. Hence $\sigma + p^d \mu \in B^*$ so $\mu \in B$. We now have that every factor of M has the form $L(\mu)$ for some $\mu \in \Lambda^+(n, q)$, and so M is polynomial of degree q. But all factors of M are of the form $L(\mu)$ for $\mu \in \Lambda^+(n, q) \cap B_\Gamma(\lambda)$ so $M \in {}_B\textbf{mod}$. Hence $F(M) \cong Y$ and (ii) follows.

Finally, (iii) is a particular case of the above. $\qquad\square$

By the comments preceding 5.1.7 we have $B_S(\lambda) = B_\Gamma(\lambda) \cap \Lambda^+(n, r)$ and 5.1.3 follows at last.

Remark. The requirement for an algebraically closed field is not necessary. For if $F' : F$ is a field extension we have a natural isomorphism $F' \otimes_F S_F(n, r) \to S_{F'}(n, r)$ (see 3.5) and the polynomial simple modules $L_F(\lambda)$ are absolutely irreducible by 3.4.2. Thus

$$F' \otimes_F \text{Ext}^1_{S_F(n, r)}(L_F(\lambda), L_F(\mu)) \cong \text{Ext}^1_{S_{F'}(n, r)}(L_{F'}(\lambda), L_{F'}(\mu)),$$

and so the $S_F(n, r)$-blocks and the $S_{F'}(n, r)$-blocks are identical. Hence we could have worked over any field of positive characteristic. Indeed by resorting to the language of the group scheme Γ_R, the main result would be valid over the local principal ideal domain R.

Exercise. Assume $r \leq n$. Use 5.1.3 and the methods of Chapter 4 to rederive the Nakayama Rule 5.1.1. (For a possible solution see Donkin [1992, §4].)

5.5 The finiteness theorem

That there are only finitely many Morita equivalence classes of blocks of a finite group with a given defect group is a long-standing conjecture of Donovan. It has been verified for $G = \Sigma_r$ by Scopes [1991]: she defined a bijection φ between two G-blocks and showed that this induces a Morita equivalence between the corresponding block algebras. The paper of Erdmann, Martin and Scopes [1993] shows that φ, defined in (5.3) below, induces a Morita equivalence between the respective S_r-blocks and hence the finiteness condition for G carries over to S_r. We have assumed up to now that K is algebraically closed; here we shall relax this condition and work over a more general ring of coefficients.

Morita equivalence for symmetric groups

We now outline the main result of Scopes [1991] and establish additional notation at the same time. One needs to be familiar with β-numbers; recall that they are a generalisation of first column hook lengths and see James and Kerber [1981, pp. 77–78].

Let B be a p-block of KG of weight $w \geq 0$, having p-core $\gamma \in \Lambda^+(r - pw)$ with $c \geq 0$ parts, say. Let $\zeta = \{\beta_1, \ldots, \beta_{c+pw}\}$ be the β-set for γ with $c+pw$ members. Display ζ on an abacus with p runners in the usual manner, that is, let the bead positions be determined by the elements of ζ. It will be convenient to identify partitions belonging to B with their corresponding $(c+pw)$ bead abacus display. Observe also that the number of beads on each runner determines γ (since it is a core).

Suppose that for some $i \in \mathbf{p}$, the ith runner has ℓ more beads on it than the $(i-1)$th, where $\ell \geq w$. Swap these two runners (formally, slide the bottom ℓ beads of column i into column $(i-1)$: the resulting abacus defines a new $(c + pw)$-element β-set, which we call $\bar{\zeta}$, representing a p-core $\bar{\gamma}$ (a partition of $r - pw - \ell$). Let $\bar{B}$ be the block of $K\Sigma_{r-\ell}$ of weight w and core $\bar{\gamma}$. The partitions belonging to the blocks B and $\bar{B}$ are related by a map

$$\varphi : \lambda \mapsto \bar{\lambda} \tag{5.3}$$

whose effect is to interchange the ith and $(i-1)$th columns in the associated abaci. It is easily seen that φ is a bijection between the set of partitions of r with p-core γ and the set of partitions of $r - \ell$ with p-core $\bar{\gamma}$; this bijection preserves p-regularity and the lexicographic ordering. Now Scopes [1991, Theorem 4.2] states that under these hypotheses

Theorem 5.5.1 (Scopes' Equivalence) B is Morita equivalent to $\bar{B}$.

The proof proceeds by finding an explicit equivalence between the categories $_B\mathbf{mod}$ and $_{\bar{B}}\mathbf{mod}$, and we now briefly recall the method. There are exact functors:

$$_B\mathbf{mod} \; \underset{F_1'}{\overset{F_1}{\underset{\longleftarrow}{\longrightarrow}}} \; KG\mathbf{mod} \; \underset{F_2'}{\overset{F_2}{\underset{\longleftarrow}{\longrightarrow}}} \; K\Sigma_{r-\ell}\times\Sigma_\ell\,\mathbf{mod} \; \underset{F_3'}{\overset{F_3}{\underset{\longleftarrow}{\longrightarrow}}} \; K\Sigma_{r-\ell}\,\mathbf{mod} \; \underset{F_4'}{\overset{F_4}{\underset{\longleftarrow}{\longrightarrow}}} \; _{\bar{B}}\mathbf{mod}$$

where F_1, F_1', F_4 and F_4' are the usual functors obtained by moving between the modules of a group algebra and a block. Define F_2 and F_2' by restriction and coinduction respectively. View $M \in {}_{K(\Sigma_{r-\ell}\times\Sigma_\ell)}\mathbf{mod}$ as a $(K\Sigma_\ell{}^{\mathrm{op}}, K\Sigma_{r-\ell})$-bimodule, and objects of $_{K\Sigma_{r-\ell}}\mathbf{mod}$ as $(K, K\Sigma_{r-\ell})$-bimodules. Let k be the trivial $(K, K\Sigma_\ell{}^{\mathrm{op}})$-bimodule. If M is a $(K\Sigma_\ell{}^{\mathrm{op}}, K\Sigma_{r-\ell})$-bimodule, and N is a $(K, K\Sigma_{r-\ell})$-bimodule, define functors

$$F_3(M) = k \otimes_{(K\Sigma_\ell{}^{\mathrm{op}})} M \qquad F_3'(N) = \mathrm{Hom}_K(k, N),$$

the latter being a $(K\Sigma_\ell{}^{\mathrm{op}}, K\Sigma_{r-\ell})$-bimodule. In fact

Theorem 5.5.2 $F = F_4 \circ F_3 \circ F_2 \circ F_1$ and $F' = F_1' \circ F_2' \circ F_3' \circ F_4'$ are equivalences of categories, and are mutually inverse.

Repeating the swapping procedure until we can proceed no further, one finally deduces the finiteness condition for blocks of $G = \Sigma_r$, as r varies:

Corollary 5.5.3 For fixed weight w, there exist only finitely many Morita equivalence classes of p-blocks of symmetric groups of weight w.

Let R be a complete d.v.r. (in particular R could be a field) with residue field $k = R/\mathcal{M}$. Donkin [1992, §5], working over R, extends Scopes' result by applying the theory of partial tilting modules in the category $\mathcal{F}(\Delta) \cap \mathcal{F}(\nabla)$ (for the latter see Donkin [1993]). We don't need this; we merely observe that for $i \in \mathbf{4}$ the functors F_i and F_i' can be defined over R and are 'liftable' in the sense that $k \otimes_R F_{i,R}(-)$ and $F_{i,k}(- \otimes_R k)$ are equivalent. This is clear.

The main theorem

Here we work over R. When considering algebras that are equivalent to $S_R(n,r)$, it is no loss to work with the *basic algebra*, Λ, associated with $S_R(n,r)$. This is the unique algebra with the two properties that (1) all of its simple modules have R-rank one, and (2) Λ is Morita equivalent to $S_R(n,r)$. For basic algebras see Erdmann [1990, I.2]; we recall here the construction. Decompose r-tensor space of the free R-module E into indecomposable, pairwise non-isomorphic summands, say $E^{\otimes r} = \oplus m_i E_i$ (the integers m_i are the multiplicities).

Definition 5.5.4 The algebra $S = \operatorname{End}_{R\Sigma_r}(\oplus E_i)$ is the *basic algebra* for $S_R(n,r)$.

Definition 5.5.5 Let B be a G-block. Define the *component* of $S_R(n,r)$ and S corresponding to B to be

$$S_R(n,r)_B = \operatorname{End}_{R\Sigma_r}\left(\bigoplus_{E_i \in B} m_i E_i\right) \quad \text{and} \quad S_B = \operatorname{End}_{R\Sigma_r}\left(\bigoplus_{E_i \in B} E_i\right).$$

Clearly, S_B is a direct sum of blocks of S, and $S = \oplus S_B$, summed over all blocks of $R\Sigma_r$. Moreover S_B is the basic algebra of $S_R(n,r)_B$ and from 5.3.2, if $r \leq n$ then S_B is actually a block. Clearly also

$$S_B = \operatorname{End}_{R\Sigma_r}\left(\bigoplus_{\lambda} Y^{\lambda}\right)$$

summed over those $\lambda \in B$ with at most n non-zero parts. Our main message (valid over R or a field) is:

Theorem 5.5.6 *For fixed weight w, there exist only finitely many Morita equivalence classes of algebras of the form $S_R(n,r)_B$ where B is a block of G of weight w, for arbitrary integers $n, r \geq 1$.*

Define R-free RG-modules M_R^{λ}, S_R^{λ} and Y_R^{λ} as in the last chapter. From 4.6.4, $Y = Y_R^{\lambda}$ has a Specht filtration

$$Y = Y_1 \supseteq Y_2 \supseteq \cdots \supseteq Y_s = \{0\}$$

with Y_i/Y_{i+1} isomorphic to some $S_R^{\lambda_i}$. Moreover S_R^{λ} occurs once, and all other $S_R^{\lambda_i}$ are such that $\lambda_i > \lambda$.

Let $\bar{\lambda}$ be a partition of $r - \ell$ belonging to $\bar{B}$. The first part of the proof involves showing that, under the equivalence 5.5.2, the induced module

$R_{\Sigma_{\bar{\lambda}}} \uparrow^{\Sigma_{r-\ell}}$ is sent to some induced module $R_{\Sigma_{\nu}} \uparrow^{\Sigma_r}$ for an appropriate Young subgroup Σ_{ν}.

Lemma 5.5.7 *Let ν be the partition of r obtained from $\bar{\lambda}$ by adjoining the part ℓ appropriately. Then $F_2' F_3'(M^{\bar{\lambda}}) \cong M^{\nu}$.*

Proof Now $F_3'(M^{\bar{\lambda}}) \cong M^{\bar{\lambda}}$ as a $(\Sigma_\ell{}^{\mathrm{op}}, \Sigma_{r-\ell})$-bimodule with trivial $R\Sigma_\ell$-action. Now $M^{\bar{\lambda}} \cong R_{\Sigma_{\bar{\lambda}}} \uparrow^{\Sigma_{r-\ell}} \otimes_R R$ as $(\Sigma_\ell{}^{\mathrm{op}}, \Sigma_{r-\ell})$-bimodules. Since $R \cong R_{\Sigma_\ell} \uparrow^{\Sigma_\ell}$, it follows that $M^{\bar{\lambda}} \cong (R_{\Sigma_{\bar{\lambda}} \times \Sigma_\ell}) \uparrow^{\Sigma_{r-\ell} \times \Sigma_\ell}$, and the group $\Sigma_{\bar{\lambda}} \times \Sigma_\ell$ is a Young subgroup for the partition ν. Thus by transitivity of induction,

$$F_2'(F_3'(M^{\bar{\lambda}})) \cong R_{\Sigma_{\nu}} \uparrow^{\Sigma_r} = M^{\nu}.$$

$\square$

We recall from 5.5.2 the functor $F' = F_R' : \mathbf{mod}\ \bar{B} \to \mathbf{mod}\ B$ and the bijection φ of (5.3). If $\bar{\lambda}$ is a partition of $r - \ell$ belonging to $\bar{B}$, let $\lambda = \varphi^{-1}(\bar{\lambda})$ be the corresponding partition of r lying in B.

Lemma 5.5.8 $F'(Y^{\bar{\lambda}}) \cong Y^{\lambda}$.

Proof The Young module $Y^{\bar{\lambda}}$, as a summand of $M^{\bar{\lambda}}$, is sent by F' to an indecomposable summand of M^{ν} by 5.5.7. We claim that this summand is actually Y^{λ}.

Now since $Y^{\bar{\lambda}}$ belongs to $\bar{B}$, the module $F_2' F_3' F_4'(Y^{\bar{\lambda}}) = F_2' F_3'(Y^{\bar{\lambda}})$ is some direct summand of M^{ν}, say $F_2' F_3'(Y^{\bar{\lambda}}) = \sum \alpha_\mu Y^{\mu}$ where $\mu > \nu$. But we have, already, that $F'(Y^{\bar{\lambda}}) = Y^{\mu}$ with $\mu \geq \nu$. Take an R-Specht filtration for $Y^{\bar{\lambda}}$ as above. Then $S^{\bar{\lambda}}$ appears with multiplicity one and all other factors $S^{\bar{\rho}}$ are such that $\bar{\rho} > \bar{\lambda}$. Since $F'(S^{\bar{\rho}}) = S^{\rho}$ (Scopes [1991, p. 453]), and since φ preserves the ordering, we have that $F'(Y^{\bar{\lambda}}) = Y^{\mu}$ has a Specht series where S^{λ} appears once and any other factors S^{ρ} are such that $\rho > \lambda$. By indecomposability, $\mu = \lambda$ as required. $\square$

From the two lemmas we have that $Y^{\bar{\lambda}}$ is sent to the summand Y^{λ} of M^{ν}. Moreover, if $\alpha \neq \beta$ then $F'(Y^{\alpha}) \not\cong F'(Y^{\beta})$, since F' is an equivalence.

Before completing the proof of 5.5.6 we will prove a combinatorial lemma. Recall again the means by which a β-set has a representation on an abacus with p runners. The *weight*, w_i of abacus runner i is the total number of places through which the beads of the runner are moved to reach their positions in the core. Thus $w = \sum_{i=1}^p w_i$.

Lemma 5.5.9 *Let B_m be the collection of partitions belonging to B that have exactly m non-zero parts. Then $\varphi(B_m) = \bar{B}_t$ where $t = m$ or $m - 1$, with only one exception: namely, if $\ell = w$, $i = 2$ and $m = c$ then $\varphi(B_c) = \bar{B}_c \cup \bar{B}_{c-1}$, and $\bar{B}_c$ consists of a unique partition.*

Proof We begin with the observation that a partition λ belongs to B_m if and only if there are m beads after the first gap. The total number of beads must add up to $c + pw$, and so $\lambda \in B_m$ if and only if the number of beads before the first gap is $m' = c + pw - m$, that is, if and only if the first gap is at position m'. Effectively, we are saying that there is a bead in the $(m' - 1)$th slot and that the first gap occurs in the m'th slot.

We need to analyse the relationship between $m' - 1$, m' and $i - 1$, i.

(1a) *Neither $m' - 1$ nor m' lies on runner $i - 1$ or i .* In this case φ fixes the first gap, and $\varphi(B_m) = \bar{B}_m$.

(1b) *$m' - 1$ is in the ith column.* As in case (1a), $\varphi(B_m) = \bar{B}_m$.

(1c) *$m' - 1$ lies on the $(i - 1)$th runner.* This situation does not arise. For if it does, then in column i one would have to use beads from the first w rows of the abacus, since by assumption there are no gaps in the first column up to the row of m'. Denoting by l_i the number of beads on the ith runner and by w_i its weight, the hypothesis gives $w_i > l_i = l_{i-1} + \ell \geq w$. But the total weight w is such that $w \geq w_i > w$, a contradiction.

(2) *m' belongs to column $i - 1$.* First of all consider the case that all $\lambda \in B_m$ have a bead at position $m' + 1$. Then φ will move the first gap one place to the right, hence $\varphi(B_m) = \bar{B}_{m-1}$. Now assume that there exists $\lambda \in B_m$ whose display has no bead in the $(m' + 1)$th slot.

We have made the universal assumption that the ith runner has ℓ more beads than runner $i - 1$, and $\ell \geq w$. Now up to position $m' + 1$ the $(i - 1)$th and ith runners are identical, hence there must be at least ℓ beads in column i below position $m' + 1$. By assumption, there is a gap at position $m' + 1$, so to reach the core γ, we must move at least ℓ beads up this runner. Therefore the runner has weight at least ℓ. Hence

$$w = \text{total weight of } \lambda \geq \text{weight of } i\text{th runner} \geq \ell \geq w,$$

forcing $\ell = w$. Moreover the total weight of λ occurs in the ith column, and each bead moves up one place to reach γ—hence λ is uniquely determined, since all other $\mu \in B_m$ will have a bead on the ith runner in position $m' + 1$.

But we have also shown that none of the first pw beads are moved, which means that λ can be represented on an abacus for γ with c beads. Hence $m = c$, and the first gap has to be in the first column. Therefore $i = 2$.

Finally, since $i = 2$, $\bar{\gamma}$ has $c - 1$ non-zero parts. Thus φ takes λ to the unique partition of $\bar{B}$ with exactly c non-zero parts, and φ sends the rest of $B_m = B_c$ to $\bar{B}_{c-1}$. $\qquad\square$

To complete the proof of 5.5.6, we fix the prime p and the weight w and let B and $\bar{B}$ be as before. Now 4.3.3 tells us that $S(N, r)$ and $S(n, r)$ are already Morita equivalent if $N \geq n \geq r$, so if we consider the Morita equivalence class of $S(n, r)_B$ we may assume $r \geq n$. We shall need a final lemma about the bijection φ.

Lemma 5.5.10 *The mapping φ induces a bijection*

$$\varphi : \bigcup_{m \leq n} B_m \to \bigcup_{s \leq t} \bar{B}_s$$

where $t = n$ or $n - 1$.

Proof We proceed by induction on n, the base case occurring when $n = c$. Details are safely left to the reader. $\qquad\square$

We deduce by the previous lemmas that the functor $F = F_R$ of 5.5.2 sends $\bigoplus_{\lambda \in B \cap \Lambda(n,r)} Y^\lambda$ to $\bigoplus_{\bar{B} \cap \Lambda(t,r-\ell)} Y^{\bar{\lambda}}$. This induces an isomorphism of endomorphism rings

$$\mathrm{End}_{R\Sigma_{r-\ell}}\left(\bigoplus_{\bar{B}} Y^{\bar{\lambda}}\right) \xrightarrow{\sim} \mathrm{End}_{R\Sigma_r}\left(\bigoplus_{B} Y^\lambda\right),$$

that is, $S(n, r)_B$ is Morita equivalent to $S(t, r-\ell)_{\bar{B}}$ where $t = n$ or $n-1$. By induction, and using the methods of 4.3.3, any block component $S(n, r)_B$ is Morita equivalent to a block component of some $S(t', r')_{B'}$ with $t' \leq r'$, where B' is one of finitely many blocks of weight w, and there are finitely many of these by 5.5.3.

5.6 Examples

This section contains examples of the basic algebras of certain Schur algebras possessing an underlying structure that is particularly simple. Let us review the necessary background.

Finite representation type

Recall from Curtis and Reiner [1987, §62C] that the K-algebra S is said
to be of *finite (representation) type* if the number of isomorphism classes
of indecomposable S-modules is finite. If $S = B$ is a block of a finite
group, then it is well known that this occurs if and only if the defect
group of B is cyclic. Moreover the PIMs of all simple B-modules have
a structure described by a certain tree called a Brauer tree.

In our case we are not going to be particularly ambitious: taking
$S = S(n,r)$, we shall concentrate only on those components S_B for
which B is a p-block of G of finite type. This has the advantage of our
knowing the decomposition matrix already and having a construction of
the Brauer tree for B.

Recently there has been much work done on blocks of Schur algebras
of finite type. Using Young modules, Erdmann [1993a] determines which
of the Schur algebras are of finite type and describes the finite type Schur
algebras up to Morita equivalence. The final classification is that $S(n,r)$
has finite type if and only if one of the following holds:

- $n = 1$;
- $n = 2$, $p = 2$ and $r = 1, 2, 3, 5$ or 7;
- $n = 2$, $p > 2$ and $r < p^2$;
- $n > 2$ and $r < 2p$.

A similar investigation appears in Xi [1991], [1992]. It is curious that
if p is odd and $r \geq 2p$ then $K\Sigma_r$ has infinite type, whereas $S(2,r)$ will
have finite type provided $r < p^2$. These results have been generalised
in two directions, as follows. Xi also considered q-Schur algebras of
finite type; and then Donkin and Reiten [1993] studied Morita types of
quasi-hereditary algebras of finite type over the algebraically closed field
K (or rather those that satisfied a certain 'duality fixing irreducibles'
condition). A by-product of their researches was a classification of the
blocks of finite type of the Schur algebras: the primitive block B of
$S(n,r)$ of weight w has finite type if and only if one of the following
holds:

- $n = 1$;
- $n = 2$ and $w < p$;
- $n \geq 3$ and $w \leq 1$.

Thus by 5.4.3 we have an exact description of all blocks of finite type
for the Schur algebras.

To commence study in this area one has to be familiar with some methods from the representations of algebras; perhaps this is now standard knowledge (otherwise, see Erdmann [1990, Chapter 4]). We also require some facts about permutation modules, for example the Green Correspondence for Young modules stated in 4.6.3(i).

Quivers for basic algebras

We recall from Erdmann [1990, I.5.1] that a *quiver* Q is a directed graph. Write module homomorphisms on the right. The *path algebra* KQ of Q is a K-algebra with K-basis consisting of all paths in Q including those of length zero; the product of two basis elements is the composition of paths in reverse order (since we are dealing with left modules) if the paths are so composable, or zero if the paths are not composable. All the quivers we consider will have only finitely many vertices and edges, so KQ is always finite-dimensional and associative with 1. We are interested in the following well-known quiver:

Definition 5.6.1 Suppose S is a finite-dimensional K-algebra, and let $S_1, \ldots, S_m$ be the isomorphism types of simple S-modules with projective covers $P_i = Se_i$. The *ordinary quiver* for S has vertex set $Q_0 = \{x_1, \ldots, x_m\}$, i.e. one vertex for each simple module, and the number of edges $x_i \to x_j$ is the integer $d_{ij} = \dim_K \operatorname{Ext}^1_S(S_i, S_j)$.

It is not hard to show that $d_{ij} = e_j J(S) e_i / e_j J^2(S) e_i$, where $J(S)$ is the Jacobson radical of S. The author has constructed many of these quivers for $S = KG$; see Martin [1989]. They give strong information about the representation theory of the algebra S, for example, Gabriel noted that if Λ is a finite-dimensional basic K-algebra, with ordinary quiver Q, then there exists a surjection of algebras $\phi : KQ \to \Lambda$ having kernel I contained in the ideal of paths of length at least two. The kernel is called the *ideal of relations*. In other words, we assign to each pair $x_i, x_j \in Q_0$ a subspace of the space of paths of length at least two from x_i to x_j (the *relations*), such that composing a relation on either side with any path produces another relation. If S is any finite-dimensional K-algebra we say that 'Q is the quiver of S', if the basic algebra of S is of the form KQ/I. Calculation of the relations is normally highly non-trivial; see for example the computation of relations for the basic algebra of the principal p-block of $K\Sigma_{2p}$ in Erdmann and Martin [1993].

Example. We are most concerned here with the algebra $\mathcal{A}_m$ $(m \geq 1)$, having the quiver

$$1 \bullet \underset{\beta_1}{\overset{\alpha_1}{\underset{\longleftarrow}{\longrightarrow}}} \bullet \underset{\beta_2}{\overset{\alpha_2}{\underset{\longleftarrow}{\longrightarrow}}} \bullet \underset{\beta_3}{\overset{\alpha_3}{\underset{\longleftarrow}{\longrightarrow}}} \bullet \cdots \bullet \underset{\beta_{m-1}}{\overset{\alpha_{m-1}}{\underset{\longleftarrow}{\longrightarrow}}} \bullet\, m$$

with ideal I generated by the relations

$$\alpha_i \alpha_{i+1} = 0, \quad \beta_{i+1}\beta_i = 0, \quad \alpha_1 \beta_1 = 0, \quad \beta_i \alpha_i = \alpha_{i+1}\beta_{i+1} \quad (1 \leq i \leq m-2).$$

By tradition maps are written on the right. It is known that this is an algebra of finite type. In particular, $\mathcal{A}_1$ is simple.

Exercise. Show that this is the quiver for any basic block of Σ_n where $n < 2p$. (Hint: try Σ_9 in characteristic 5.)

Proposition 5.6.2 *Suppose S is a finite-dimensional K-algebra, with m simple modules $S_1, \ldots, S_m$, with $S_m = 0$ possibly. Assume further that there are modules for S of the form*

$$M_1 = S_1, \quad M_2 = \begin{matrix} S_1 \\ S_2 \\ S_1 \end{matrix}, \quad M_i = \begin{matrix} S_{i-1} \\ S_i \quad S_{i-2} \\ S_{i-1} \end{matrix}$$

for $3 \leq i \leq m$. Suppose that $\dim_K \mathrm{Hom}_S(M_i, M_j) = 1$ for $|i - j| = 1$. Then

$$\mathrm{End}_S \left(\bigoplus_{i=1}^m M_i \right) \cong \mathcal{A}_m.$$

Proof Let $M = \bigoplus_{i=1}^m M_i$, and let e_i be the projection onto the ith summand $(i \in \mathbf{m})$. The e_i are to be the vertices of the quiver of $\mathrm{End}(M)$. Choose and fix non-zero homomorphisms $\alpha_i : M_i \to M_{i+1}$ and $\beta_i : M_{i+1} \to M_i$. Thus $\mathrm{Im}\,\alpha_1$ and $\mathrm{Im}\,\beta_1$ are simple, whereas all other images have composition length two. We now check the relations.

Now $\mathrm{Im}\,\alpha_1 = \mathrm{Soc}\,M_2 \subseteq \mathrm{Rad}(M_2) = \mathrm{Ker}\,\beta_1$ and so $\alpha_1\beta_1 = 0$. Also $\mathrm{Hom}(M_i, M_j) = 0$ for $|i - j| \geq 2$, so $\alpha_i\alpha_{i+1} = 0 = \beta_{i+1}\beta_i$ independently of the choice of α_i, β_i.

Finally, choose $i \leq m - 2$ and suppose we have, by induction, the commutativity relations for e_j $(j \leq i - 1)$. But $\beta_i\alpha_i$ and $\alpha_{i+1}\beta_{i+1}$ are both elements of $\mathrm{End}(M_i)$ having simple image. Thus we can find $\kappa \in K^*$ such that $\kappa(\alpha_{i+1}\beta_{i+1}) = \beta_i\alpha_i$, and replacing α_{i+1} by $\kappa\alpha_{i+1}$ does not affect any earlier relations.

We have now an epimorphism $\mathcal{A}_m \to \mathrm{End}(M)$. It is an isomorphism, by dimension considerations.   □

The following theorem was stated for the first time in Martin [1991] and then in Erdmann [1993a]. We recall the definition of the components of $S = S(n, r)$ from 5.5.5.

Theorem 5.6.3 *Suppose B is a block of KG of finite type. Then S_B is of finite type. Moreover, if B contains m partitions with at most n parts then S_B is Morita equivalent to $\mathcal{A}_m$.*

Proof Any block of KG of finite type may be described by a Brauer tree, which is an open polygon with p vertices none of which is exceptional; see Chung [1951] or prove it as an exercise. The block has $p-1$ modular irreducible modules; for each $\ell \in \{0, 1, \ldots, p-2\}$ there is a unique such module occurring as a factor with multiplicity one in Specht modules indexed by partitions having skew p-hooks with leg lengths ℓ and $\ell + 1$. We can take the decomposition matrix for B in the form

$$
\begin{array}{c}
\lambda_0 \\
\lambda_1 \\
\lambda_2 \\
\vdots \\
\vdots \\
\lambda_{p-2} \\
\lambda_{p-1}
\end{array}
\left[
\begin{array}{cccccc}
1 & & & & & \\
1 & 1 & & & & \\
0 & 1 & 1 & & & \\
& & & & & \\
& & & & & \\
0 & & & & 1 & 1 \\
0 & & & & 0 & 1
\end{array}
\right]
\qquad (5.4)
$$

It follows that we can obtain the decomposition matrix for S_B from that for G by adjoining a single 1 in the bottom right-hand corner. The matrix for S_B is therefore the $p \times p$ matrix

$$
\left[
\begin{array}{cccccc}
1 & & & & & 0 \\
1 & 1 & & & & \\
0 & 1 & 1 & & & \\
& & & & & \\
0 & & & 0 & 1 & 1
\end{array}
\right]
\qquad (5.5)
$$

where the $\lambda_i \in B$ have been placed in increasing lexicographic order. (To obtain the decomposition matrix (5.4) for G, ignore the bottom-right 1, replace partitions by their conjugates and read the (λ, μ)th entry as $[S^{\lambda} : D^{\mu}]$.)

We want to apply 5.6.2, so we would like to show that Y^{λ_i} has the same shape as the M_i occurring in that proposition. This is an easy task with the filtration theorem of 4.6.4 at hand. In the process we obtain the characters (see below) and the homomorphism condition follows from this using the proof of 4.6.2(i).

Now, by the dual Specht filtration, $Y^{\lambda_p} = S_{\lambda_{p-1}}$ and is (modularly) irreducible. Also $Y^{\lambda_{p-1}}$ has two factors, namely $S_{\lambda_{p-2}}$ and $S_{\lambda_{p-1}}$. Reducing modulo p we see that $Y^{\lambda_{p-1}}$ is uniserial with factors from the top $D_{\lambda_{p-1}}, D_{\lambda_{p-2}}, D_{\lambda_{p-1}}$, since it is self-dual. Similarly for $i \le p-3$ the Y^{λ_i} have the shape of the M_i.

Finally, the number of parts of λ_i is non-increasing with i, so there exists an m for which the last m partitions in B have at most n parts. By 5.6.2, S_B is Morita equivalent to $\mathcal{A}_m$.   □

For our final example we explicitly record some information, uncovered in the proof of 5.6.3, about certain Young modules .

Lemma 5.6.4 *Take a block B of G having decomposition matrix of the form (5.4), and order the partitions in B in decreasing lexicographic order, $\lambda_0 > \lambda_1 > \cdots$. For $\lambda \in \Lambda^+(r)$, write χ^{λ} for the ordinary character of the Specht module S^{λ}. Then we have the following.*

    (i) *Y^{λ_0} is simple.*

    (ii) *For the first two partitions Y^{λ_1} is uniserial of length three with factors; from the top these are D^{λ_0}, D^{λ_1}, D^{λ_0}.*

    (iii) *If $M \in B$ is a summand of some M^{α} ($\alpha \in B$) and the character of M has inner product of value 2 with itself, then $M \cong Y^{\lambda}$ for some $\lambda \ge \alpha$. Moreover if we have three consecutive partitions $\nu > \mu > \lambda$ then Y^{λ} has a simple socle and top isomorphic to D^{μ} and a semisimple heart isomorphic to $D^{\nu} \oplus D^{\lambda}$, i.e. Y^{λ} is a module 'of diamond shape'.*

    (iv) *If $\lambda \in B$ and Y^{λ} has an irreducible character, then χ^{λ} must be in the first row of (5.4).*

Proof   This is left as a very easy exercise that may be done either directly by using basic properties of Young modules, or indirectly by adapting the proof of 5.6.3. For part (iv), just note that if case (i) is excluded then the reduction mod p of χ^{λ} has two distinct factors, so cannot give rise to a self-dual indecomposable module.   □

With these preliminaries complete, we concentrate on odd characteristic and study the algebra $S(2, r)$ where $r = wp + t$ $(0 \leq w, t \leq p - 1)$. These Schur algebras possess a rather transparent structure.

Theorem 5.6.5 *Let B be the principal p-block of $K\Sigma_r$ and let S_r be the Schur algebra $S(2, r)$, as above.*

   (i) *If $t = p - 1$ then $(S_r)_B$ is semisimple, with $w/2 + 1$ or $(w-1)/2 + 1$ simple components.*

  (ii) *If $t \neq p - 1$ then $(S_r)_B$ is a block of S_r, and $(S_r)_B$ is Morita equivalent to $\mathcal{A}_{w+1}$.*

Proof There are published algorithms for computing decomposition numbers for Specht modules labelled by two-part partitions (here called pairings) due originally to James [1981]. Another proof is given in Erdmann [1993b, §6]. Let D be the matrix for the pairings in B.

For (i), $D = I$, so all permutation modules $M^{\lambda(a)}$ indexed by the pairings $\lambda(a) = (r - ap, ap)$ are simple for $0 \leq a \leq w/2$ or $(w - 1)/2$. So the component $(S_r)_B$ is semisimple with the same number of simple components as there are pairings.

For (ii), the pairings are either of the form $\lambda(a)$ as in (i) or of the form $\mu(b) = (r - bp - t - 1, bp + t + 1)$ with $0 \leq b \leq w/2 - 1$ or $(w - 1)/2$. The James algorithm demonstrates that D is a $(w + 1) \times (w + 1)$ matrix with the same form as (5.4). We now execute a less trivial application of 5.6.2. Let $B = B(\varepsilon)$, where ε is the idempotent defining the block, and recall the Green Correspondence for Young modules from 4.6.3.

(1) $M^{\lambda(a)}\varepsilon$ has $a + 1$ components with Young vertex $V = \Sigma_{(p^w)}$, for $0 < a < (w/2)$ or $(w - 1)/2$; one component is trivial. We put $N = N_G(V)/V$ and $N_1 = N_{\Sigma_\lambda}(V)/V$. It is enough to show that $K_{N_1} \uparrow^N$ has $a + 1$ summands having Green Correspondents in the principal block. But

$$K_{N_1} \uparrow^N = \left(K_{\Sigma_{w-a} \times \Sigma_a \times \Sigma_t} \right) \uparrow^{\Sigma_w \times \Sigma_t} \cong \left(K_{\Sigma_{w-a} \times \Sigma_a} \right) \uparrow^{\Sigma_w} \otimes K,$$

which is semisimple, affording a character with $a + 1$ simple constituents (e.g. by the Littlewood-Richardson Rule). The correspondent of any summand must lie in the principal block, since each summand is of the form $M \otimes K$ with M having maximal vertex as an $N_G(V)$-module.

(2) $M^{\mu(b)}\varepsilon$ has $b + 1$ components with Young vertex $V = \Sigma_{p^{w-1}}$ for $0 < b < w/2 - 1$ or $(w - 1)/2$. Define N and N_1 by analogy with (1) above. This time we have

$$K_{N_1} \uparrow^N = (K_{\Sigma_{w-b-1} \times \Sigma_b \times \Sigma_{p-1} \times \Sigma_{t+1}}) \uparrow^{\Sigma_{w-1} \times \Sigma_{p+t}}$$

$$\cong (K_{\Sigma_{w-b-1} \times \Sigma_b}) \uparrow^{\Sigma_{w-1}} \otimes (K_{\Sigma_{p-1} \times \Sigma_{t+1}}) \uparrow^{\Sigma_{p+1}}.$$

As above, the first factor is semisimple with $b+1$ summands. The second summand is isomorphic to the Σ_{p+1}-module $M^{(p-1,t+1)}$. The part of the character lying in the principal block of Σ_{p+1} has two irreducible constituents, and the module is indecomposable.

(3) *For M_{2i+1} and M_{2i+2} in 5.6.2 we may take $Y^{\lambda(i)}$ and $Y^{\mu(i)}$ respectively.* Using 5.6.4 we have that $Y^{\lambda(0)}$ and $Y^{\mu(0)}$ have the desired shape. Now $\operatorname{ch} M^{\lambda(a)} \varepsilon$ has $2a+1$ simple constituents, by the Littlewood-Richardson Rule, and by (1) above there are at least $a+1$ components. Any non-trivial summand has a character with at least two constituents. Hence there are exactly $a+1$ summands, and any non-trivial summand has two constituents in its character. This is true in particular for $Y^{\lambda(a)}$; similarly for $Y^{\mu(a)}$. $\square$

Exercise. As an (easy) exercise show that $S(2,2)$ is Morita equivalent to $\mathcal{A}_2$ and $S(2,7)$ is Morita equivalent to $\mathcal{A}_2 \oplus K \oplus K$.

6

The q-Schur algebra

The representation theory of the finite general linear groups was investigated throughout the nineteen-eighties, the cases of the describing and non-describing characteristic being attacked with equal vigour. Fong and Srinivasan [1982] produced an elegant classification of the p-block structure of $GL_n(q)$ where $(p, q) = 1$. Their labours produced a rule similar to the Nakayama Rule, 5.1.1, thus demonstrating that the case of positive non-describing characteristic could be as interesting as the case of the natural characteristic. James [1984] unearthed plentiful evidence for the assertion that the modular representation theory of Σ_n is 'the case $q = 1$' of this coprime theory.

Following this line, Dipper and James (independently and also in collaboration) unleashed the q-Schur algebra, $S_q(n, r)$, on an unsuspecting world. The q-Schur algebras are related to the Hecke algebras associated to $G = \Sigma_r$ in much the same way as are Schur algebras and group algebras of symmetric groups. Moreover, $_{S_q(n,r)}\mathbf{mod}$ embeds into $_{GL_n(q)}\mathbf{mod}$ if we take q to be a prime power and this enables us to handle the representations of the finite reductive group. It is quite astonishing that both the representation theory of symmetric groups and the representation theory of the infinite general linear groups over K, discussed in previous chapters, are special cases of the representation theory of $KGL_n(q)$, where n any integer such that $n \geq 2$ and where q is a prime power coprime to the characteristic of K (so that q is a root of unity mod p). As revealed in the next chapter there exists a quantum group, Γ_q, whose representation theory provides an elegant description of these phenomena.

Nowadays, there are several quite distinct treatments in the literature, each with a different set of goals and *ad hoc* notation. This chapter and the next form an attempt to present an account covering various

159

aspects of the work of Dipper and James, together with that covered in the fundamental paper of Dipper and Donkin [1991]. The approach in the next two chapters is elementary, in the sense that we use little of the quantum group theory developed by Parshall and Wang [1991] and others. Yet another angle on this area can be found in Du [1991], [1992a], [1992b]. The purpose here is to give a straightforward, leisurely but substantial survey of the current situation, the presentation being chosen so as to bring into sharp relief the analogies with ideas in previous chapters. Certainly, no claim is made for completeness, indeed we often omit either routine or overly complex arguments that are available in the literature; we recall the Aristotelian aphorism that

it is the mark of the educated mind to use for each subject the degree of exactness which admits it.

Let R be a commutative domain, and $q \in R$ a unit. The basic idea is to construct a continuous q-deformation, $A_{R,q}(n)$, of the commutative Hopf algebra $A_R(n) = R[c_{ij}]$ (the coordinate ring of the algebraic variety $\mathrm{Mat}_n(R)$). Roughly, this means we require some bialgebra involving n^2 parameters in non-commuting variables and a 'quantum' parameter q. The point is that the bialgebra so constructed is *not* in general commutative, but it should yield $A_1(n)$ when we put $q = 1$. In the special case where $R = S[[t]]$, with S a commutative domain and $q = e^t = 1 + t + t^2/2 + \cdots$, we can consider the coalgebra $A_{R,q}(n)$ as a quantisation of $A_R(n)$ and whose quantum space is the spectrum of $A_{R,q}(n)$.† Let me also point out that if q is not a root of unity, then the quantised objects, $A_q(n)$ and $S_q(n,r)$ are not very exciting: they behave like their algebraic group counterpart in characteristic zero.

Initially, the q-Schur algebra is defined as the R-dual of the coalgebra $A_{R,q}(n,r)$; this is an obvious analogue of 2.3.5 in which the usual Schur algebra is identified with $A_{R,1}(n,r)^*$. Actually this was not the original definition of Dipper and James. It was known at the time that the symmetric group algebra KG has a q-analogue, namely a certain Hecke algebra $\mathcal{H}_q$ with Coxeter group G. We recall from (4.12) that the permutation module $E^{\otimes r}$ is a direct sum of transitive permutation modules

† The notion of replacing commutative algebras by a non-commutative 'deformation' is familiar from quantum mechanics, and indeed this was the motivation for the recent interest in Hopf algebras. In classical mechanics 'observables' are functions on a manifold hence generate a commutative associative algebra; in quantum mechanics the observables are operators on a Hilbert space, so form a non-commutative algebra. For more fascinating background on the physics behind quantisation see Drinfel'd [1986].

with multiplicities: actually these have q-analogues too. In the paper of
Dipper and James [1989], the q-Schur algebra is defined to be the endomorphism algebra of the direct sum of these q-permutation modules,
each of multiplicity one. In a sequel Dipper and James [1991] redefine
the q-Schur algebra by upping the multiplicities to coincide with those
of tensor space, the new algebra being Morita equivalent to the first.
The main result of this chapter is the assertion that this second algebra is canonically isomorphic to $A_q(n,r)^*$; see 6.6.6 below. The proof
proceeds by obtaining a satisfactory definition for the action of $\mathcal{H}_q$ on
r-tensor space and provides another strong analogy with the classical
case. It is interesting to note that the Hecke algebra and an action of it
on tensor space had been given by Jimbo [1986] in a different context:
it specialised to the usual G-action, twisted by sign.

Some notational points: if C is a coalgebra, we will write (M, ω_M) for
the right C-comodule M with comodule structure map ω_M (see Sweedler
[1969] or Abe [1980] for details). In this chapter there are various spaces
being acted upon by two quite different rings, one on the left and one
on the right. To avoid confusion between elements and maps we have
written homomorphisms on the side opposite to the ring action.

6.1 Quantum matrix space

Throughout, R will be a commutative integral domain with 1.

The basic ingredient in constructing $A_{R,q}(n)$ is the free R-algebra
$F(n)$ in the n^2 non-commuting variables X_{ij} $(1 \le i, j \le n)$.

Definition 6.1.1 Let $A_{R,q}(n) = A_q(n) = F(n)/J$ where J is the ideal
generated by elements of the form

$$
\begin{array}{lll}
X_{ik}X_{jl} & - \quad qX_{jl}X_{ik} & (i > j,\ k \le l) \\
X_{ik}X_{jl} & - \quad X_{jl}X_{ik} - (q-1)X_{jk}X_{il} & (i > j,\ k > l) \\
X_{ik}X_{il} & - \quad X_{il}X_{ik} & (\text{all } i, l, k)
\end{array}
$$

$(i, j, k, l \in \mathbf{n})$. Write the image $X_{ij} + J$ of X_{ij} in $A_q(n)$ by c_{ij} $(i, j \in \mathbf{n})$,
so that $A_q(n)$ is an affine R-algebra generated by $\{c_{ij} : i, j \in \mathbf{n}\}$, with
relations

$$
\begin{array}{llll}
(A) & c_{ik}c_{jl} = qc_{jl}c_{ik} & (i > j,\ k \le l) & \\
(B) & c_{ik}c_{jl} = c_{jl}c_{ik} + (q-1)c_{jk}c_{il} & (i > j,\ k > l) & (6.1) \\
(C) & c_{ik}c_{il} = c_{il}c_{ik} & (\text{all } i, l, k). &
\end{array}
$$

$(i, j, k, l \in \mathbf{n})$.

Now $F(n)$ is graded by the $F(n,r) = \langle X_{i_1 j_1}, \ldots, X_{i_r j_r} : i, j \in I(n,r) \rangle$ ($r \geq 0$). The ideal J is homogeneous and generated by degree two elements, so $A_q(n)$ is graded by $A_q(n,r) = \langle c_{i,j} : (i,j) \in I^2 \rangle$. We seek a basis for $A_q(n,r)$ similar to the one obtained in 1.3.4 for A_r. This should be a sensible strategy, given that $A_{K,1}(n,r)$ is precisely the Schur coalgebra A_r of 1.3.4.

Combinatorics

We recall that the *length* $\ell(\pi)$ of a permutation $\pi \in \Sigma_r$ is defined as the cardinality of $\{(i,j) : i < j \text{ and } i\pi > j\pi\}$; equivalently, if $\pi = v_1 \ldots v_k$ is a product of basic transpositions with k minimal, then $k = \ell(\pi)$ ($v \in G$ is said to be a *basic transposition* if $v = (a, a+1)$ for $1 \leq a < r$). Each right coset $\Sigma_\lambda \pi$ contains a unique shortest element: we call this the *distinguished coset representative* of Σ_λ in $\Sigma_\lambda \pi$, and the set of all such is denoted by $\mathcal{D}_\lambda$. To think of this combinatorially, consider the λ-tableau τ^λ having as entries the numbers 1 up to r appearing in order along successive rows (so that the row stabiliser of τ^λ is Σ_λ). Then clearly

$$\mathcal{D}_\lambda = \{d \in G : \tau^\lambda d \text{ is row semistandard}\}.$$

We will take $r \leq n$ and $\lambda, \mu \in \Lambda(n,r)$. Let $\mathcal{D}_\mu^{-1}$ be a set of distinguished left coset representatives of Σ_μ in G; then $\mathcal{D}_{\lambda\mu} = \mathcal{D}_\lambda \cap \mathcal{D}_\mu^{-1}$ is a set of distinguished $(\Sigma_\lambda, \Sigma_\mu)$-double coset representatives. This set also has a simple combinatorial description. Namely, let $\mathcal{T}(\mu, \lambda)$ be the set of all μ-tableaux of type λ, and take $T \in \mathcal{T}(\mu, \lambda)$. If $\sigma \in G$ let σ_T be that element of $\mathcal{D}_\lambda$ such that $\tau^\lambda \sigma_T$ is a row semistandard λ-tableau, and where p is in the sth row if the place occupied by $p \in \tau^\mu \sigma$ is occupied by s in T ($p \in \mathbf{r}$).

Example. With $\mu = (3,2)$ and $\lambda = (3,1^2)$,

$$T = \begin{matrix} 1 & 2 & 1 \\ 1 & 3 \end{matrix} \quad , \quad \tau^\mu \sigma = \begin{matrix} 1 & 5 & 3 \\ 2 & 4 \end{matrix} \quad , \quad \tau^\lambda \sigma_T = \begin{matrix} 1 & 2 & 3 \\ 5 \\ 4 \end{matrix}$$

Then the map

$$\mathcal{T}(\mu, \lambda) \rightarrow \mathcal{D}_\lambda$$
$$T \mapsto 1_T \qquad\qquad (6.2)$$

is well defined and is easily seen to be a bijection (see Dipper and James

[1986, 1.7]—indeed its inverse will be given explicitly in 7.1). On restriction we have a bijection between elements of $\mathcal{T}(\mu, \lambda)$ that are row semi-standard, and $\mathcal{D}_{\lambda\mu}$.

Order I lexicographically and define a linear order of $I \times I$ by

$$(i, j) < (k, l) \text{ if } i < k \text{ or } i = k \text{ and } j < l$$

for $i, j, k, l \in I$. This induces an order on the indeterminates $X_{ij} \in F(n, r)$. If $\lambda \in \Lambda(n, r)$, consider the index $i_\lambda = (i_1, \ldots, i_r) \in I$ of weight λ such that $i_1 \leq \ldots \leq i_r$: this *initial index* is the smallest element in its G-orbit $\{i_\lambda d : d \in \mathcal{D}_\lambda\}$. Similarly, for $i, j \in I$ the smallest element in the G-orbit $\{(i\pi, j\pi) \in I^2 : \pi \in G\}$ will be called the *initial double index* $(i, j)_0 = (k, l)$, say. The parts satisfy

$$k_1 \leq \ldots \leq k_r \text{ and } l_\alpha \leq l_{\alpha+1} \text{ if } k_\alpha = k_{\alpha+1} \ (1 \leq \alpha < r).$$

We denote the set of initial double indices in $I \times I$ by I_0^2.

A basis for $A_q(n)$

We find $\pi \in G$ of smallest length such that $(k\pi, l\pi) = (i, j)$. Then $\pi = uv$ for some $u \in \Sigma_\lambda \cap \mathcal{D}_\mu$, $v \in \mathcal{D}_\lambda$, with i and j of weight λ and μ respectively. Also by Dipper and James [1986, 1.6], $\ell(\pi) = \ell(u) + \ell(v)$. Let $\{\bar{v}\} = \Sigma_\mu v \cap \mathcal{D}_\mu$ and let $\varepsilon(i, j) = \ell(\bar{v}) \in \mathbf{N}_0$. The following comes from the relations (6.1) by induction on $\ell(\pi)$: it will be needed later when we are looking for a basis for a certain quantum space.

Lemma 6.1.2 *In the notation of the last paragraph, there exists the relation*

$$c_{i,j} = q^{\varepsilon(i,j)} c_{k,l} + \text{terms involving } c_{s,t}$$

in $A_{R,q}(n, r)$, where (s, t) is an initial double index not in the G-orbit of (i, j), and $(s, t) > (k, l) = (i, j)_0$.

The basis theorem, which follows, is proved in Dipper and Donkin [1991, 1.1.8] by ring-theoretic methods. For a proof involving Yang-Baxter operators, see Parshall and Wang [1991, (3.5.1)].

Theorem 6.1.3 $A_q(n) = \bigoplus_{r \geq 0} A_q(n, r)$ *is free as an R-module. The rth graded piece has R-basis $\mathcal{B}_r = \{c_{i,j} : (i, j) \in I_0^2\}$, so that*

$$\mathcal{B} = \bigcup_{r \geq 0} \mathcal{B}_r = \{c_{i,j} : (i, j) \in I_0^2, \ r \geq 0\}$$

is a basis of $A_q(n)$. (By convention, $I_0^2(n,0) = \emptyset$ and $c_\emptyset = 1_R$.) In particular,

$$\dim_R A_{R,q}(n,r) = \binom{n^2+r-1}{r}.$$

A total ordering on monomials in $\mathcal{B}$ now suggests itself. We say that $c_{i,j} \preceq c_{k,l}$ for $(i,j) \in I_0^2(n,r)$ and $(k,l) \in I_0^2(n,s)$, if $r \leq s$, or if $r = s$ and $i \geq k$, or if $r = s$, $i = k$ and $j \geq l$. Here $\leq$ is the usual dictionary order on I.

The bialgebra structure on $A_q(n)$

The quantisation $A_q(n)$ is also a bialgebra for any $q \in R$. For, the diagonal map $\Delta(X_{ij}) = \sum X_{ik} \otimes X_{kj}$ can be extended to an R-algebra homomorphism $\Delta : F(n) \to F(n) \otimes F(n)$ by universality. Δ will be coassociative and defines a diagonalisation on $F(n)$. Similarly, $\varepsilon : F(n) \to R$ induced by $\varepsilon : X_{ij} \mapsto \delta_{ij}$ defines a co-unit on $F(n)$. Then $F(n)$ satisfies the axioms for being a bialgebra. An axiom crunch, given explicitly in Dipper and Donkin [1991, 1.4.1], demonstrates that $\varepsilon(J) = 0$ and $\Delta(J) \subseteq J \otimes F(n) + F(n) \otimes J$, so

Lemma 6.1.4 J is a bi-ideal of the bialgebra $F(n)$.

Consequently (cf. 1.3.6),

Theorem 6.1.5 $A_q(n)$ is a bialgebra with diagonalisation $\Delta : A_q(n) \to A_q(n) \otimes A_q(n)$ defined by $\Delta(c_{ij}) = \sum_k c_{ik} \otimes c_{kj}$ and co-unit $\varepsilon : A_q(n) \to R$ defined by $\varepsilon(c_{ij}) = \delta_{ij}$. Moreover, for all $r \geq 0$, $A_q(n,r)$ is a sub-coalgebra of $A_q(n)$.

6.2 The q-Schur algebra, first visit

For $r \geq 0$, consider the R-dual $A_q(n,r)^* = \mathrm{Hom}_R(A_q(n,r), R)$ of the coalgebra $A_q(n,r)$. In the usual way, $A_q(n,r)$ has the structure of an R-algebra with multiplication Δ^*. The first manifestation of $A_q(n,r)^*$ as a q-Schur algebra appeared in Dipper and Donkin [1991, 1.4.4].

Definition 6.2.1 $A_q(n,r)^*$ is called the q-Schur algebra.

So, a basis for $A_q(n,r)^*$ is the dual basis $\{\xi_{i,j} : (i,j) \in I_0^2\}$ to $A_q(n,r)$.

Remark. Of course, if $R = K$ and $q = 1$ we obtain the Schur algebra

$S_K(n, r)$ of 2.3 (and the above basis reduces to that of 2.2.6). Nevertheless, we have refrained from denoting $A_q(n, r)^*$ by anything emotional like $S_q(n, r)$: we reserve this notation for another algebra that, in any case, turns out to be canonically isomorphic to $A_q(n, r)^*$. This identification will be the primary goal for the remainder of the chapter.

Tensor space and tensor algebras

Abusing previous notation, we will write E for the free R-module of dimension n with basis $\{e_1, \ldots, e_n\}$. Then once again the rth tensor power, $E^{\otimes r}$ ($r \in \mathbf{N}_0$), of E plays a pivotal rôle. For a start, $E^{\otimes r}$ is a right $A_q(n, r)$-comodule: the structure map $\tau_r : E^{\otimes r} \to E^{\otimes r} \otimes A_q(n, r)$ is

$$\tau_r(e_j) = \sum_{i \in I} e_i \otimes c_{i,j}$$

for $j \in I$. By embedding $E^{\otimes r} \otimes A_q(n, r)$ into $E^{\otimes r} \otimes A_q(n)$ using the embedding of $A_q(n, r)$ into $A_q(n)$ and composing with τ_r, one makes $E^{\otimes r}$ into an $A_q(n)$-comodule: let τ_r again denote the structure map.

One can also form the tensor algebra in the normal way, that is, $T(E) = \bigoplus_{r \geq 0} E^{\otimes r}$ is a right $A_q(n)$-comodule with structure map given by the R-algebra homomorphism $\tau = \sum_r \tau_r : e_i \mapsto \sum_{j \in I} e_j \otimes c_{j,i}$ for $i \in I$ (see Dipper and Donkin [1991, 2.1.2]).

Now we look for exterior powers. Omitting tensor signs, let us consider the ideal

$$J_q = \langle e_i^2, \ e_j e_k + q e_k e_j : i \in \mathbf{n}, \ 1 \leq j < k \leq n \rangle$$

of $T(E)$ ($q \in R$). We check using (6.1) that J_q is in fact a sub-comodule of $T(E)$.

Definition 6.2.2 The *q-exterior power* $\Lambda_q E$ is the $A_q(n)$-comodule $T(E)/J_q$. As an R-algebra it is generated by $\{\bar{e}_i : i \in \mathbf{n}\}$, where $\bar{e}_i$ is the image of e_i in $\Lambda_q E$.

J_q is a homogeneous ideal in the graded algebra $T(E)$, and so $\Lambda_q E = \bigoplus_{r \geq 0} \Lambda_q^r E$ is graded. In fact, as for the usual exterior power, $\Lambda_q^r E = 0$ whenever $r > n$. Even more is true: $\Lambda_q^r E$ is a free R-module.

An inductive proof of the next fact is set out in Dipper and Donkin [1991, 2.1.5].

Theorem 6.2.3 *For $r \leq n$, $\Lambda_q^r E$ is a free R-module with free basis $\{\bar{e}_{i_1} \ldots \bar{e}_{i_r} : i_1 > \ldots > i_r, i_\alpha \in \mathbf{n}\}$.*

Remark. The basis given in 6.2.3 is independent of the choice of R or q.

There is also a natural left comodule structure on the tensor space: let V be R-free with basis $\{v_1, \ldots, v_n\}$ and turn $V^{\otimes r}$ into a left $A_q(n,r)$-comodule by means of the R-algebra homomorphism

$$\sigma_r : V^{\otimes r} \rightarrow A_q(n,r) \otimes V^{\otimes r}$$
$$v_i \mapsto \sum_{j \in I} c_{i,j} \otimes v_j$$

for each $i \in I$. Endow $V^{\otimes r}$ with the structure of an $A_q(n)$-comodule as above, and sum up to obtain the left $A_q(n)$-comodule $T(V)$, with structure map $\sigma = \sum \sigma_r$. Define $\Lambda_q V = T(V)/I_q$ where I_q is the subcomodule

$$I_q = \langle v_i^2, \; v_j v_k + v_k v_j : i \in \mathbf{n}, \; 1 \leq j < k \leq n \rangle$$

of $T(V)$ (so $\Lambda_q V$ is the exterior algebra we know and love). We grade $\Lambda_q V$ by the spaces $\Lambda_q^r V$, to get a basis $\{\bar{v}_{i_1} \ldots \bar{v}_{i_r} : i_1 < \ldots < i_r, i_\alpha \in \mathbf{n}\}$ where $\bar{v}_i$ is the image of $v_i \in T(V)$ in $\Lambda_q V$ (see Dipper and Donkin [1991, 2.1.6]).

Remark. In some sense the exterior powers lead a double life: respectable and familiar on one side, but with a 'q-alter ego' on the other. We shall see that this characteristic is shared by the q-determinant of 7.2. Similar definitions are available for q-symmetric powers.

6.3 Weights and polynomial modules

We recall the discussion of weights in 1.4. Let $Z = \mathbf{N}_0^n$. There is an obvious grading of $F(n)$ by $Z \times Z$, namely, for $\lambda, \mu \in Z$, let $F^{\lambda,\mu}(n) = \langle X_{i,j} : i \in \lambda, \; j \in \mu \rangle$. Thus $F(n) = \bigoplus_{\lambda,\mu \in Z} F^{\lambda,\mu}(n)$. The (λ, μ)th piece is 0, if λ and μ are not compositions of the same integer and clearly $F^{\lambda,\mu}(n).F^{\alpha,\beta}(n) \subseteq F^{\lambda+\alpha,\mu+\beta}(n)$ where $\lambda, \mu, \alpha, \beta \in Z$. Since J is homogeneous, we have a $Z \times Z$-grading on

$$A_q(n) = \bigoplus_{\lambda,\mu \in Z} A_q^{\lambda,\mu}(n)$$

where the (λ, μ)th piece is the span of all $c_{i,j} : (i,j) \in I_0^2$ with $i \in \lambda$ and $j \in \mu$. By the comment above, if this piece is non-zero, λ, μ must be compositions of the same integer. Hence

$$A_q(n, r) = \bigoplus_{\lambda, \mu} A_q^{\lambda, \mu}(n, r)$$

summed over all compositions λ, μ of r. For a weight μ, define

$$A_q^{\mu}(n) = \bigoplus_{\lambda \in \Lambda(n,r)} A_q^{\lambda, \mu}(n) \qquad (6.3)$$

Example. The generators c_{ij} are members of $A_q^{\omega_i, \omega_j}(n)$, where $\omega_i \in Z$ has 1 in the ith slot and zeros elsewhere $(i, j \in \mathbf{n})$.

Definition 6.3.1 Any right $A_q(n)$-comodule (M, ω) that is free of finite rank will be called *polynomial*. For $r \geq 0$, the rth *homogeneous component* M_r of M is

$$M_r = \{m \in M : \omega(m) \in M \otimes A_q(n, r)\}.$$

Call the integer r the *degree*. The $A_q(n, r)$-comodule M will be called *homogeneous* if $M = M_r$.

We shall say that M is an $A_q(n)$-*colattice* if it is an $A_q(n)$-comodule that is R-free of finite rank.

Proposition 6.3.2 *Any $A_q(n)$-colattice is a direct sum of its homogeneous components. Hence any indecomposable comodule is homogeneous. Subcomodules and quotients of homogeneous $A_q(n)$-comodules are themselves homogeneous of the same degree.*

Proof One way would be to mimic the classical case of 1.3.7 by using Green [1976, 1.6c]. □

Example. *Tensor products.* Let us take two $A_q(n)$-colattices M and N, with structure maps ω_1 and ω_2. Then $M \otimes_R N$ is a $A_q(n)$-comodule with structure map

$$\omega_3 : M \otimes_R N \quad \rightarrow \quad M \otimes_R N \otimes_R A_q(n)$$
$$m \otimes n \quad \mapsto \quad \sum_{i,j} m_i \otimes n_j \otimes r_i s_j$$

where $\omega_1(m) = \sum_i m_i \otimes r_i$ and $\omega_2(n) = \sum_j n_j \otimes s_j$. Clearly,

Lemma 6.3.3 *If M and N are right $A_q(n)$-colattices, homogeneous of degree r and degree s respectively, then $M \otimes N$ is a homogeneous polynomial module of degree $r + s$.*

Dually, we can consider the direct sum of rings $S_q(n) = \bigoplus_{r \geq 0} A_q(n,r)^*$, a subalgebra (without 1) of $A_q(n)^*$. Given a $A_q(n)$-colattice (M, ω_M), the dual may be made into a right $A_q(n)^*$-module by dualising ω_M and restricting the action of $A_q(n)^*$ to $S_q(n)$. One can also identify M and M^{**} (because of finite rank) to give a *left* action of $S_q(n)$ on M. Thus $M_r = 1_{A_q(n,r)^*} M$.

By (6.3) there is a decomposition of the left comodule $(A_q(n,r), \Delta)$,

$$A_q(n,r) = \bigoplus_{\lambda \in \Lambda(n,r)} A_q^\lambda(n,r),$$

as a direct sum of left sub-comodules. We dualise to obtain a decomposition of the q-Schur algebra of 6.2.1 as a direct sum of right ideals, i.e. projective right $A_q(n,r)^*$-modules, each generated by an idempotent $\xi_\lambda \in A_q(n,r)^*$. But $1 = \sum_{\lambda \in \Lambda(n,r)} \xi_\lambda$, so that every right $A_q(n,r)^*$-lattice M can be decomposed as a direct sum of *weight spaces*

$$M^\lambda = M\xi_\lambda.$$

Starting rather with an $A_q(n)$-colattice M, let us decompose it into its homogeneous components and view the piece of degree r, M_r, as a left $A_q(n,r)^*$-module (cf. the discussion after 6.3.3) to obtain a decomposition of M into its weight spaces M^λ ($\lambda \in \Lambda(n,r)$, $r \geq 0$). The final lemma of the section ties this together:

Lemma 6.3.4 *Let (M, ω_M) be an $A_q(n)$-colattice. For $\lambda \in Z$,*

$$M^\lambda = \{ m \in M : \omega_M(m) \in M \otimes A_q^\lambda(n) \},$$

and for $r \geq 0$, $M_r = \bigoplus_{\lambda \in \Lambda(n,r)} M^\lambda$, so $M = \sum_{\lambda \in Z} M^\lambda$, as R-modules.

6.4 Characters and irreducible $A_q(n)$-modules

In this section we assume that R is a principal ideal domain. Thus far all our modules have a well-defined dimension, being R-free, and so it makes sense to define their formal character, by analogy with 1.5.3. Thus, given an $A_{R,q}(n)$-colattice M, we define the formal character of M as the polynomial

$$\operatorname{ch}(M)(\mathbf{X}) = \sum_{\lambda \in Z}(\dim_R M^\lambda)\mathbf{X}^\lambda.$$

As usual the character function is exact with respect to short exact sequences, multiplicative with respect to tensoring, and isomorphic modules have the same character. In particular the ordinary character of M can be defined to be ch $M \otimes_R Q$ where Q is the quotient field of R.

Having plundered 1.5, let us see what we can use from the methods of 1.6; there, to construct the simple polynomial modules, we took suitable exterior powers of the natural module and separated off composition factors with the 'correct' character. We recall that E is the right colattice of dimension n and $\Lambda_q^r E$ is its rth q-exterior power. By the description of $\Lambda_q E$ in 6.2, we have

Lemma 6.4.1 *Let $0 \le r \le n$. Choose $\lambda = (\lambda_1, \ldots, \lambda_n) \in \Lambda(n, r)$ such that the parts λ_i are either 0 or 1 for all i. Then $(\Lambda_q^r E)^\lambda$ is 1-dimensional over R with generator $\bar{e}_i$, where $i \in I$ is the unique decreasing element $i_1 > \cdots > i_r$ in the G-orbit on I with weight λ. In particular the character is the rth elementary symmetric function*

$$\operatorname{ch} \Lambda_q^r E = \mathbf{X}^{\omega_r} + \sum_\mu a_\mu \mathbf{X}^\mu$$

summed over all μ dominated by the Schur weight ω_r,

We write the conjugate to the arbitrary $\lambda \in \Lambda(n, r)$ as $\lambda' = (1^{a_1}, \ldots, n^{a_n}) \in \Lambda^+(m, r)$, say. The next definition may be compared with the product of exterior powers V mentioned in the proof of 1.6.1.

Definition 6.4.2 Define the $A_q(n)$-colattice

$$M_q(\lambda) = E^{\otimes a_1} \otimes (\Lambda_q^2 E)^{\otimes a_2} \otimes \cdots \otimes (\Lambda_q^n E)^{\otimes a_n}.$$

It is easy to compute the a_i: $a_i = \lambda_i - \lambda_{i+1}$ $(i < n)$, $a_n = \lambda_n$, and $\lambda = a_1\omega_1 + \cdots + a_n\omega_n$ where $\omega_i = (1, 1, \ldots, 1, 0, \ldots, 0) \in \Lambda(n, i)$.

The main result, 6.4.4, is a consequence of

Theorem 6.4.3 *For $\lambda \in \Lambda^+(n, r)$,*

$$\operatorname{ch} M_q(\lambda) = \mathbf{X}^\lambda + \sum_\mu a_{\mu\lambda}\mathbf{X}^\mu$$

over all $\mu \lhd \lambda$ and for some $a_{\lambda\mu} \in \mathbf{Z}$.

Proof From 6.4.1, with $r = 1$,

$$\operatorname{ch} M_q(\lambda) = e_1^{a_1} \cdots e_n^{a_n} = \mathbf{X}^\lambda + \sum_\mu \alpha_{\mu\lambda} \mathbf{X}^\mu$$

over all $\mu \lhd \lambda$, and for certain $\alpha_{\mu\lambda} \in \mathbf{Z}$.   □

We conclude that for $\lambda \in \Lambda^+(n, r)$, the highest weight occurring in $M_q(\lambda)$ is λ. Choose $R = K$. Then for any weight $\lambda \in \Lambda(n, r)$, $M_q(\lambda)$ is a finite-dimensional $A_q(n)$-comodule, so has a composition series. By 6.4.3, and the fact that isomorphic modules have the same character,

Theorem 6.4.4 *There exists a unique composition factor $L_{K,q}(\lambda)$ of $M_q(\lambda)$ with the property that*

$$\operatorname{ch} L_{K,q}(\lambda) = \mathbf{X}^\lambda + \sum_\mu b_{\mu\lambda} \mathbf{X}^\mu$$

for $\mu \lhd \lambda$ and $b_{\mu\lambda} \in \mathbf{Z}$. Moreover $L_{K,q}(\lambda) \cong L_{K,q}(\mu)$ if and only if $\lambda' = \mu'$. So we have a set $\{L_{K,q}(\lambda) : \lambda \in \Lambda^+(n, r), \ r \geq 0\}$ of non-isomorphic irreducible $A_{K,q}(n)$-comodules.

Remarks.
(1) Since $M_q(\lambda)$ is r-homogeneous, so too is $L_{K,q}(\lambda)$ provided $\lambda \in \Lambda(n, r)$.
(2) We say more later as to why the set is complete. We will also show that the L_q mentioned in 6.4.4 have an explicit description as the simple heads of the q-Weyl modules.
(3) The L_q are absolutely irreducible. We must show that if V is an irreducible $A_{K,q}(n)$-comodule with $\dim_K V^\lambda = 1$ then $\operatorname{End}_{A_{K,q}(n,r)}(V) = K$. Now, any $A_{K,q}(n, r)$-linear endomorphism θ of V sends V^λ to V^λ. Thus by hypothesis we can find $c \in K$ such that $\theta(v) = cv$ for all $v \in V$. So $(\theta - c1_V)(v) = 0$ and $v \in \operatorname{Ker}(\theta - c1_V)$. If V is simple then θ is the homothety $c1_V$ as required.

6.5 R-forms for q-Schur algebras

Suppose $\theta : R \to R'$ is a homomorphism between the commutative rings R and R'. Let $q \in R$ and $q' = \theta(q)$. The following assertions all follow from the definition of $A_{R,q}(n)$ in 6.1.1. First of all we have that θ induces a homomorphism of rings

$$A_{R,q}(n) \;\rightarrow\; A_{R',q'}(n)$$
$$c_{ij}^{R,q} \;\mapsto\; c_{ij}^{R',q'}$$

for $i, j \in \mathbf{n}$. Viewing R' as an R-module via θ, there exists a canonical isomorphism of bialgebras $\psi : R' \otimes_R A_{R,q}(n) \rightarrow A_{R',q'}(n)$. For $r \geq 0$ and $\alpha \in \Lambda(n, r)$, restrict ψ to obtain an isomorphism of coalgebras $\psi_r : R' \otimes_R A_{R,q}(n, r) \rightarrow A_{R',q'}(n, r)$, and an isomorphism of R-modules $\psi_\alpha : R' \otimes_R (A_{R,q}(n, r))^\alpha \rightarrow (A_{R',q'}(n, r))^\alpha$. Dualising, we obtain a ring isomorphism

$$A_{R',q'}(n, r)^* \rightarrow R' \otimes_R A_{R,q}(n, r)^*.$$

As usual let R be local, complete principal ideal domain with field of fractions Q and residue class field k. Let $q' \in k$ be the image of $q \in R$ under the canonical epimorphism. We pick a finite-dimensional $A_{Q,q}(n, r)$-comodule, M_Q say, and view it as a right $A_{Q,q}(n, r)^*$-module under the structure duality of 6.3. Now we find an R-form M_R contained in it in the usual way (this, you will remember from Curtis and Reiner [1981, §16C] and 3.5, means finding some M_R that is finitely-generated and R-free such that $Q \otimes_R M_R = M_Q$). Define the $A_{k,q'}(n, r)$-comodule $M_k = k \otimes_R M_R$. The next result is standard.

Lemma 6.5.1 *For $\mu \in \Lambda(n, r)$,*

  (i) $(M_R)^\mu = M_R \cap (M_Q)^\mu$;

  (ii) $(M_k)^\mu = k \otimes_R (M_R)^\mu$;

  (iii) $\dim_Q (M_Q)^\mu = \mathrm{rank}_R (M_R)^\mu = \dim_k (M_k)^\mu$ *and* $\mathrm{ch}\, M_Q = \mathrm{ch}\, M_R$
  $= \mathrm{ch}\, M_k$.

Simple modules in the generic case

This allows us to treat a particular instance in characteristic zero in greater detail. On display is also a general principle we shall continually exploit: first understand the generic case in characteristic 0, and then apply a 'base change'. I shall explain these terms as we proceed.

The *generic case* is associated with the system $(\mathbf{Q}, \mathbf{Q}[t], \mathbf{Q}(t))$, where $\mathcal{M} = \langle t - 1 \rangle$ and $q = t$. We are looking for the simple $A_{\mathbf{Q}(t),t}(n)$-comodules, consequently, by 6.3.2, we need only construct the simple $A_{\mathbf{Q}(t),q}(n, r)$-comodules. A (not necessarily complete) set of such simple comodules was exhibited in 6.4.4. Now, for any $\lambda \in \Lambda^+(n, r)$ choose an

$A_{\mathbf{Q}[t],t}(n,r)$-colattice $L_{\mathbf{Q}[t],t}(\lambda)$ inside $L_{\mathbf{Q}(t),t}(\lambda)$, whose character (from 6.5.1) is such that

$$\mathrm{ch}(\mathbf{Q} \otimes_{\mathbf{Q}[t]} L_{\mathbf{Q}[t],t}(\lambda)) = \mathrm{ch}\, L_{\mathbf{Q}[t],t}(\lambda) = \mathrm{ch}\, L_{\mathbf{Q}(t),t}(\lambda),$$

where we view $\mathbf{Q}$ as a $\mathbf{Q}[t]$-module under the specialisation $q \to 1$. Note here that $q' = t' = 1 \in \mathbf{Q} = \mathbf{Q}[t]/\mathcal{M}$.

Let $F(\lambda)$ be the unique composition factor of the finite-dimensional $A_{\mathbf{Q},1}(n,r)$-comodule $\mathbf{Q} \otimes_{\mathbf{Q}[t]} L_{\mathbf{Q}[t],t}(\lambda)$ of highest weight λ.

Theorem 6.5.2 *Given $\lambda \in \Lambda^+(n,r)$, the $A_{\mathbf{Q},1}(n,r)$-comodules $F(\lambda)$ and $\mathbf{Q} \otimes_{\mathbf{Q}[t]} L_{\mathbf{Q}[t],t}(\lambda)$ coincide. Therefore $L_{\mathbf{Q},1}(\lambda) \cong \mathbf{Q} \otimes_{\mathbf{Q}[t]} L_{\mathbf{Q}[t],1}(\lambda)$. Hence we have a complete set*

$$\{L_{\mathbf{Q},1}(\lambda) : \lambda \in \Lambda^+(n,r)\}$$

of non-isomorphic absolutely irreducible $A_{\mathbf{Q},1}(n,r)$-comodules.

Proof From the preamble we have a set $\{F(\lambda) : \lambda \in \Lambda^+(n,r)\}$ of non-isomorphic absolutely irreducible $A_{\mathbf{Q},1}(n,r)$-comodules. In other words we have a set of absolutely irreducible $S_{\mathbf{Q},1}(n,r)$-modules, i.e. a set of $|\Lambda^+(n,r)|$ irreducible modules for the classical Schur algebra $S_{\mathbf{Q}}(n,r)$. So the set is complete by 3.4.2.

Now from 2.2.8, $S_{\mathbf{Q}}(n,r)$ is semisimple, and so Wedderburn's theorem implies that

$$\dim_{\mathbf{Q}} S_{\mathbf{Q}}(n,r) = \sum_{\lambda \in \Lambda^+(n,r)} (\dim_{\mathbf{Q}} F(\lambda))^2.$$

We deduce, since we already have the set $\{L_{\mathbf{Q}(t),t}(\lambda) : \lambda \in \Lambda^+(n,r)\}$ of non-isomorphic absolutely irreducible $A_{\mathbf{Q}(t),t}(n,r)^*$-modules by 6.4.4, Remark (3) in 6.4 and the first paragraph of 6.5:

$$\sum_{\lambda}(\dim_{\mathbf{Q}(t)} L_{\mathbf{Q}(t),t}(\lambda))^2 \;\leq\; \dim_{\mathbf{Q}(t)} A_{\mathbf{Q}(t),t}(n,r)$$

$$= \dim_{\mathbf{Q}} A_{\mathbf{Q},1}(n,r)$$

$$= \sum_{\lambda}(\dim_{\mathbf{Q}} F(\lambda))^2$$

$$\leq \sum_{\lambda}(\dim_{\mathbf{Q}}(\mathbf{Q} \otimes_{\mathbf{Q}[t]} L_{\mathbf{Q}[t],t}(\lambda)))^2$$

$$= \sum_{\lambda}(\dim_{\mathbf{Q}(t)} L_{\mathbf{Q}(t),t}(\lambda))^2$$

where all sums are over $\Lambda^+(n,r)$. Hence equality must hold everywhere; since

$$\dim_{\mathbf{Q}} F(\lambda) \leq \dim_{\mathbf{Q}}(\mathbf{Q} \otimes_{\mathbf{Q}[t]} L_{\mathbf{Q}[t],t}(\lambda)) = \dim_{\mathbf{Q}(t)} L_{\mathbf{Q}(t),t}(\lambda),$$

we have equality here too. $\square$

From the above, Wedderburn's theorem, and the definition of the Schur function (Remark (2) in 1.6),

Theorem 6.5.3 *The t-Schur algebra $A_{\mathbf{Q}(t),t}(n,r)^*$ is semisimple and the collection*

$$\{L_{\mathbf{Q}(t),t}(\lambda) : \lambda \in \Lambda^+\}$$

is a complete set of inequivalent absolutely irreducible $A_{\mathbf{Q}(t),t}(n,r)^*$-modules such that

$$\operatorname{ch} L_{\mathbf{Q}(t),t}(\lambda) = s_\lambda.$$

Example. We can use exterior powers to compute some simple comodules explicitly. We recall that $\Lambda_t^n(E) = \mathbf{Q}(t)\bar{e}$, where $\bar{e} = \bar{e}_n \cdots \bar{e}_1$, and choose $\lambda = (1) \in \Lambda^+(n,1)$. Then $L_t(1)$ is n-dimensional and simple, since its character is $s_{(1)} = \sum_1^n X_i$. But it is also a quotient of E by construction, which implies $L_t(1) = E$.

Now $E \otimes \Lambda_t^n E$ is an irreducible $A_t(n)$-comodule (since $\Lambda_t^n E$ is 1-dimensional), so it must be isomorphic to $L_t(\lambda)$ for some $\lambda \in \Lambda^+(n,r)$ and $r \geq 0$ by 6.5.3. But $\operatorname{ch}(E \otimes \Lambda_t^n E) = \operatorname{ch}(E)\operatorname{ch}(\Lambda_t^n E) = (\sum X_i)(\prod X_i)$ $= s_{(2,1^{n-1})}$ and we have an irreducible comodule of highest weight $(2, 1^{n-1}) \in \Lambda(n, n+1)$. After 6.5.3, we find that $\theta : E \otimes \Lambda_t^n E \to L_t(2, 1^{n-2})$ is an isomorphism of $A_t(n)$-comodules.

Exercises.

(1) Show that $\Lambda_t^n E \otimes E \cong L_q(2, 1^{n-2})$.

(2) Let $\lambda_i \in \Lambda(n, n+1)$ be the weight having all 1s except a 2 in the ith slot. Show that the weight spaces $(E \otimes \Lambda_t^n(E))^{\lambda_i}$ and $(\Lambda_t^n(E) \otimes E)^{\lambda_i}$ are $\mathbf{Q}(t)(e_i \otimes \bar{e})$ and $\mathbf{Q}(t)(\bar{e} \otimes e_i)$, respectively.

6.6 The q-Schur algebra, second visit

One of the highlights of Chapter 2 is undoubtedly Schur's Commutation Theorem, 2.1.3: take the natural module E for Γ and consider the permutation action of Σ_r on $E^{\otimes r}$; the Schur algebra is then characterised as the centralising algebra of this action. The present section introduces a Hecke algebra $\mathcal{H}_q$ with Coxeter group Σ_r, defines an action of $\mathcal{H}_q$ on r-tensor space and then proves the assertion that the q-Schur algebra of 6.2 is precisely the centralising algebra of this action of $\mathcal{H}_q$ on r-tensor space. This fundamental result provides the strongest analogy yet with the classical ($q = 1$) case.

Motivation and definitions

Before defining $\mathcal{H}_q$, let us pause to recall the permutation module M on the cosets of the upper triangular matrices in $\mathrm{GL}_n(q)$. This module is important because over K its composition factors provide the unipotent representations of $K\mathrm{GL}_n(q)$. A classical theorem of Iwahori [1964], (or see Curtis and Reiner [1981, §67]) proves that $\mathcal{H}_q = \mathrm{End}_{K\mathrm{GL}_n(q)}(M)$, where the characteristic of K does not divide q, is an endomorphism algebra with basis $\{T_\pi : \pi \in G\}$, whose multiplication is defined in 6.6.1 below. Actually we could easily have given the definition for the case where q a unit in our ring R.

Definition 6.6.1 The *Hecke algebra* $\mathcal{H}_q = \mathcal{H}_{R,q}(G)$ over R with Coxeter group G and distinguished generators the basic transpositions (i.e. transpositions of length one) is defined as follows: as an R-module it is free with basis $\{T_\pi : \pi \in G\}$; the multiplication in $\mathcal{H}_q$ satisfies the relations

$$T_\pi T_b = \begin{cases} T_{\pi b} & \text{if } \ell(\pi b) > \ell(\pi) \\ qT_{\pi b} + (q-1)T_\pi & \text{otherwise} \end{cases}$$

for $\pi \in G$ and for a basic transposition, b.

Here $q = q1 \in \mathcal{H}_q$, and we note that $T_1 = 1_{\mathcal{H}_q}$. The fundamental properties of this algebra are listed in Dipper and James [1986, 2.1].

Remarks.
(1) A characteristic zero forerunner of $\mathcal{H}_q$ appears in the thesis of Hoefsmit [1974] using the language of Young's seminormal form.
(2) There is a vast literature on Hecke algebras having any desired Coxeter system. Good accounts of these are to be found in Curtis and Reiner [1987, §67] or Humphreys [1992]. We shall be interested only in the case where the Weyl group is associated to a root system of type A, as is the case in 6.6.1. A recent account of Hecke algebras of type B appears in Dipper and James [1992].

If $q \in K^*$ then we can think of $\mathcal{H}_q$ as a q-analogue of KG as follows:

start with any K-space V with basis $\{T_\pi : \pi \in G\}$; define a multiplication on V as in 6.6.1 to obtain a K-algebra. If $q = 1$ we are back to KG. If q is the power of the prime in $\mathrm{GL}_n(q)$ then we have $\mathrm{End}_{K\mathrm{GL}_n(q)}(M)$ where M is as above.

It is as well to formalise the procedure of 'putting $q = 1$'.

Definition 6.6.2 Let $S = R[t, t^{-1}]$ be the ring of Laurent polynomials in the indeterminate t. Call $\mathcal{H}_{S,t}(G)$ the *generic algebra* of G over R. The epimorphism $\mathcal{H}_{S,t}(G) \to \mathcal{H}_{R,q}(G)$ induced by $h_q(t) = q \in R$ is called the *specialisation* of $\mathcal{H}_{S,t}$ with respect to q.

Now comes the clever bit: for $\lambda \in \Lambda^+(n, r)$, consider the element

$$x_\lambda = \sum_{\pi \in \Sigma_\lambda} T_\pi. \tag{6.4}$$

For example, if $r = 2$, $x_{(2)} = T_1 + T_{(12)}$ and $x_{(1^2)} = T_1$.

Remark. If H is an arbitrary subgroup of Σ_r then the R-module $\sum_{\pi \in H} RT_\pi$ is not necessarily a subalgebra, though if H is a Young subgroup this is guaranteed. If $H = \Sigma_\lambda$ we denote this *Young subalgebra* by $\mathcal{H}_\lambda$.

Returning to the special case where M is the permutation module of $A = K\mathrm{GL}_n(q)$ on B (in the non-describing characteristic), most readers will recognise $M_q^\lambda = x_\lambda M$ as the permutation module on the parabolic subgroup

$$\begin{bmatrix} * & & * \\ & * & \\ 0 & & * \end{bmatrix},$$

where the diagonal $*$s stand for blocks having dimensions $\lambda_1 \times \lambda_1, \ldots,$. It is a non-trivial fact from Dipper and James [1989, 2.24] that there is a natural isomorphism

$$\mathrm{End}_A \left(\bigoplus_{\Lambda^+(r)} x_\lambda M \right) \cong \mathrm{End}_{\mathcal{H}_q} \left(\bigoplus_{\Lambda^+(r)} x_\lambda \mathcal{H}_q \right);$$

the algebra appearing on the right is a version of the q-Schur algebra! This is a strong motivation for studying Hecke algebras.

Notice that Rx_λ is a cyclic right ideal of $\mathcal{H}_\lambda$, since $T_\pi x_\lambda = q^{\ell(\pi)} x_\lambda$ for $\pi \in \Sigma_\lambda$. Hence, choosing $q = 1$ we obtain the trivial character of Σ_λ: thus Rx_λ is called the trivial representation of $\mathcal{H}_\lambda$ and one can think of $M_q^\lambda = x_\lambda \mathcal{H}_q$ as a q-version of our old friend M^λ, defined at the start of 4.2; see especially Remark (1) in 4.1.

Define the q-Schur algebra as the endomorphism algebra of an external direct sum of the the R-free $\mathcal{H}_q$-modules

$$M_q^\lambda = x_\lambda \mathcal{H}_q;$$

i.e. we define the q-Schur algebra to be

$$S_{R,q}(n,r) = \mathrm{End}_{\mathcal{H}_q} \left(\bigoplus_{\lambda \in \Lambda^+(r)} M_q^\lambda \right).$$

This turns out to be canonically isomorphic to the q-Schur algebra of 6.2.1.

Now $\lambda' = \mu'$ implies $d^{-1} \Sigma_\lambda d = \Sigma_\mu$ for some $d \in \mathcal{D}_{\lambda\mu}$, so $T_d^{-1} \mathcal{H}_\lambda T_d = \mathcal{H}_\mu$; hence $x_\lambda T_d = T_d x_\mu$ and there is an isomorphism of right ideals

$$\theta : x_\lambda \mathcal{H}_q \rightarrow x_\mu \mathcal{H}_q$$

given by $(x_\lambda h)\theta = T_d x_\lambda h = x_\mu T_d h$ for all $h \in \mathcal{H}_q$. Hence it is immaterial whether we take the sum over all partitions of r or over compositions of r. In fact we normally index over the latter set: if $r \leq n$ then $\Lambda^+(r) \subseteq \Lambda(n,r)$. So in this case

Lemma 6.6.3 $S_q(n,r)$ *is Morita equivalent to* $\mathrm{End}_{\mathcal{H}_q} \left(\bigoplus_{\lambda \in \Lambda(n,r)} x_\lambda \mathcal{H}_q \right)$.

Notice that

$$S_q(n,r) = \bigoplus_\lambda \bigoplus_\mu \mathrm{Hom}_{\mathcal{H}_q}(M_q^\lambda, M_q^\mu).$$

These spaces are easy to work with for one can show that each element is left $\mathcal{H}_q$-multiplication, i.e. $\varphi x_\lambda h = h_\varphi x_\lambda h$. The aim in the next few sections is to show that $A_{R,q}(n,r)^*$ of 6.2.1 and the endomorphism ring $S_q(n,r)$ of 6.6.3 are canonically isomorphic.

A basis

The reader may be reassured by the sight of a basis for the new q-Schur algebra. Given compositions λ, μ of r and $\sigma \in G$, one defines an element $\varphi_{\lambda\mu}^\sigma \in \mathrm{Hom}_{\mathcal{H}_q}(M_q^\mu, M_q^\lambda)$ by

$$\varphi_{\lambda\mu}^\sigma : x_\mu h \mapsto \sum_{\pi \in \Sigma_\lambda \sigma \Sigma_\mu} T_\pi h$$

for all $h \in \mathcal{H}_q$. Extend these maps to be elements of $S_q(n, r)$ in the usual way. Use the same notation for this extended map. The following basis theorem is proved in Dipper and James [1986, 3.4] in a straightforward manner.

Theorem 6.6.4 $S_{R,q}(n, r)$ *is free as an R-module with basis*

$$\{\varphi_{\lambda\mu}^d : \lambda, \mu \in \Lambda(n, r), \ d \in \mathcal{D}_{\lambda\mu}\}.$$

Remark. What about an analogue of the Product Rule 2.2.11? Suppose we are given $\lambda, \mu, \alpha, \beta \in \Lambda(r)$, $c \in \mathcal{D}_{\alpha\beta}$ and $d \in \mathcal{D}_{\lambda\mu}$. It is clear that $\varphi_{\lambda\mu}^d$ sends $x_\mu\mathcal{H}$ into $x_\lambda\mathcal{H}$ and kills off $x_\alpha\mathcal{H}$ if $\alpha \neq \mu$. Thus $\varphi_{\alpha\beta}^c \varphi_{\lambda\mu}^d = 0$ unless $\lambda = \beta$. If $\lambda = \beta$ then $\varphi_{\alpha\beta}^c \varphi_{\lambda\mu}^d$ is some (probably complicated) combination of $\varphi_{\alpha\mu}^e$ where $e \in \mathcal{D}_{\alpha\mu}$. Du [1992a] gets round this by finding another basis for $S_q(n, r)$ (in terms of a so-called relative norm) and analysing the product of any two elements.

We note also that $\varphi_{\lambda\lambda}^1 = 1_{x_\lambda\mathcal{H}_q}$ and $\varphi_{\lambda\mu}^d = \varphi_{\lambda\lambda}^1 \varphi_{\lambda\mu}^d = \varphi_{\lambda\mu}^d \varphi_{\mu\mu}^1$. So $\varphi_{\lambda\lambda}^1 S_q \varphi_{\mu\mu}^1 = \mathrm{Hom}_{\mathcal{H}_q}(M_q^\mu, M_q^\lambda)$, and $\{\varphi_{\lambda\lambda}^1 : \lambda \in \Lambda(n, r)\}$ is a set of orthogonal idempotents whose sum is $1_{S_q(n,r)}$.

Contravariant duals

Let S be any R-free R-algebra possessing an anti-automorphism $a \mapsto a^\circ$ such that $a^{\circ\circ} = a$ for all $a \in S$. If M is a finite-dimensional right S-lattice, define the contravariant dual M° as in 3.4.4 (so that S acts on the left). If $\phi \in \mathrm{End}_S(M)$ let $\phi^\sharp \in \mathrm{End}_S(M^\circ)$ be defined by $\phi^\sharp f = f\phi$ for all $f \in M^\circ$. So $\phi \mapsto \phi^\sharp$ is an anti-isomorphism $\mathrm{End}_S(M) \to \mathrm{End}_S(M^\circ)$. If in addition M is self-dual, there is an S-isomorphism $\theta : M \to M^\circ$ inducing an isomorphism $\hat{\theta} : \mathrm{End}_S(M) \to \mathrm{End}_S(M^\circ)$. Composing $\sharp$ with $\hat{\theta}^{-1}$ we get an anti-automorphism '$*$' of $\mathrm{End}_S(M)$ such that

$$\phi^*(m) = \hat{\theta}^{-1}\phi^\sharp(m) = \theta^{-1}\phi^\sharp\theta(m) = \theta^{-1}(\theta(m) \circ \phi)$$

for all $\phi \in \mathrm{End}_A(M)$ and $m \in M$.

From 3.4.5, $\theta(m)n = \langle m, n \rangle$ defines a correspondence between the set of S-isomorphisms $M \to M^\circ$ and the non-degenerate bilinear forms on M such that $\langle ma, n \rangle = \langle m, na^* \rangle$ for all $m, n \in M$, $a \in S$. Since $\langle \phi^* m, n \rangle = \langle m, \phi n \rangle$, not only is M self-dual as a right S-module, but also as a left $\mathrm{End}_S(M)$-module.

Let us apply this theory directly to our situation. Define $T_\pi^\circ = T_{\pi^{-1}}$ ($\pi \in G$), and extend this to be an R-linear map $\circ$ of the R-module $\mathcal{H}_q$; it can be seen that $\circ$ is an anti-automorphism of $\mathcal{H}_q$ of order 2. We now define a symmetric bilinear form on M_q^λ by decreeing that

$$\langle x_\lambda T_u, x_\lambda T_v \rangle_\lambda = \begin{cases} q^{\ell(u)} & \text{if } u = v \\ 0 & \text{otherwise} \end{cases}$$

for $u, v \in \mathcal{D}_\lambda$. This extends to a form (,) on $M = \displaystyle\bigoplus_{\lambda \in \Lambda(n,r)} M_q^\lambda$ by requiring M_q^λ and M_q^μ to be orthogonal when $\lambda \neq \mu$. By Dipper and James [1986, 4.4], for $h \in \mathcal{H}_q$, $m, n \in M$,

$$(mh, n) = (m, nh^*). \tag{6.5}$$

Example. Taking $\lambda = (1^r)$, $M_q^\lambda = \mathcal{H}_q$ and we thus obtain a symmetric, bilinear form on $\mathcal{H}_q$. Setting $f(h_1, h_2) = (h_1, h_2^\circ)$, f is an associative, symmetric, non-degenerate bilinear form, and hence for $0 \neq q \in K$, $\mathcal{H}_{K,q}$ is a symmetric K-algebra.

By (6.5) M is self-dual as a $\mathcal{H}_q$-module and so, by the above, the q-Schur algebra $\mathrm{End}_{\mathcal{H}_q}(M)$ possesses an anti-automorphism $*$. Hence the theory of contravariant duals as expounded in 3.4 holds in the present situation.

Remarks.
(1) If $\lambda, \mu \in \Lambda(n,r)$ and $d \in \mathcal{D}_{\lambda\mu}$ then $(\varphi_{\lambda\mu}^d)^* = \varphi_{\mu\lambda}^{d^{-1}}$ —for this see Dipper and James [1991, 1.11].
(2) In fact, if $q = 1$ then $*$ reduces to the map J appearing in the discussion before 3.4.6.

Action of $\mathcal{H}_q$ on M_q^λ and on M

A free R-basis for M_q^λ is given by $\{x_\lambda T_d : \lambda \in \Lambda(n,r), \ d \in \mathcal{D}_\lambda\}$, and if $d \in \mathcal{D}_\lambda$, the action of the basic generator T_b is given by

$$x_\lambda T_d T_b = \begin{cases} qx_\lambda T_d & \text{if } \ell(db) = \ell(d) + 1 \text{ and } db \notin \mathcal{D}_\lambda \\ x_\lambda T_{db} & \text{if } \ell(db) = \ell(d) + 1 \text{ and } db \in \mathcal{D}_\lambda \\ qx_\lambda T_{db} + (q-1)x_\lambda T_d & \text{if } \ell(db) = \ell(d) - 1 \end{cases}$$

$$(6.6)$$

Recalling the sub-section on combinatorics in 6.1 we see that the first case occurs if $i, i+1$ lie in the same row of $\tau^\lambda d$, whereas the second case occurs if the row index of i in $\tau^\lambda d$ is less than that of $i+1$. Now the T_b generate $\mathcal{H}_q$ as an R-algebra, so this determines the action of $\mathcal{H}_q$ on M_q^λ completely. Hence we obtain a basis of $M = \bigoplus M_q^\lambda$.

Example. To avoid being overwhelmed by so much symbolism, we shall provide a numerical example. The simple case of $\lambda = (2^2)$ will suffice. Let $d \in \mathcal{D}_{(2^2)}$. Using the correspondence of (6.2), a basis of $M_q^{(2^2)}$ is the collection of the $6(= |\mathcal{D}_{(2^2)}|)$ $\tau^{(2^2)}d$ terms, i.e. $\{\tau^\lambda d\} = x_\lambda T_d$ is one of

1 2	1 3	1 4	2 3	2 4	3 4
3 4	2 4	2 3	1 4	1 3	1 2

and the corresponding lengths of d are 0, 1, 2, 2, 3 and 4 respectively. As is standard practice, tabloids are denoted by braces; the horizontal lines denote 'unordered row entries'.

From (6.6) we have, for example,

$$\frac{\overline{1\ \ 4}}{2\ \ 3}T_{(2,3)} = q\frac{\overline{1\ \ 4}}{2\ \ 3}, \qquad \frac{\overline{1\ \ 4}}{2\ \ 3}T_{(1,2)} = \frac{\overline{2\ \ 4}}{1\ \ 3},$$

$$\frac{\overline{1\ \ 4}}{2\ \ 3}T_{(3,4)} = q\frac{\overline{1\ \ 3}}{2\ \ 4} + (q-1)\frac{\overline{1\ \ 4}}{2\ \ 3}.$$

Endow $M^* = \operatorname{Hom}_R(M, R)$ with a left $\mathcal{H}_q$-module structure as usual. Let $\{x^*_{\lambda,d} : \lambda \in \Lambda(n,r), d \in \mathcal{D}_\lambda\}$ be the basis of M^* dual to (6.6). Clearly, $x^*_{\lambda,d}(x_\mu T_e) = \delta_{\lambda\mu}\delta_{de}$ for $\lambda, \mu \in \Lambda(n,r)$, $d \in \mathcal{D}_\lambda$, $e \in \mathcal{D}_\mu$. Using this we have, for b basic, $\lambda \in \Lambda(n,r)$ and $d \in \mathcal{D}_\lambda$,

$$T_b x^*_{\lambda,d} = \begin{cases} qx^*_{\lambda,db} & \text{if } \ell(db) = \ell(d) + 1 \\ qx^*_{\lambda,db} + (q-1)x^*_{\lambda,d} & \text{otherwise} \end{cases}$$

$$(6.7)$$

Action of $\mathcal{H}_q$ on tensor space

We need to make various identifications in order to proceed. Let $N_q^\lambda = \mathcal{H}_q x_\lambda$. Then there is a canonical isomorphism between

$$\mathrm{End}_{\mathcal{H}_q}\left(\bigoplus_{\lambda\in\Lambda(n,r)} M_q^\lambda\right) \simeq \mathrm{End}_{\mathcal{H}_q}\left(\bigoplus_{\lambda\in\Lambda(n,r)} N_q^\lambda\right).$$

For, suppose we have some $\mathcal{H}_q$-map of M_q^μ to M_q^λ, that is given as usual via left multiplication by an element of $h \in \mathcal{H}_q$. Right multiplication by h gives an $\mathcal{H}_q$-map from N_q^λ to N_q^μ. Thus, the map that sends left h-multiplication to right h-multiplication, when composed with the anti-automorphism $*$ of $S_q(n,r)$ produces the required canonical identification.

Next, we identify the left $\mathcal{H}_q$-modules M^* and $\bigoplus_{\Lambda(n,r)} N_q^\lambda$ by the map

$$x_{\lambda,d}^* \mapsto q^{-\ell(d)} T_{d^{-1}} x_\lambda.$$

But there is an R-module isomorphism between $E^{\otimes r}$ and M^* induced by $e_i \mapsto x_{\lambda,d}^*$ where $i = i_\lambda d$. So we identify these two spaces. Carrying over the $\mathcal{H}_q$-action to $E^{\otimes r}$, we have, from (6.7) with $b = (a, a+1)$ $(a \leq r-1)$,

$$T_b e_i = \begin{cases} q e_{ib} & i_a \leq i_{a+1} \\ e_{ib} + (q-1)e_i & i_a > i_{a+1} \end{cases} \tag{6.8}$$

for $i = (i_1, \ldots, i_r) \in I$. Hence tensor space is canonically a left $\mathcal{H}_q$-module with the given action. With all these identifications, we have an isomorphism

$$E^{\otimes r} \cong \bigoplus_{\Lambda(n,r)} N_q^\lambda \tag{6.9}$$

via $e_{id} \mapsto q^{-\ell(d)} T_{d^{-1}} x_\lambda$ where $i \in \lambda$ is weakly increasing. Recalling the $A_q(n,r)$-comodule structure on $E^{\otimes r}$ at the end of 6.2, we have:

Theorem 6.6.5 *Multiplication by elements of $\mathcal{H}_q$ is a comodule endomorphism of the $A_q(n,r)$-comodule $E^{\otimes r}$.*

Proof The complete argument is set out in Dipper and Donkin [1991, 3.1.6]. Without loss of generality one can take $r = 2$, since $\mathcal{H}_q$ is generated by basic transpositions. Now (6.8) gives the action of $\mathcal{H}_q$ on $E^{\otimes 2}$; explicitly,

$$\begin{aligned} T_{(12)}(e_i \otimes e_j) &= q e_j \otimes e_i \quad (i \leq j) \\ T_{(12)}(e_j \otimes e_i) &= e_i \otimes e_j + (q-1)e_j \otimes e_i \quad (i < j). \end{aligned}$$

$E^{\otimes 2}$ has structure map $\tau_2(e_i \otimes e_j) = \sum_{r,s} e_r \otimes e_s \otimes c_{ri} c_{sj}$. Finally, we must check that $\tau_2 \circ T_{(12)} = (T_{(12)} \otimes 1)\tau_2$. Let us do the special case $i \leq j$. Write $T = T_{(12)}$ and $\tau = \tau_2$. We have

$$
\begin{aligned}
(T \otimes 1)\tau(e_i \otimes e_j) &= (T \otimes 1)\Big(\sum_{r,s} e_r \otimes e_s \otimes c_{ri} c_{sj}\Big)\\
&= q \sum_{r,s} e_s \otimes e_r \otimes c_{ri} c_{sj}\\
&\quad + \sum_{r>s} e_s \otimes e_r \otimes c_{ri} c_{sj}\\
&\quad +(q-1) \sum_{r>s} e_r \otimes e_s \otimes c_{ri} c_{sj}.
\end{aligned}
$$

But

$$
\tau T(e_i \otimes e_j) = q\tau(e_j \otimes e_i) = q \sum_{r,s} e_r \otimes e_s \otimes c_{rj} c_{si}.
$$

Now compare coefficients of $e_r \otimes e_s$ for $s > r$. In the first equation we have $c_{si} c_{rj}$ while in the second we have $q c_{rj} c_{si}$. But these are equal by (6.1)(A)! The reader is left to complete the case $r > s$ for $i \leq j$, and then to do $i > j$. □

Remarks.

(1) We could have used the canonical $F(n)$-comodule structure of tensor space, together with the relations imposed by equating coefficients on both sides of the structure theorem in the above proof, to derive the equations in 6.1.1. If this approach is taken, a check has to made that the relations do indeed generate a bi-ideal of $F(n)$.

(2) Unfortunately the procedure in Remark (1) is not invariant under base change: the tensor structure of $E^{\otimes r}$ has to go into the calculations. Carrying out a base change of E will induce a base change in $E^{\otimes r}$, producing new relations but *isomorphic* bialgebras; however, base changes in $E^{\otimes r}$ not arising from a base change in E may give new relations and may produce non-isomorphic bialgebras. This question is discussed fully in Dipper and Donkin [1991, 4.4]. In particular Dipper and Donkin extend Manin's example (Manin [1988, 1,6], Jimbo [1986]) to produce a quantum deformation of $n \times n$ matrices and hence obtain another quantum GL$_n$ by localising at a certain central q-determinant; see the next chapter. To make matters worse, the $\mathcal{H}_q$-endomorphism ring of Manin's tensor space is canonically isomorphic to the Schur alge-

bra! But the $A_q(n)$ of 6.1.1 and the version of $A_q(n)$ obtained by Manin are not isomorphic.

Now we know that $E^{\otimes r}$ is a comodule for $A_q(n, r)$; it is therefore a left $A_q(n, r)^*$-module via

$$\xi e_i = \sum_{j \in I} \xi(c_{j,i}) e_j \tag{6.10}$$

for $\xi \in A_q^*$ ($i \in I$). We know from (6.9) that the left $\mathcal{H}_q$-module $E^{\otimes r}$ is isomorphic to $N = \bigoplus_{\lambda \in \Lambda(n,r)} \mathcal{H}_q x_\lambda$ and from the start of this subsection that $\text{End}_{\mathcal{H}_q} N$ is isomorphic to $S_q(n, r)$. N is a left $\mathcal{H}_q$- and a left $A_q(n, r)^*$-module since $A_q(n, r)^*$ acts on the left. Now $S_q(n, r) \cong S_q(n, r)^{\text{op}}$ since it possesses $*$. Hence $\text{End}_{\mathcal{H}_q} N \cong S_q(n, r)$ canonically (writing endomorphisms on the left). Finally, by 6.6.5 the $A_q(n, r)^*$- and $\mathcal{H}_q$-action on $E^{\otimes r}$ centralise each other, and so we have induced a canonical R-algebra homomorphism

$$f : A_{R,q}(n, r)^* \to \text{End}_{\mathcal{H}_{R,q}} \left(\bigoplus_{\lambda \in \Lambda(n,r)} N_q^\lambda \right) \cong S_{R,q}(n, r). \tag{6.11}$$

Theorem 6.6.6 f, as defined in (6.11), is an isomorphism.

Proof We have split the justification of this into three short sections.

(1) *Let* $(i, j) \in I_0^2$ *be such that* $i \in \lambda$ *and* $j \in \mu$. *Then for* $\xi_{i,j} \in A_q(n, r)^*$, $f(\xi_{i,j}) = \phi_{i,j}$ *where* $\phi_{i,j}(h x_\rho) = 0$ *for all* $h \in \mathcal{H}_q$ *and* $\mu \neq \rho \in \Lambda(n, r)$; *also* $\text{Im}\phi_{i,j} \subseteq N_q^\lambda$. Consider the basis $\mathcal{B}_r = \{\xi_{i,j} : (i, j) \in I_0^2\}$ of $A_q(n, r)^*$ that is dual to the R-basis $\{c_{i,j} : (i, j) \in I_0^2\}$ of $A_q(n, r)$. Now for $(i, j) \in I_0^2$ and any $s \in I(n, r)$,

$$\xi_{i,j}(e_s) = \sum_{t \in I} \xi_{i,j}(c_{t,s}) e_t = \sum_{d \in \mathcal{D}_\lambda} \xi_{i,j}(c_{id,s}) e_{id}$$

where $i \in \lambda$. Under the homomorphism of (6.11), this translates as

$$
\begin{aligned}
f(\xi_{i,j})(h x_\rho) &= 0 \\
f(\xi_{i,j})(h x_\mu) &= h(f(\xi_{i,j}) x_\mu) \\
&= h \sum_{\mathcal{D}_\lambda} q^{-\ell(d)} \xi_{i,j}(c_{id,u_\mu}) T_{d^{-1}} x_\lambda
\end{aligned}
\tag{6.12}
$$

where $\mu \neq \rho \in \Lambda(n, r)$, $h \in \mathcal{H}$ and $u_\mu \in \mu$ is increasing.

(2) *The $\mathcal{H}_q$-map $\phi_{i,j} : N_q^\mu \to N_q^\lambda$ is specified by the rule*

$$x_\mu \mapsto \sum_{w \in \mathcal{D}_\nu \cap \Sigma_\mu} T_{w^{-1}} T_{d^{-1}} x_\lambda = \sum_{w \in \Sigma_\mu d^{-1} \Sigma_\lambda} T_w$$

where ν is a composition of r corresponding to $d^{-1}\Sigma_\lambda d \cap \Sigma_\mu$ and d is the distinguished double coset representative in $\Sigma_\lambda \tilde{d} \Sigma_\mu$ with $\Sigma_\mu \tilde{d} = \{\pi \in G : u_\mu \pi = j\}$. Now we express the monomial c_{id,u_μ} in terms of the basis elements $\mathcal{B}_r$ of $A_q(n,r)$. The hypothesis on u_μ means that the reordering process does not require the use of rule (B) in (6.1). So $c_{id,u_\mu} = q^{\ell(d)} c_{i,k}$ with $k = u_\mu d^{-1}$. Now find an initial double index using $(6.1)(C)$, that is, find $d' \in \Sigma_\lambda$ such that $(i, kd') = (id', kd') = (i, k)d' \in I_0^2$. This yields

$$\xi_{i,j}(c_{id,u_\mu}) = \begin{cases} q^{\ell(d)} & \text{if } u_\mu d^{-1} d' = j \text{ for some } d' \in \Sigma_\lambda \\ 0 & \text{otherwise} \end{cases}.$$

Now, $\{\pi \in G : u_\mu \pi = j\}$ is a left coset of Σ_μ with distinguished coset representative $\tilde{d}$, say. Let d be the unique distinguised double coset representative in $\Sigma_\lambda \tilde{d} \Sigma_\mu$. Then

$$\begin{aligned} \Sigma_\lambda \tilde{d} \Sigma_\mu \cap \mathcal{D}_{\lambda\mu} &= \{g \in \mathcal{D}_\lambda : \exists\, d' \in \Sigma_\lambda : u_\mu g^{-1} d' = j\} \\ &= \{d'd : d' \in \mathcal{D}_\nu \cap \Sigma_\mu\} \end{aligned}$$

where ν is a composition of r corresponding to the Young subgroup $d^{-1}\Sigma_\lambda d \cap \Sigma_\mu \leq G$.

(3) *Completion of proof.* Paragraph (2) implies that $\phi_{i,j}$ acts on the generator $x_\mu \in \mathcal{H}_q x_\mu$ in the same way as does the basis element $\varphi_{\lambda\mu}^d \in S_q(n,r)$ of 6.6.4. We have shown that f maps a basis of $A_q(n,r)^*$ to a basis of $S_q(n,r)$ and hence is an isomorphism. The proof is complete.

□

Theorem 6.6.6 implies that the canonical basis of $A_q(n,r)^*$ is precisely the canonical basis of the q-Schur algebra derived from the spaces

$$\text{Hom}_{\mathcal{H}_q}(N_q^\mu, N_q^\lambda).$$

The latter basis is calculable via the 'Nakayama Rules' for Hecke algebras from Dipper and James [1986, 2.8].

Identifications

Let $\lambda \in \Lambda(n,r)$. In 6.3 we used the idempotent ξ_λ to define λ-weight spaces of comodules. Now identify $\xi_\lambda = \xi_{i,i} \in A_q(n,r)^*$ where $i \in \lambda$ and is weakly increasing. We can use the isomorphism f to translate between the comodule viewpoint and the module viewpoint followed by Dipper and James [1991]. Thus we can use the basis $\{\xi_{i,j} : (i,j) \in I_0^2\}$ if we are using 6.2.1 and the basis given in 6.6.4 if we are thinking in terms of Definition 6.6.3. For example, if M is a left $S_q(n,r)$-module the λ-weight is $M\varphi_{\lambda\lambda}^1$, whereas if one prefers to view it as a right $A_q(n,r)$-comodule the weight space is $\xi_\lambda M$.

We pursue this reference: Dipper and James [1991] define the *q-tensor space*. Formally, $E_{R,q}(n,r)$ is the ω-weight space $S_q(n,r)\varphi_{\omega\omega}^1$ of $S_q(n,r)$ (in the classical case cf. (4.3)). This curiosity is nothing other than a version of familiar tensor space: $E_{R,q}(n,r)$ is isomorphic to $(E^{\otimes r})^*$ as a $(S_q(n,r), \mathcal{H}_q)$-bimodule. Here $(E^{\otimes r})^*$ has the usual right $\mathcal{H}_q$ action; the canonical right $S_q(n,r)$-action on $(E^{\otimes r})^*$ is moved to the left using the anti-automorphism $*$ appearing in Remark (1) after (6.5).

7

Representation theory of $S_q(n, r)$

One task remaining from the previous chapter is to obtain a complete list of irreducible $S_q(n, r)$-modules, thus completing the job begun for characteristic zero in 6.5.3. Unfortunately the situation for finite characteristic is more problematic: whichever approach is adopted there seems to be no quick way of showing that the irreducible $S_q(n, r)$-modules are indexed by $\Lambda^+(n, r)$. One approach would be to bring out the quantum GL_n and use quantum group theory, a method employed by Parshall and Wang [1991, §8]. The Dipper and James [1991] approach is to define a certain q-analogue of the Weyl module. Over K the similarity with the classical case is striking: the q-Weyl module $V_q(\lambda)$ has a unique maximal submodule and therefore a simple top factor $F_q(\lambda)$. They show that, as λ runs through the dominant weights, the $F_q(\lambda)$ will then give a complete set of simple $S_q(n, r)$-modules. The analogy with the situation studied in previous chapters (referred to henceforth as the 'classical case') is a persuasive one: for example in 7.5 we show that $S_q(n, r)$ is quasi-hereditary having as standard modules the q-Weyl modules.

Returning to the q-universe constructed in Chapter 6, we already have at our disposal several q-analogues of structures in the classical situation. We have, for example, a q-analogue of the polynomial coalgebra, namely $A_q(n, r)$, and of the Schur algebra, namely $S_q(n, r)$, but as yet there is nothing even remotely resembling a q-version of GL_n. One might suppose that there is no obvious q-counterpart of the statement that representations of Schur algebras and representations of $\mathrm{GL}_n(K)$ are equivalent. My purpose in the middle sections of this chapter is to describe a certain quantum group, Γ_q, the so-called quantum GL_n that serves as such a counterpart, and at the same time explains most of the phenomena encountered in Chapter 6.

We take R to be a field or a complete valuation ring and recall that

185

the ring of regular functions $R[\Gamma]$ on $\Gamma = \mathrm{GL}_n(R)$ is the polynomial ring $A_R(n) = A_{R,1}(n)$ localised at the set of powers of the determinant function. The idea is to construct a q-deformation of $R[\Gamma]$ by suitably deforming $A_R(n)$. The word 'suitably' means that our deformation is tailored in such a way that $\mathcal{H}_q$-multiplication on tensor space is a co-module map. The q-Schur algebra then plays the rôle of the centralising algebra of this action. Upon localisation at the powers of a certain 'quantum determinant' we obtain an object that is a non-commutative non-cocommutative Hopf algebra with antipode. Taking the prime spectrum we obtain a quantum GL_n, Γ_q. Note that if q is not a root of unity, $R[\Gamma_q]$ behaves like the coordinate ring in characteristic zero. In general we construct a free R-basis for Γ_q and in 7.4 we observe that the r-homogeneous polynomial representations of Γ_q are equivalent to representations of $S_{R,q}(n,r)$.

In 7.6 we discuss the main theorem of Dipper and James [1989]. This should be viewed as the climax of this text and incorporates James' theorem 4.4.3 as a special case. The quantum GL_n explains all the formal similarities between the classical case ($q = 1$) for Σ_n and $\mathrm{GL}_n(R)$, and the quantum case (q a prime power coprime to $p =$char R) for $\mathcal{H}_q$ and $\mathrm{GL}_n(q)$. For example if q is a prime power, the module category for $S_q(n,n)$ embeds into the category of representations of $\mathrm{GL}_n(K)$, and from this it follows that the decomposition matrix of $S_q(n,n)$ is a submatrix of the p-modular decomposition matrix of $\mathrm{GL}_n(q)$. At the end we give an extended example to illustrate how all these ideas work out in practice.

7.1 q-Weyl modules

The combinatorial path to the q-Weyl module requires us to build on some of the ideas contained in 6.1. Take the element $w_\lambda \in \Sigma_r$ that makes $1, 2, \ldots, r$ appear in order down the columns of $\tau^\lambda w_\lambda$. Now given $d \in \mathcal{D}_\lambda$ let $\sigma_d \in \mathcal{T}(\mu, \lambda)$ be such that for all i, j, if the (i,j)th entry of $\tau^\mu w_\mu$ is p then the (i,j)th entry of σ_d is the row index of p in $\tau^\lambda d$.

Example. Take $\lambda = (3, 2^2, 1)$ and $\mu = (4, 2^2)$. We recall that $d \in \mathcal{D}_\lambda$ is such that $\tau^\lambda d$ is row semistandard. So, we have

$$\tau^\lambda d = \begin{matrix} 1 & 2 & 7 \\ 3 & 5 & \\ 6 & 8 & \\ 4 & & \end{matrix} \qquad \tau^\mu w_\mu = \begin{matrix} 1 & 4 & 7 & 8 \\ 2 & 5 & & \\ 3 & 6 & & \end{matrix} \ .$$

Replacing each entry of $\tau^\mu w_\mu$ by its row index in $\tau^\lambda d$ gives

$$\sigma_d = \begin{array}{cccc} 1 & 4 & 1 & 3 \\ 1 & 2 & & \\ 2 & 3 & & \end{array}$$

If we denote by $C_{\lambda\mu}$ the set $\{d \in \mathcal{D}_{\lambda\mu} : \Sigma_\lambda \cap d\Sigma_\mu d^{-1} = \{1\}\}$, then the map

$$\begin{aligned}
\mathcal{D}_\lambda &\to \mathcal{T}(\mu, \lambda) \\
d &\mapsto \sigma_d
\end{aligned} \tag{7.1}$$

is a bijection (the inverse to (6.2) to be precise), whose restriction to $C_{\lambda\mu'}$ gives a bijection between $C_{\lambda\mu'}$ and the strictly column standard μ-tableau of type λ (Dipper and James [1991, 7.6, 7.9]). By the above, w_λ is the unique distinguished double coset representative in $\mathcal{D}_{\lambda\lambda'}$ with the property that $w_\lambda \Sigma_\lambda w_\lambda^{-1} \cap \Sigma_{\lambda'} = \{1\}$.

For $\lambda \in \Lambda^+(n, r)$ we define the element

$$y_\lambda = \sum_{\pi \in \Sigma_\lambda} (-q)^{-\ell(\pi)} T_\pi \tag{7.2}$$

of $\mathcal{H}_q$.

Remark. The ideals $x_\lambda \mathcal{H}$ and $y_\lambda \mathcal{H}$ are closely related: by Dipper and James [1989, 2.1] these two modules are conjugate with respect to a certain R-algebra automorphism of $\mathcal{H}_q$. One should not then be surprised to hear that it would have been perfectly possible to use the ideal $y_\lambda \mathcal{H}_q$ in place of M_q^λ at many points in the previous sections; indeed there is a natural isomorphism between algebras, as follows:

$$S_q(n, r) = \mathrm{End}_{\mathcal{H}_q} \left(\bigoplus_{\lambda \in \Lambda(n,r)} x_\lambda \mathcal{H}_q \right) \cong \mathrm{End}_{\mathcal{H}_q} \left(\bigoplus_{\lambda \in \Lambda(n,r)} y_\lambda \mathcal{H}_q \right). \tag{7.3}$$

A sizeable number of arguments below require weight space theory, and we have seen at the end of Chapter 6 that there are two possible approaches for this via right comodules for $A_q(n, r)$ or left modules for the algebra $S_q(n, r)$ of (7.3). Both versions are employed below as the circumstances demand.

Consider first the right $\mathcal{H}_q$-module $X_q^\mu = \varphi_{\mu w}^1 \mathcal{H}_q$. Both X_q^μ and

$\varphi^1_{\mu\mu}S_q(n,r)\varphi^1_{\omega\omega}$ have basis $\{\varphi^d_{\mu\omega} : d \in \mathcal{D}_\mu\}$ by the discussion just after 6.6.4, hence they are equal (this therefore characterises X^μ_q as the double (ω,μ)-weight space of $S_q(n,r)$). In turn, they are also equal to $\mathrm{Hom}_{\mathcal{H}_q}(\mathcal{H}_q, M^\mu_q)$. Note also there is an isomorphism between M^μ_q and X^μ_q induced from the map of generators:

$$x_\mu h \mapsto \varphi^1_{\mu\omega}h \quad (h \in \mathcal{H}_q). \tag{7.4}$$

Two general facts are also at hand from Dipper and James [1986, 4.1]. For given weights λ, μ

(1) $x_\lambda \mathcal{H}_q y_\mu \neq \{0\}$ only if $\mu' \unrhd \lambda$;

(2) $x_\lambda \mathcal{H}_q y_{\lambda'}$ is a rank one R-free module with basis $x_\lambda T_{w_\lambda} y_{\lambda'}$. (Indeed one shows that if $\mu'' = \lambda'$ and $x_\lambda T_\pi y_\mu \neq 0$ ($\pi \in G$), then $x_\lambda T_\pi y_\mu$ is plus or minus some non-negative q-power of $x_\lambda T_{w_\lambda} y_\mu$.)

Now $\varphi^1_{\lambda\lambda}S_q(n,r)y_\mu \subseteq \varphi^1_{\lambda\lambda}S_q(n,r)\varphi^1_{\omega\omega} = X^\lambda_q$. Using the isomorphism (7.4) between the right $\mathcal{H}_q$-modules M^λ_q and X^λ_q, deduce

Lemma 7.1.1 *If $\lambda, \mu \in \Lambda(n,r)$,*

     (i) *$\varphi^1_{\lambda\lambda}S_q(n,r)y_\mu \neq 0$ only if $\mu' \unrhd \lambda$.*

     (ii) *$\varphi^1_{\lambda\lambda}S_q y_{\lambda'}$ is a rank one R-free module with basis $\varphi^1_{\lambda\omega}T_{w_\lambda}y_{\lambda'} \neq 0$.*

We are now in a position to define the q-Weyl module.

Definition 7.1.2 For $\lambda \in \Lambda(n,r)$, consider the element

$$z_\lambda = x_\lambda T_{w_\lambda} y_{\lambda'} = \sum_{\pi \in \Sigma_{\lambda'}} (-q)^{-\ell(\lambda)} x_\lambda T_{w_\lambda \pi} \tag{7.5}$$

of $\mathcal{H}_q$. Under the isomorphism (7.4) this element maps to $\varphi^1_{\lambda\omega}T_{w_\lambda}y_{\lambda'}$. Call the left $S_q(n,r)$-submodule $V_q(\lambda) = S_q(n,r)z_\lambda$ of $\mathcal{H}_q$ the *q-Weyl module*.

Remark. Application of (6.11), (6.12) allows one to obtain the z_λ used in Dipper and James [1989, 3.3]; namely, the element

$$\sum_{\pi \in \Sigma_{\lambda'}} (-q)^{-\ell(\pi)} T_w T_{w_{\lambda'}} e_i = \sum_{\pi \in \Sigma_{\lambda'}} (-1)^{\ell(\pi)} e_{iw_\lambda \pi}$$

of $E^{\otimes r}$. Here $i = \ell(\lambda)$. Thus we have defined an $A_q(n,r)$-comodule

$$V_q(\lambda) = \{v \in E^{\otimes r} : \tau_r(v) = z_\lambda \otimes c \text{ for some } c \in A_q(n,r)\}.$$

Example. With $\lambda = (2^2)$ we compute

$$x_\lambda = \frac{\boxed{\begin{array}{cc} 1 & 2 \\ \hline 3 & 4 \end{array}}}{} \qquad x_\lambda T_{w_\lambda} = \frac{\boxed{\begin{array}{cc} 1 & 3 \\ \hline 2 & 4 \end{array}}}{}$$

and the generator of $V_q(\lambda)$ is given by (7.5):

$$z_\lambda = x_\lambda T_{w_\lambda}(1 - q^{-1}T_{(12)})(1 - q^{-1}T_{(34)})$$

$$= \frac{\boxed{\begin{array}{cc} 1 & 3 \\ \hline 2 & 4 \end{array}}}{} - q^{-1}\frac{\boxed{\begin{array}{cc} 2 & 3 \\ \hline 1 & 4 \end{array}}}{} - q^{-1}\frac{\boxed{\begin{array}{cc} 1 & 4 \\ \hline 2 & 3 \end{array}}}{} + q^{-2}\frac{\boxed{\begin{array}{cc} 2 & 4 \\ \hline 1 & 3 \end{array}}}{}.$$

The q-Weyl modules may be indexed by partitions, rather than weights:

Lemma 7.1.3 *If $\lambda' = \mu'$ then $V_q(\lambda) = V_q(\mu)$.*

Proof This is immediate from the alternative definition of z_λ in the last remark. For variety, let us also prove this using the generator of 7.1.2.

Pick $d \in \mathcal{D}_{\lambda\mu}$ such that $d^{-1}\Sigma_\lambda d = \Sigma_\mu$. The idea is to see what happens to the generator z_λ under the non-singular basis element $\varphi_{\lambda\mu}^d$. Now, since $\lambda' = \mu'$,

$$\begin{aligned} \varphi_{\lambda\mu}^d z_\mu &= \varphi_{\lambda\mu}^d \varphi_{\mu\omega}^1 T_{w_\mu} y_{\mu'} \in \varphi_{\lambda\lambda}^1 S_q(n,r) y_{\lambda'} \\ &= r\varphi_{\lambda\omega}^1 T_{w_\lambda} y_{\lambda'} \\ &= r z_\lambda \end{aligned}$$

for some $r \in R$ by 7.1.1(ii). Thus $r z_\lambda \in V_q(\mu)$, and since the $\varphi_{\lambda\mu}^d$ are non-singular, $V_q(\lambda) = V_q(\mu)$. $\qquad\square$

Henceforth let $R = K$. We have a collection of q-Weyl modules $\{V_{K,q}(\lambda) : \lambda \in \Lambda^+(n,r)\}$ labelled by dominant weights. Also $V_q(\lambda)$ is a highest weight module of highest weight λ, and is the universal object of $_{S_q(n,r)}\mathbf{mod}$ with this property. Now the q-Weyl modules have simple tops, which can be shown to provide a complete set of irreducible $S_q(n,r)$-modules. The proof of completeness is involved whichever approach is adopted and we merely outline the procedure.

We refer back to (6.5) for the non-degenerate form defined on the direct sum $M = \bigoplus M_q^\lambda$. If we identify M^* with tensor space using (6.9) and the remarks preceding (6.8), we have a form $\langle \, , \, \rangle$ on tensor space. (Observe that if $q = 1$ we have the form of Exercise (1) after 3.4.6.) We aim again to define a contracted form on a certain submodule of tensor space containing $V_q(\mu)$ and then restrict this form to produce a contravariant form on $V_q(\mu)$. Thus, for $\mu \in \Lambda(n,r)$, we consider the $S_q(n,r)$-module

$$W_q(\mu) = S_q(n,r)y_\mu(e_1 \otimes \cdots \otimes e_r) = S_q(n,r)\tilde{y}_\mu$$

where

$$\tilde{y}_\mu = y_\mu(e_1 \otimes \cdots \otimes e_r) = \sum_{\pi \in \Sigma_\mu} (-1)^{\ell(\pi)} e_{\pi(1)} \otimes \cdots \otimes e_{\pi(r)}.$$

Then $V_q(\mu)$ is contained in the submodule $W_q(\mu')$ of tensor space; I should point out that $W_q(\mu)$ is the module called L^α in Dipper and James [1991], where $\alpha'' = \mu'$.

We define the symmetric form

$$\langle\langle u\tilde{y}_\mu, v\tilde{y}_\mu\rangle\rangle = \langle\langle u, vy_\mu\rangle\rangle$$

for $u, v \in E^{\otimes r}$. A simple check, employing the fact that $\tilde{y}_\mu^* = \tilde{y}_\mu$, shows that this is well defined and, moreover, that for all $u, v \in W_q(\mu)$ and $s \in S_q(n,r)$, $\langle\langle s^*u, v\rangle\rangle = \langle u, sv\rangle$. It is easy to verify that this form is also non-degenerate, for $\langle\langle z_\mu, z_\mu\rangle\rangle = q^{\ell(w_\mu)} \neq 0$. Denote in the usual way the perp space $U^\perp = \{v \in W_q(\mu) : \langle\langle v, u\rangle\rangle = 0 \,\forall u \in U\}$, where U is any subset of $W_q(\mu)$. The next result is reminiscent of the good old Submodule Theorem from James [1978a].

Theorem 7.1.4 (The Submodule Theorem) *If U is an $S_q(n,r)$-submodule of $W_q(\mu')$ then either $V_q(\mu)$ is a submodule of U or U is a submodule of $V_q(\mu)^\perp$.*

Proof The trick here is to consider μ-weight spaces of U. For, by 7.1.1,

$$\varphi_{\mu\mu}^1 U \neq 0 \quad \Rightarrow \quad \varphi_{\mu\omega}^1 T_{w_\mu} y_{\mu'} \in U$$
$$\Rightarrow \quad V_q(\mu) \subseteq U.$$

Otherwise, for all $u \in U$,

$$\varphi_{\mu\mu}^1 U = 0 \quad \Rightarrow \quad \forall u \in U, \xi \in S_q(n,r),$$
$$0 = \langle\langle \varphi_{\mu\mu}^1 \xi u, \varphi_{\mu\omega}^1 T_{w_\mu} y_{\mu'}\rangle\rangle$$
$$= \langle\langle u, \xi^* \varphi_{\mu\omega}^1 T_{w_\mu} y_{\mu'}\rangle\rangle$$
$$\Rightarrow \quad \langle\langle u, \xi^* z_\mu\rangle\rangle = 0$$

$$\Rightarrow \quad U \subseteq V_q(\mu)^{\perp},$$

since we have $\langle\langle u, \xi^* z_\mu \rangle\rangle = 0$ for all $u \in V_q(\mu)$. We have used the fact that $(\varphi^1_{\mu\mu})^* = \varphi^1_{\mu\mu}$, from Remark (1) after (6.5). $\qquad\square$

Theorem 7.1.5 *If $\mu \in \Lambda^+(n,r)$, $V_q(\mu) \cap V_q(\mu)^{\perp}$ is the unique maximal submodule of $V_q(\mu)$, and $F_q(\mu) = V_q(\mu)/(V_q(\mu) \cap V_q(\mu)^{\perp})$ is an absolutely irreducible self-dual $S_q(n,r)$-module. Since $F_q(\lambda) = F_q(\mu)$ provided $\lambda' = \mu'$, the irreducible top factors may be indexed by dominant weights.*

Now the $V_q(\lambda)$ and the $F_q(\lambda)$ are highest weight modules of highest weight λ by 7.1.1(i), and the λ-weight space is generated by z_λ. How do the Fs relate to the Ls which were defined in 6.4.4? The $L_q(\lambda)$ have been shown to be absolutely irreducible pairwise non-isomorphic $A_q(n)$-comodules, of highest weight λ (λ dominant) and by a deep result of Dipper and James [1991, §8], we have actually constructed a complete set. One need now only compare highest weights of $F_q(\lambda)$ and $L_q(\lambda)$ to see that these modules are identical. Hence

Theorem 7.1.6 *Let $q \in K^*$. Then*

$$\{L_{K,q}(\lambda) : \lambda \in \Lambda^+(n,r),\ r \geq 0\}$$

is a complete set of inequivalent absolutely irreducible $A_q(n)$-comodules. If λ is dominant, $L_{K,q}(\lambda)$ is the unique top factor of the q-Weyl comodule $V_q(\lambda)$, and is of highest weight λ.

Remarks.
(1) The classification of the simple $S_q(n,r)$-modules (or, equivalently, of the composition factors of the q-tensor space $E_q(n,r)$) follows a circuitous route leading in Dipper and James [1991, 8.8] to the expected conclusion that the partitions of r are in one-to-one correspondence with the isomorphism types of irreducible $S_q(n,r)$-modules, under $\lambda \leftrightarrow L_q(\lambda)$. A new proof of this classification appears in Dipper and Du [1993]. They develop a q-version of 4.6.3(ii) to obtain a Green Correspondence for Hecke algebras of type A; see also Du [1991], [1992b].

Of course, the completeness would not be a problem in cases where the q-Schur algebra was known to be semisimple. A Hecke algebra result in Dipper and James [1987, 4.3] gives necessary and sufficient conditions for this to be so (one case of this was given in 6.5.3; and another in the classical case 2.2.8). Precisely, if $R = K$ and $q \in K^*$ then $\mathcal{H}_q$ is not semisimple if and only if either $q = 1$ and char $K \leq r$, or $q \neq 1$ and

q is a primitive ℓth root of unity for some $2 \leq \ell \leq r$. In particular, if $1 + q + \cdots + q^{i-1} \neq 0$ for all $i \in \mathbf{r}$ then $\mathcal{H}_q$ is semisimple and one knows that there will be as many pairwise inequivalent irreducible modules as there are partitions of r. Hence $S_q(n,r)$ has $|\Lambda^+(r)|$ irreducibles. So, by 7.1.5, $L_q(\lambda) = F_q(\lambda) = V_q(\lambda)$, and $\{V_q(\lambda) : \lambda \in \Lambda^+(r)\}$ is a complete set of irreducible $S_{K,q}(n,r)$-modules.

(2) There are also q-Schur modules as well; they may be defined as the contravariant duals $M_q(\lambda) = V_q(\lambda)^\circ$. Such modules may be characterised as induced modules for the quantum GL_n defined in 7.3.

(3) Some deep properties of $V_q(\lambda)$ are proved by Dipper and James [1991]. For instance, the q-Weyl module $V_q(\lambda)$ is R-free, with a basis indexed by the semistandard λ-tableaux of type $\mu \in \Lambda(n,r)$. They also show that $V_q(\lambda)$ is the intersection of the kernels of homomorphisms $W_q(\lambda) \to W_q(\mu)$ where $\mu \in \Lambda(n,r)$ is not dominated by λ, cf. James [1978a, 26.5].

(4) There are obvious open questions here: can one construct q-versions of bideterminants and codeterminants that give bases of the q-Schur modules and q-Weyl module? It appears that the main problems here are notational.

7.2 The q-determinant in $A_q(n,r)$

To proceed, we need something to play the rôle of determinant. We recall that an element g of a coalgebra (C, Δ, ε) is called *group-like* if $\varepsilon(g) = 1$ and $\Delta(g) = g \otimes g$.

Lemma 7.2.1 *There is a unique group-like element, d say, in the coalgebra $A_{\mathbf{Q}(t),t}(n,n)$.*

Proof By the classification of 6.5.3, for $\lambda \in \Lambda^+(n,r)$, $\mathrm{ch}\, L_{\mathbf{Q}(t),t}(\lambda) = s_\lambda$. In particular, $\mathrm{ch}\, L_{\mathbf{Q}(t),t}(1^n)(\mathbf{X}) = X_1 \ldots X_n$ so is 1-dimensional. Now, the Schur function s_λ has leading term $\mathbf{X}^\lambda$, so involves $\mathbf{X}^{\lambda\sigma}$ for all $\sigma \in \Sigma_n$. We conclude that if $L_{\mathbf{Q}(t),t}(\lambda)$ is 1-dimensional, λ is Σ_n-invariant, hence $\lambda = (1^n)$. So we have, up to isomorphism, a unique 1-dimensional $A_{\mathbf{Q}(t),t}(n,n)$-comodule $L = L_{\mathbf{Q}(t),t}(1^n)$.

Let $0 \neq v$ be an element of the comodule (L, ω_L). Since $L = \mathbf{Q}(t)v$, there must exist $d \in A_{\mathbf{Q}(t),t}(n,n)$ such that $\omega_L(v) = v \otimes d$. But $d \neq 0$ since ω_L is non-trivial, and consequently $\Delta(d) = d \otimes d$ by the comodule laws. This proves existence.

For uniqueness, choose any $v' \in A_{\mathbf{Q}(t),t}(n,n)$ satisfying $\Delta(v') = v' \otimes v'$. Now $L' = \mathbf{Q}(t)v'$ is a 1-dimensional right comodule, with structure map $\omega_{L'}$ defined by

$$\omega_{L'}(cv') = cv' \otimes v' \in L' \otimes A_{\mathbf{Q}(t),t}(n,n).$$

Thus, $L' \cong L$ so that L' and L have the same coefficient space, which implies that $\mathbf{Q}(t)v = \mathrm{cf}(L') = \mathbf{Q}(t)v'$. We deduce the existence of some $0 \neq c \in \mathbf{Q}(t)$ such that $v' = cd$. Since both v' and d are group-like, $c = 1$. $\square$

Definition 7.2.2 Define the *q-determinant* to be the element

$$\det{}_{R,q} = \sum_{\sigma \in \Sigma_n} (-1)^{\ell(\sigma)} c_{1,1\sigma} \cdots c_{n,n\sigma} \tag{7.6}$$

of $A_{R,q}(n,n)$.

So, formally $\det_{R,q}$ resembles the usual determinant function of $A_{R,1}(n)$. Actually, it leads a double life: the familiar classical 'det' on the one hand, the q-version on the other (as with the Λ_q^n operator). Classically the determinant function appears as a co-coefficient in the structure map of the 1-dimensional comodule $\Lambda_1^n(E)$. In the quantum case, we recall that $\Lambda_q^n E$ is 1-dimensional. We shall compute the image of the generator under the structure map.

Lemma 7.2.3 *The t-determinant* $\det_{\mathbf{Q}(t),t}$ *is the element d of 7.2.1.*

Proof Consider the left $A_{\mathbf{Q}(t),t}(n,n)$-comodule $(\Lambda_t^n(V), \sigma_n)$, which we know to be cyclic, generated by $\bar{v} = \bar{v}_1 \ldots \bar{v}_n$ (see the remarks at the end of section 6.2). One checks that $\sigma_n(\bar{v}) = d_{\mathbf{Q}(t),t} \otimes \bar{v}$. Since $\mathrm{cf}(\Lambda_t^n(V)) \neq 0$, $d_{\mathbf{Q}(t),t} \neq 0$, and since $(\Delta \otimes 1)\sigma_n = (1 \otimes \sigma_n)\sigma_n$, $\Delta(d_{\mathbf{Q}(t),t}) = d_{\mathbf{Q}(t),t}$, as required. $\square$

Now, similar computations with the right $A_{\mathbf{Q}(t),t}(n,r)$-comodule, $\Lambda_t^r(E) = \mathbf{Q}(t)\bar{e}$ having structure map τ_r $(r \geq 0)$ yield that $\tau_r(\bar{e}) = \bar{e} \otimes b$, where $\bar{e} = \bar{e}_n \ldots \bar{e}_1$ and

$$b = \sum_{\rho \in \Sigma_n} (-t)^{\ell(\rho)} c_{n\rho,n} \cdots c_{1\rho,1}.$$

Again, the q-determinant resembles the usual det $\in R[c_{ij}]$, the 'q-parameter' entering the game on the left.

Lemma 7.2.4 *The t-determinant,* $\det_{\mathbf{Q}(t),t}$, *is the unique non-zero group-like element* $d \in A_{\mathbf{Q}(t),t}(n,n)$ *and is given by*

$$
\begin{aligned}
\det_{\mathbf{Q}(t),t} &= \sum_{\sigma \in \Sigma_n} (-1)^{\ell(\sigma)} c_{1,1\sigma} \cdots c_{n,n\sigma} \\
&= \sum_{\rho \in \Sigma_n} (-q)^{\ell(\rho)} c_{n\rho,n} \cdots c_{1\rho,1}.
\end{aligned}
$$

Now we have to consider the general case. First of all, if we have arbitrary $q \in R$ we note that $A_{R,q}(n) \cong R \otimes_{\mathbf{Z}[t]} A_{\mathbf{Z}[t],t}(n)$, where we think of R as a $\mathbf{Z}[t]$-module, by the specialisation $t \to q$. The statement of 7.3.4 is valid inside $A_{\mathbf{Z}[t],t}(n)$. In addition if q is a unit in R, we can actually view R as a $\mathbf{Z}[t,t^{-1}]$-module, by the specialisation $t \to q$, to get an isomorphism $A_{R,q}(n) \cong R \otimes A_{\mathbf{Z}[t,t^{-1}],t}(n)$. Thus the statement of 7.3.4 here applies inside $A_{\mathbf{Z}[t,t^{-1}],t}(n)$. Hence, in both cases we need only tensor with R to obtain:

Theorem 7.2.5 *For any ring R, and arbitrary $q \in R$*

$$
\det_q = \sum_{\sigma \in \Sigma_n} (-1)^{\ell(\sigma)} c_{1,1\sigma} \cdots c_{n,n\sigma} = \sum_{\rho \in \Sigma_n} (-q)^{\ell(\rho)} c_{n\rho,n} \cdots c_{1\rho,1};
$$

while if q is a unit,

$$
\det_q = \sum_{\pi \in \Sigma_n} (-q)^{-\ell(\pi)} c_{1\pi,1} \cdots c_{n\pi,n}.
$$

The major difference between $\det_q$ and the element D of $K[M_q(n)]$ in Parshall and Wang [1991, §4] is that $\det_q$ defined in 7.2.5 is *not* central in $A_{R,q}(n)$, whereas D is central in the Parshall-Wang q-matrix space $K[M_q(n)]$. Failure to be central is not, in fact, a problem when we come to localise, as we shall see later.

Theorem 7.2.6 *For the generic case $R = \mathbf{Q}(t)$, $q = t$, we have, for $i,j \in \mathbf{n}$, $c_{ij}d_{\mathbf{Q}(t),t} = t^{i-j}d_{\mathbf{Q}(t),t}c_{ij}$. Hence by base change, for arbitrary $q \in R$,*

$$
\begin{aligned}
c_{ij}\det_q &= q^{i-j}\det_q c_{ij} \quad (1 \le j \le i \le n) \\
q^{j-i}c_{ij}\det_q &= \det_q c_{ij} \quad (1 \le i < j \le n)
\end{aligned}.
$$

Note that t is a unit, so t^{i-j} makes sense in the case $i < j$.

Proof We use the notation established in the example after 6.5.3. We have proved that there is a comodule isomorphism $\theta : E \otimes \Lambda_t^n E \to \Lambda_t^n E \otimes E$, and since comodule maps preserve weight spaces, we find $0 \neq f_i \in \mathbf{Q}(t)$ such that $\theta(e_i \otimes \bar{e}) = f_i(\bar{e} \otimes e_i)$. But

$$\omega_1(e_j \otimes \bar{e}) = \sum_i e_i \otimes \bar{e} \otimes c_{ij}d,$$

$$\omega_2(\bar{e} \otimes e_j) = \sum_i \bar{e} \otimes e_i \otimes dc_{ij}.$$

Since θ is a comodule isomorphism, $(\theta \otimes 1)\omega_1 = \omega_2\theta$ and so $c_{ij}d = f_{ij}dc_{ij}$ for some $0 \neq f_{ij} \in \mathbf{Q}(t)$.

Recall now the ordering $\preceq$ defined just after 6.1.3. We have

$$
\begin{aligned}
c_{ij}d &= c_{ij}c_{11}\cdots c_{nn} + \text{smaller terms} \\
&= t^m c_{11}\cdots c_{nn} + \text{smaller terms} \\
&= f_{ij}dc_{ij}
\end{aligned}
$$

where $m = m_1 - m_2$ is the difference of the integers

$$
\begin{aligned}
m_1 &= |\{k \in \mathbf{n} : c_{ij}c_{kk} = tc_{kk}c_{ij}\}| \\
m_2 &= |\{k \in \mathbf{n} : c_{ij}c_{kk} = t^{-1}c_{kk}c_{ij}\}|.
\end{aligned}
$$

Using 6.1.1, $i \leq j$ implies $m_1 = 0$ and $m_2 = j - i$, whereas $i > j$ implies $m_1 = i - j$ and $m_2 = 0$. Hence $m = i - j$. Since $f_{ij} \in \mathbf{Q}(t)$ we have $f_{ij} = t^{i-j}$ as wanted. $\qquad\square$

7.3 A quantum GL_n

All along, our goal has been to exhibit a Hopf algebra with antipode, dependent on a ring parameter q, whose spectrum is what will be called the quantum GL_n. We proceed as follows.

Let q be an element of the commutative ring R. Let $d = \det_q \in A_q(n)$ be the q-determinant, which, we observe, generates an ideal of $A_q(n)$. Thus the collection of powers $S = \{1, d, d^2, \ldots\}$ is a multiplicative set in $A_q(n)$ and satisfies the left and right Ore condition. By standard ring theory, this means that an Ore localisation exists.

Definition 7.3.1 The Ore localisation of $A_q(n)$ at S, $A_q(n)_S$, is an R-algebra denoted by $R[\Gamma_q]$, and called the *quantum $R[\Gamma]$*.

Remarks.

(1) In Dipper and Donkin [1991, 4.2], $R[\Gamma_q]$ is termed *the quantum* GL_n, but that conflicts with Drinfel'd [1986, p. 800] who reserves this term for the prime spectrum of $R[\Gamma_q]$. It's hard to understand quite what Γ_q, standing alone, can be!

(2) The first person to study a q-analogue of the reciprocity between Σ_n and GL_n was Jimbo [1986]: he started with the quantum group and exhibited the Hecke algebra as its 'partner'. The current presentation follows the opposite path: we start with the Hecke algebra and find its reciprocal partner. The latter approach, however, has a drawback. Recalling Remark (2) before (6.10), another (non-isomorphic) quantum GL_n exists; a good reference here is Parshall and Wang [1991]. This memoir, developing some themes in Manin [1988], constructs a quantum deformation of the coordinate ring but then localises at a central determinant. So there are now two possible ways of producing a quantum GL_n. Curiously, though, taking the dual of the coordinate algebra of the $n \times n$ matrices in both cases produces isomorphic q-Schur algebras! An explanation of this phenomenon is offered in Du *et al.* [1991], which describes the notion of 'hyperbolic invariance'. Other comments on this difficult area, and further references, are contained in Jimbo [1986], Drinfel'd [1986] and Dipper and Donkin [1991, 4.4]. There are further complications: in the classical case the hyperalgebra $\mathcal{U}$ and the coordinate ring of semisimple algebraic groups are in Hopf duality. Lusztig has produced certain q-hyperalgebras $\mathcal{U}_q$ (see for example Andersen [1989] or Andersen *et al.* [1991]), but these are not in Hopf duality with $R[\Gamma_q]$ because there is no generic way of constructing a deformation of $R[\Gamma]$. Lusztig's algebras are, however, related to the coordinate rings of Manin's quantum SL_n by Hopf duality over fields of characteristic zero. See Dipper [1991] for more details.

Now, Δ extends (canonically) to a map

$$\Delta : A_q(n)_S \;\rightarrow\; A_q(n)_S \otimes A_q(n)_S$$
$$d^{-i} \;\mapsto\; d^{-i} \otimes d^{-i}$$

($i \in \mathbf{N}$). By 7.2.3, this is the only way to extend Δ. Similarly, extend ε to $A_q(n)_S$. Since $\varepsilon(d^{-i}) = 1$ by 6.1.5, we are forced to put $\varepsilon(d^{-i}) = 1$. Now we check that we have a bialgebra, containing $A_q(n)$ as a sub-bialgebra. Actually much more is true: serious effort is expended in Dipper and Donkin [1991, 4.2.21] in proving the following important result.

Theorem 7.3.2 *Let R be a commutative ring, and let $q \in R$ be a unit. Then $R[\Gamma_q]$ is a Hopf algebra with antipode.*

Proof Let me outline the motivation behind the Dipper and Donkin proof. Classically, inversion in Γ induces the antipode γ on the Hopf algebra $R[\Gamma]$, and γ has the following effects: if $c \in R[\Gamma]$ is such that $\Delta(c) = \sum u_i \otimes v_i$ then $\sum u_i \gamma(v_i) = \sum \gamma(u_i)v_i = \varepsilon(c)1_{R[\Gamma]}$. Thus $\gamma(\det) = \det^{-1}$, and for $i, j \in \mathbf{n}$, $\gamma(c_{ij}) = h_{ij}$, where h_{ij} is the (i,j)th entry in the adjugate matrix of $[c_{ab}]$, i.e. $h_{ij} = \det(X_{ji})$, and X_{ji} is obtained by deleting row j and column i from $[c_{ab}]$, multiplied by $\det^{-1}$. There is nothing magical about these terms: they are simply the co-coefficients of $\tau(e_{i_1} \cdots e_{i_{n-1}})$ in the usual $(n-1)$th exterior power, $\Lambda^{n-1}(E)$, regarded as a comodule for $A_R(n)$ (here τ is its structure map).

Now, $\det_q$ will provide us with a way of computing determinants of $n \times n$ matrices with entries in $A_q(n)$, and hence of square matrices of arbitrary size. In particular the q-determinant d_{ij} of the $(n-1) \times (n-1)$ minor C_{ij} of $[c_{ab}]$ over $A_q(n)$ is the co-coefficient of

$$\tau(e_1 \cdots e_{i-1}e_{i+1} \cdots e_n) \in \Lambda_q^{n-1} E \otimes A_q(n).$$

Let $h_{ij} = (-1)^{i+j} \det_q^{-1} d_{ji} \in R[\Gamma_q]$. The map $\gamma : R[\Gamma_q] \to R[\Gamma_q]$ such that $\gamma(c_{ij}) = h_{ij}$ and $\gamma(d^r) = d^{-r}$ is an algebra anti-endomorphism It turns out that this produces the required antipode, though the detailed check in Dipper and Donkin [1991, 4.2.21] is very extensive. They also go on to show that the antipode γ has order greater than two, provided $q \neq 1$ (more precisely, γ^2 is the inner automorphism of $R[\Gamma_q]$ defined by conjugation by $\det_q$). $\qquad \square$

A basis for $R[\Gamma_q]$

This completes the construction phase. The next round is to look for a basis for $R[\Gamma_q]$ in the case when q is a unit in the commutative ring R.

Let $a = [a_{ij}] \in \mathrm{Mat}_n(\mathbf{Z})$ and define $c^a = c_{11}^{a_{11}}c_{12}^{a_{12}} \cdots c_{1n}^{a_{1n}}c_{21}^{a_{21}} \cdots c_{nn}^{a_{nn}}$. We show that

$$\mathcal{A} = \{a \in \mathrm{Mat}_n(\mathbf{N}_0) : a_{ii} = 0 \text{ for some } i \in \mathbf{n}\}$$

provides another indexing set for a basis of $A_q(n)$.

Proposition 7.3.3 $\mathcal{B}' = \{c^a d^r : a \in \mathcal{A}, r \in \mathbf{N}_0\}$ *is a basis for $A_q(n)$ over R.*

Proof We recall the ordering of monomials in the usual basis $\mathcal{B}$ of $A = A_q(n)$ from 6.1.2. We will say that $c_{i,j} \preceq c_{k,l}$ if $r \leq s$, and $r = s$ implies $i \geq k$, respectively $j \geq l$ if also $i = k$. [Here '$\geq$' is the dictionary order on $I(n,r)$].

(1) *If $a, b \in \mathrm{Mat}_n(\mathbf{N}_0)$, there exists $z \in \mathbf{N}_0$ such that $c^a c^b = q^z c^{a+b} +$ smaller terms.* Of course $q^z \neq 0$ as q is a unit. This is just a rerun of 6.1.2.

(2) *Linear independence.* Suppose that $\sum \alpha_{a,r} c^a d^r = 0$ where the pair (a, r) runs through some finite subset B of $\mathcal{A} \times \mathbf{N}_0$ and $\alpha_{a,r} \neq 0$ for each pair $(a, r) \in B$. By definition $d = \prod c_{ii} +$ smaller terms, so from (1) we have $\sum \alpha_{a,r} q_{a,r} c^{a+re} +$ smaller terms $= 0$, where $q_{a,r}$ is a q-power for $(a, r) \in B$ and e is the identity in $\mathrm{Mat}_n(\mathbf{Z})$. As (a, r) varies over B, there will be a largest monomial C of the c^{a+re}. Let $B_0 \subseteq B$ consist of all pairs (a, r) such that $c^{a+re} = C$. By 6.1.3

$$\sum_{(a,r) \in B_0} \alpha_{a,r} q_{a,r} = 0. \tag{7.7}$$

If $b \in B_0$, since $b \in \mathcal{A}$ we can find some i such that $b_{ii} = 0$. If $g = b + re$, then r is the least of the diagonal entries of g. Now $C = c^g$, so that g, and hence r and $b = g - re$, are specified uniquely once we know C. So B_0 is the singleton $\{(a_0, r_0)\}$, hence (7.7) becomes

$$\alpha_{a_0,r_0} q_{a_0,r_0} = 0.$$

Since q_{a_0,r_0} is a unit (being a q-power), $\alpha_{a_0,r_0} = 0$, a contradiction.

(3) *Let A' be the R-span of $\mathcal{B}'$. Then $A' = A_q$.* If not, find the smallest monomial c^a not contained in A', where $a \in \mathrm{Mat}_n(\mathbf{N}_0)$. As $c \notin \mathcal{B}'$, $a_{ii} > 0$ for all $i \in \mathbf{n}$. Define $b \in \mathrm{Mat}_n(\mathbf{Z})$ by putting $b_{ij} = a_{ij}$ if $i \neq j$ and $b_{ii} = a_{ii} - 1$ if $i \in \mathbf{n}$. Put $c_1 = c^b$, and note $c_1 \preceq c$. By minimality of c_1, the latter must then lie in A', as must $c_1 d$, since A' is closed to multiplication by d. Using (1) again,

$$c_1 d = c^b c_{11} \cdots c_{nn} + \text{other terms} = q^z c^{b+e} + \text{other terms}$$
$$= q^z c^a + \text{other terms} \in A'$$

where the 'other terms' are smaller in the $\preceq$ order. All these other terms in $c_1 d$ lie in A', by minimality of c^a, forcing $c^a \in A'$ (since q^z is a unit). This contradiction implies $A = A'$.   □

Localising at $\det_q$ yields

Theorem 7.3.4 $R[\Gamma_q]$ *is R-free with basis* $\hat{\mathcal{B}} = \{c^a d^r : a \in \mathcal{A}, r \in \mathbf{Z}\}$.

Remark. The rational representation theory of the quantum GL_n corresponds almost word for word with the classical theory of rational representations of linear algebraic groups, as outlined in the Appendix, though the proofs in this new situation often require considerable technical adjustments. For the theory of q-tori (which are exactly as for GL_n), q-weights, root systems etc. see Parshall and Wang [1991, §§2, 6, 8].

7.4 The category $\mathcal{P}_q(n,r)$

In keeping with an accepted perversion of language I shall call right comodules for $R[\Gamma_q]$ *left rational* Γ_q-*modules*. Such (co)modules M come equipped with a map $\tau : M \to M \otimes K[\Gamma_q]$. Let us assume henceforth that all (co)modules are R-free of finite R-rank.

Weights and polynomial modules

We take the weight space decomposition (6.3). Now $d = \det_q$ lies in the weight space $A_q^{\omega,\omega}(n)$ where $\omega = (1^n) \in \Lambda(n,n)$, hence we may extend the decomposition to the localisation at d, i.e.

$$R[\Gamma_q] = \bigoplus_{\lambda,\mu \in \mathbf{Z}^n} A_q^{\lambda,\mu}(n)_S$$

where by 7.3.4 the elements of $A_q^{\lambda,\mu}(n)_S \subseteq A_q(n)_S$ are of the form $f d^{-r}$ ($r \in \mathbf{Z}$), where $f \in A^{\lambda+r\omega,\mu+r\omega}(n)$ and $\lambda + r\omega, \mu + r\omega \in \Lambda(n,r)$.

For $\lambda \in \mathbf{Z}^n$, consider the subspace of $R[\Gamma_q]$ defined by

$$A_d^{\lambda}(n) = \bigoplus_{\mu \in \mathbf{Z}^n} A_q^{\lambda,\mu}(n)_S.$$

Definition 7.4.1 If (M, ω_M) is a $R[\Gamma_q]$-comodule, the λ-*weight space* of M is

$$M^\lambda = \{m \in M : \omega_M(m) \in M \otimes A_d^{\lambda}(n)\}.$$

Thus we obtain a decomposition $M = \bigoplus_{\lambda \in \mathbf{Z}^n} M^\lambda$ as in 6.3.4. With the assumption that M is R-free of finite R-rank, then one can define its formal character in the Laurent ring, namely

$$\mathrm{ch}\, M = \sum_{\lambda \in \mathbf{Z}^n} (\dim_R M^\lambda) \mathbf{X}^\lambda \in \mathbf{Z}[\mathbf{X}, \mathbf{X}^{-1}]$$

where $\mathbf{X}^\lambda = X_1^{\lambda_1} \cdots X_n^{\lambda_n}$ for $(\lambda_1, \ldots, \lambda_n) \in \mathbf{Z}^n$. Proofs of these assertions exactly parallel those in 6.3. The important fact is that if R is a principal ideal domain and M is an R-free $R[\Gamma_q]$-comodule of finite R-rank then M^λ is also R-free, so $\dim_R M^\lambda$ is defined $(\lambda \in \mathbf{Z}^n)$. The formal characters here enjoy all the usual properties.

Definition 7.4.2 For some R-free $R[\Gamma_q]$-comodule (M, ω_M) with R-basis B, define the *coefficient space* cf(M) to be the R-subspace of $R[\Gamma_q]$ generated by

$$\{x \in R[\Gamma_q] : \omega_M(m) = m_1 \otimes x + \text{other terms, for some } m, m_1 \in B\}.$$

The right $R[\Gamma_q]$-comodule M is *polynomial* if cf$(M) \subseteq A_q(n)$. M is an r-homogeneous polynomial comodule if cf $(M) \subseteq A_q(n, r)$.

Switching instead to the language of modules, one sees that submodules and factor modules of homogeneous polynomial Γ_q-modules are again homogeneous polynomial modules of the same degree. As for 6.3.2, any polynomial module is a direct sum of homogeneous polynomial modules.

Let $\mathcal{P}_q(n)$ be the category of left rational polynomial Γ_q-comodules and $\mathcal{P}_q(n, r)$ the full subcategory of homogeneous polynomial Γ_q-modules of degree r. Now the category $_{S_q(n,r)}\mathbf{mod}$ is of course naturally isomorphic to the category of right $A_q(n, r)$-comodules. But every R-free $A_q(n)$-comodule M is canonically a $R[\Gamma_q]$-comodule via the embedding $M \otimes A_q(n) \hookrightarrow M \otimes R[\Gamma_q]$. Conversely, polynomial $R[\Gamma_q]$-comodules are precisely the $A_q(n)$-comodules extended to $R[\Gamma_q]$. The next theorem (whose proof is now obvious) generalises the equivalence of categories in 2.2.7.

Theorem 7.4.3 *There is a natural isomorphism between the categories* $\mathcal{P}_q(n, r)$ *and* $_{S_q(n,r)}\mathbf{mod}$.

7.5 $\mathcal{P}_q(n, r)$ is a highest weight category

Take $R = K$ algebraically closed in this section. That $S_q(n, r)$ is quasi-hereditary was proved by Parshall and Wang [1991, 11.5] by exhibiting explicit defining sequences of ideals in $S_q(n, r)$ but with some restrictions on q. As usual we order $\Lambda^+(n, r)$ according to the prescription in (3.3). Let $S = S_q(n, r)$.

For each $p \in \mathbf{t}$ where $|\Lambda^+(n, r)|$ has cardinality t, consider the idempotent $e_i = \xi_{\lambda_1} + \cdots + \xi_{\lambda_i} \in S$. Let us define a sequence of ideals of S

as follows: put $J_0 = 0$, and $J_i = Se_iS$ ($i \in \mathbf{t}$). Write $\bar{e}_i$ for the image of e_i in $\bar{S}_i = S/J_{i-1}$.

Theorem 7.5.1 $S_q(n,r)$ *is a quasi-hereditary algebra with weight poset* $\Lambda^+(n,r)$; *in fact*

$$\{0\} = J_0 \subseteq J_1 \subseteq \cdots \subseteq J_t = S$$

is a defining sequence for S.

Proof We may as well assume that q is a root of unity; if not then the q-Schur algebra is semisimple, the ξ_{λ_i} correspond to the $V_q(\lambda_i)$, and the result follows trivially.

(1) $\bar{e}_i$ *is a primitive idempotent in* $\bar{S}_i$. To see this, one must first identify $_{\bar{S}_i}\mathbf{mod}$ with the full subcategory of $_S\mathbf{mod}$ all of whose objects have composition factors of the form $L_q(\lambda_j)$ with $j \geq i$. This done, we notice that $\bar{e}_i$ is precisely $\bar{\xi}_{\lambda_i}$ (i.e. the image of ξ_{λ_i} in $\bar{S}_i$). Thus $\dim \bar{e}_i L_q(\lambda_j)$ is 1 if $i = j$ and is 0 if $j > i$ (since $M \in {}_{\bar{S}_i}\mathbf{mod} \Rightarrow \bar{e}_i M = M^{\lambda_i}$).

To show primitivity, write $\bar{e}_i = f_1 + f_2$. Choose any $\bar{S}_i$-module M. Since a weight space decomposition is direct, then, taking $v \in M$, it follows that if $f_a v \neq 0$ ($a = 1, 2$) $f_1 v$, $f_2 v$ are independent; since also $\bar{e}_i f_a = f_a$, we can assume without loss of generality that $\dim f_1 L_q(\lambda_j) = \delta_{ij}$, while $f_2 L_q(\lambda_j) = 0$ for all $j \geq i$. The Hom functor $_{\bar{S}_i}\mathbf{mod} \to {}_K\mathbf{mod}$ defined by f_2 is exact, and so $f_2 M = 0$ for any $\bar{S}_i$-module M. This forces $f_2 \in f_2\bar{S}_i = \{0\}$ and then it follows that $\bar{e}_i$ is primitive.

(2) *We have a defining sequence.* By (1), $\bar{P}_i = \bar{S}_i\bar{e}_i$ is a PIM for $\bar{S}_i$ with generator $\bar{e}_i$. If $0 \neq M$ is a quotient of $\bar{P}_i$ then $\bar{e}_i M \neq 0$ and this establishes $\bar{P}_i$ as the projective cover of $L_q(\lambda_i)$. By assumption λ_i is maximal in the poset $\{\lambda_j : j \geq i\}$ and so $\bar{P}_i$ is a highest weight module with highest weight λ_i. By the universal property of q-Weyl modules, $\bar{P}_i \cong V_q(\lambda_i)$. But also $\bar{e}_i\bar{S}_i\bar{e}_i \cong \mathrm{End}_{\bar{S}_i}(V_q(\lambda_i)) \cong K$ which implies $\bar{e}_i \operatorname{Rad}\bar{S}_i\bar{e}_i = 0$. This establishes that $\bar{J}_i = J_i/J_{i-1}$ satisfies properties (ii′) and (iii) of 3.3.1.

We consider condition 3.3.1(i′), namely that the multiplication map is an isomorphism. Now

$$\dim \bar{J}_i \leq (\dim V_q(\lambda_i))^2 = \dim(\bar{S}_i\bar{e}_i \otimes_{\bar{e}_i\bar{S}_i\bar{e}_i} \bar{e}_i\bar{S}_i).$$

We recall that for any q, $V_q(\lambda)$ has dimension equal to the number of semistandard λ-tableaux, hence

$$\dim S = \begin{pmatrix} n^2 + r - 1 \\ r \end{pmatrix} = \sum (\dim V_q(\lambda_i))^2$$

(cf. the proof of 6.5.2). Thus

$$\dim S = \sum (\dim V_q(\lambda_i))^2 \geq \sum \dim \bar{J}_i = \dim S,$$

forcing equality as desired.   □

Recalling the results of 3.3.8 and 3.3.4, we now deduce that $S_q(n,r)$ has finite global dimension (of at most $2t - 2$) and that $\mathcal{P}_q(n,r)$ is a highest weight category.

Corollary 7.5.2 $\mathcal{P}_q(n,r)$, equivalently $_{S_q(n,r)}\mathbf{mod}$, is a highest weight category with finite poset $\Lambda^+(n,r)$. For $\lambda_i \in \Lambda^+(n,r)$, $V_i = S\bar{e}_i$ is isomorphic to the q-Weyl module $V_q(\lambda_i)$. Therefore $V_q(\lambda_i) = \Delta(\lambda_i)$ is the standard module and $M_q(\lambda_i) = \nabla(\lambda_i)$ is the costandard module in $\mathcal{P}_q(n,r)$.

Finally, we have a q-version of (4.9):

Corollary 7.5.3 For $M, N \in {}_{S_q(n,r)}\mathbf{mod}$ and any $i \geq 0$ there is an isomorphism

$$\mathrm{Ext}^i_{S_q(n,r)}(M,N) \cong \mathrm{Ext}^i_{\Gamma_q}(M,N). \tag{7.8}$$

Proof   The argument is standard. For, 7.5.1 and the results of 3.3 imply that every injective $S_q(n,r)$-module I has a ∇_q-good filtration. Thus, given $\lambda \in \Lambda^+(n,r)$, I is acyclic with respect to the functor $\mathrm{Hom}_{\Gamma_q}(V_q(\lambda), -)$, (cf. Jantzen [1987, pp. 56, 239]). We now deduce that $\mathrm{Ext}^i_{S_q(n,r)}(V_q(\lambda), N) \cong \mathrm{Ext}^i_{\Gamma_q}(V_q(\lambda), N)$ for any i. Take any projective $S_q(n,r)$-module P; then P° 'is an I', hence it has a ∇_q-good filtration and P has a Δ_q-good filtration. Since now $\mathrm{Ext}^i_{S_q}(P,N) \cong \mathrm{Ext}^i_{\Gamma_q}(P,N) = 0$ for $i > 0$, dimension shifting gives the result.   □

7.6 Representations of $\mathrm{GL}_n(q)$ and the q-Young modules

This section provides some examples of the q-theory in action. In particular we discuss in outline the work of Dipper and James on the problem of computing decomposition numbers for the q-Schur algebra and related structures over K. The main result of Dipper and James [1989] connects the decomposition matrix of the q-Schur algebra with the decomposition matrices of $\mathcal{H}_q$, $K\Sigma_n$, $K\mathrm{GL}_n(q)$ where (char $K, q) = 1$, and $K\mathrm{GL}_n(K)$. The ideas shadow those presented in 4.5. There, certain trivial source modules (the Young modules) for G were used to study decomposition numbers in the case $q = 1$, and the approach now will be to find similar q-Young modules for $\mathcal{H}_q$. It is enough to know the multiplicities of these modules as direct summands of tensor space in order to calculate decomposition numbers.

Representations of $\mathcal{H}_q$

Many of the well-known results for Σ_r have analogues for $\mathcal{H}_q$. The starting point is to construct a certain submodule of $M_q^\lambda = x_\lambda \mathcal{H}_q$, where $\lambda \in \Lambda(n, r)$. Recalling observations (1) and (2) preceding 7.1.1 and the various manifestations of z_λ in (7.5), we make the following definition and identifications:

Definition 7.6.1 Let $\lambda \in \Lambda^+(r)$. Define the *q-Specht module* to be the $\mathcal{H}_q$-submodule

$$
\begin{aligned}
S_q^\lambda &= z_\lambda \mathcal{H}_q = x_\lambda T_{w_\lambda} y_{\lambda'} \mathcal{H}_q \\
&= \varphi_{\lambda\omega}^1 T_{w_\lambda} y_{\lambda'} \mathcal{H}_q = \sum_{\pi \in \Sigma_{\lambda'}} (-q)^{-\ell(\pi)} x_\lambda T_{w_\lambda \pi}
\end{aligned}
\tag{7.9}
$$

of M_q^λ. The dual q-Specht module is the submodule

$$
\tilde{S}_q^\lambda = y_\lambda T_{w_\lambda} x_{\lambda'} \mathcal{H}_q
$$

of $y_\lambda \mathcal{H}_q$.

Exercise. Show that $\tilde{S}_q^\lambda$ is conjugate to S_q^λ under the automorphism $\sharp$ of $\mathcal{H}_q$ induced by $T_b^\sharp = (q - 1)T_1 - T_b$ (b basic). Since it is true that $(\tilde{S}_q^{\lambda'})^\circ \cong S_q^\lambda$ by Dipper and James [1987, 3.5], we see that $(S_q^\lambda)^\sharp \cong (S_q^{\lambda'})^\circ$.

Clearly these become the objects of 4.2.1 and 4.2.2 when $q = 1$. It is an easy matter to show that associated partitions define isomorphic q-Specht modules: for, suppose that $\lambda' = \mu'$, and let $\theta : M_q^\lambda \to M_q^\mu$ be the isomorphism appearing just before 6.6.3. Then

$$
0 \neq T_d z_\lambda = z_\lambda \theta = x_\mu T_d T_{w_\lambda} y_{\lambda'} = x_\mu T_d T_{w_\lambda} y_{\mu'}
$$

and so $z_\lambda \theta$ is plus or minus some q-power multiple of z_μ, by fact (2) preceding 7.1.1. This implies that θ restricts to an isomorphism $S_q^\lambda \to$

S_q^μ. The upshot is that one can index q-Specht modules by partitions of r.

A thorough investigation of these modules was conducted by Dipper and James [1986], [1987]: in particular a free R-basis for S_q^λ (for an arbitrary domain R) was found.

Before proceeding, we return to the example following (6.6).

Exercise. Check that $S_q^{(2^2)}$ is 2-dimensional with basis

$$z_\lambda, \quad z_\lambda T_{(12)} - qz_\lambda = q\,\frac{\boxed{\begin{array}{cc}1&2\\3&4\end{array}}}{} - \frac{\boxed{\begin{array}{cc}1&3\\2&4\end{array}}}{} - q^{-1}\frac{\boxed{\begin{array}{cc}2&4\\1&3\end{array}}}{} + q^{-2}\frac{\boxed{\begin{array}{cc}3&4\\1&2\end{array}}}{}.$$

Take $R = K$ to be a field and let $1 \neq q \in K$ be a primitive eth root of unity. The representation theory of $\mathcal{H}_q$ is akin to the 'e-modular representation theory of the symmetric groups'. For example S_q^λ has a unique top factor D_q^λ provided λ is e-regular, that is self-dual and absolutely irreducible, and the set

$$\{D_q^\lambda : \lambda \in \Lambda^+(r), \ \lambda \ e\text{--regular}\}$$

gives a complete set of non-isomorphic simple $\mathcal{H}_{K,q}$-modules (Dipper and James [1986, 7.6]). A degenerate case occurs when $\mathcal{H}_{K,q}$ is semisimple (i.e. when $e > r$ or q is not a root of unity): as λ runs through $\Lambda^+(r)$ the q-Specht modules provide a complete set of simple $\mathcal{H}_q$-modules. It is also true that the Nakayama Rule, 5.1.1, can be generalised: blocks of $\mathcal{H}_q$ are parametrised by the e-cores of partitions.

By convention one puts $e = \text{char } R$ if $q = 1$, while in general $e \geq 2$ is the least number i such that

$$1 + q + q^2 + \cdots + q^{i-1} = 0 \qquad (7.10)$$

(or $e = \infty$ if there is no such i).

Analogues of Young modules. We have just seen in the semisimple case that $\mathcal{H}_{K,q}$ has $|\Lambda^+(r)|$ non-isomorphic components. There will be, in general, fewer components over K, and over the domain R. However, it turns out that, over R, when q is a prime power, the external direct sum $M = \bigoplus_{\lambda \in \Lambda^+(r)} M_q^\lambda$ has $|\Lambda^+(r)|$ non-isomorphic components, (Dipper/James [1989, 3.11]). Amongst the components of M lurks an important subclass:

Definition 7.6.2 Let $M_q^\lambda = \bigoplus_{i=1}^m Y_i$ be a decomposition of $M_q^\lambda = x_\lambda \mathcal{H}_q$ into components (the indecomposable direct summands). Then there is a unique index i such that $S_q^\lambda \subseteq Y_i$, and the summand Y_i is unique up to isomorphism. By analogy with 4.6.1 we call Y_i the *q-Young module*, and denote it by Y_q^λ.

To check the assertions implicit in 7.6.2, apply statements (1) and (2) preceding 7.1.1. Such observations also demonstrate that $\lambda \neq \mu$ if and only if $Y_q^\lambda \not\cong Y_q^\mu$ and that

$$M_q^\lambda = \bigoplus m_\mu Y_q^\mu \text{ and } m_\mu \neq 0 \Rightarrow \mu \trianglerighteq \lambda,$$

with $m_\lambda = 1$.

Remarks.

(1) Since Y_q^λ and Y_q^μ are isomorphic if λ and μ are associated, we can index the q-Young modules by elements of $\Lambda^+(n, r)$.

(2) For the semisimple $\mathcal{H}_q$ over the field K, it is clear by construction that $Y_q^\lambda = S_q^\lambda$.

(3) In the non-semisimple case, we know that $M = \bigoplus M_q^\lambda$ has at least $|\Lambda^+(r)|$ non-isomorphic components, hence, by Fitting's theorem, $S_q(n, r)$ has at least this many. It should come as no surprise that in fact the q-Young modules provide an exhaustive list of such summands. The next theorem was proved in Dipper and James [1989, 3.11] under the hypothesis that q is a prime power; the general case, however, follows at once from Fitting's theorem and Dipper and James [1991, 8.8]:

Theorem 7.6.3 *There are precisely $|\Lambda^+(r)|$ pairwise non-isomorphic components of M.*

Exercise. Take $r = 2$. In $\mathcal{H}_{R,q}(\Sigma_2)$ we have $x_{(2)} = T_1 + T_{(12)}$, $x_{(1^2)} = T_1$, $y_{(2)} = T_1 - q^{-1}T_{(12)}$ and $y_{(1^2)} = T_1$. Each of $x_{(2)}\mathcal{H}_q = Rx_{(2)}$ and $y_{(2)}\mathcal{H}_q = Ry_{(2)}$ is indecomposable. Now $(q+1)T_1 = qx_{(2)} + y_{(2)}$, so there is a decomposition $\mathcal{H}_q = x_{(2)}\mathcal{H}_q \oplus y_{(2)}\mathcal{H}_q$ if $(p, q+1) = 1$, whereas $\mathcal{H}_q$ is indecomposable if $p|(q+1)$. In the second case $\mathcal{H}_q = Y_q^{(1^2)}$ is the q-Young module for (1^2).

By considering the $RGL_2(q)$-module, M, defined by the upper Borel, show that if $(p, q+1) = 1$ the q-Young modules are $x_{(2)}M$ and $y_{(2)}M$. Of course if $p|(q+1)$ then $x_{(2)} = y_{(2)}$, M is indecomposable and $M = Y_q^{(1^2)}$.

Remark. It is impossible to resist a return to Mullineux's Conjecture (see the discussions after 4.2.9 and near the end of 4.6). Let $\lambda \in \Lambda^+(r)$ be e-regular. Since $\sharp$ preserves irreducibility, $(D_q^\lambda)^\sharp \cong D_q^{\lambda^*}$ for some e-regular λ^*. By the Exercise after 7.6.1 we have that $[S_q^\alpha : D_q^\lambda] = [S_q^{\alpha'} :$

$D_q^{\lambda^*}$]. It is clear, now, that the definition of the map $*$ agrees, in the case $q = 1$, with that given in 4.6 . Is this significant?

Exercise. Show that the map $*$ is a bijection on e-regular partitions depending only on e, not on p or q.

The Dipper-James Theorem

As stated at the beginning of the chapter, the thrust of this theorem is the assertion that the task of computing the decomposition matrices of the symmetric groups Σ_n and of the finite $GL_n(q)$ (q a prime power, $n \in \mathbf{N}$) is reducible to the task of determining decomposition numbers of the q-Schur algebras. Before stating their theorem, we briefly recall the main points of Brauer theory as they apply to the various structures considered above.

We work over the p-modular system (k, R, Q), with R complete. Let $q \in \mathbf{N}$ and suppose $(p, q) = 1$; abusing notation we write q for $q1_k \in k$. Now, $\mathcal{H}_{Q,q}$ is a semisimple finite-dimensional Q-algebra. One can view $\mathcal{H}_{R,q}$ as an R-order inside $\mathcal{H}_{Q,q}$, and moreover $\mathcal{H}_{k,q} = \mathcal{H}_{R,q} \otimes_R k$ is the reduction mod p of $\mathcal{H}_{Q,q}$. Given an $\mathcal{H}_{Q,q}$-module M, general theory guarantees the existence of an $\mathcal{H}_{R,q}$-lattice $M_R \subseteq M$ which we can also reduce mod p, i.e. we can form $\overline{M} = k \otimes_R M_R$. For example, $S_{k,q}^{\lambda}$ is the reduction mod p of $S_{Q,q}^{\lambda}$ ($\lambda \in \Lambda^+(r)$). We have therefore defined a decomposition map $\mathcal{H}_{Q,q} \to \mathcal{H}_{k,q}$ via $\mathcal{H}_{R,q}$ and the usual theory of Green [1976] to which we have appealed many times in previous chapters, applies. The reader is left to formulate an analogous modular theory for $S_q(n,r)$.

Suppose $R = K$ and again write q for $q1_K$; we assume in addition that (k, R, Q) is a splitting system for $GL_n(q)$. Consider the integral matrix Δ consisting of the multiplicities $[V_q(\lambda) : L_q(\mu)]$. Lemma 7.1.1(ii) shows that the matrix Δ is lower unitriangular, provided we order $\Lambda^+(n,r)$ compatibly with the dominance order going downwards. Now q is not a root of unity in Q, but it is a root of unity in k, so we can apply the theory of 6.5 to $A_q(n)$ over (k, R, Q). Hence Δ is the decomposition matrix of the R-order $S_{R,q}(n,r)$ in the semisimple algebra $S_{Q,q}(n,r)$.

The decomposition triangle. We now describe various functors that are used to relate the three decomposition matrices defined above. No proofs are offered here; see Dipper and James [1989], James [1990a] and Dipper [1990], [1991] for details.

First, there exists a functor H_K from $_{KGL_n(q)}\mathbf{mod}$ to $_{\mathcal{H}_q}\mathbf{mod}$, with a right inverse taking irreducibles to irreducibles, and such that $H_K(L)$ is irreducible or zero if L is irreducible. The method of construction of this functor is a little involved, but suffice it to say that it is a natural generalisation of the method for Hom functors. This method can be used to manufacture irreducible $GL_n(q)$-modules, and reduces a part of the calculation of the decomposition matrix for the finite group to a calculation for the Hecke algebra.

Next, define a *q-Schur functor* $f_q : _{S_q(n,n)}\mathbf{mod} \rightarrow _{\mathcal{H}_q}\mathbf{mod}$: let $e : \oplus_\lambda M_q^\lambda \rightarrow M_q^\omega$ be projection. By Fitting's theorem $eS_q(n,n)e \cong \mathcal{H}_q$ (cf. (4.1)), so we have a functor $f_q : M \mapsto Me$ that reduces to the Hom functor of 4.1 when $q = 1$. This too has a right inverse.

It is possible now to 'complete the triangle' and produce a functor $S_K : _{KGL_n(q)}\mathbf{mod} \rightarrow _{S_q(n,n)}\mathbf{mod}$ (and a right inverse) such that the following triangle commutes:

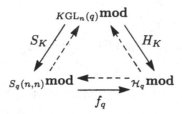

In any case the category of $S_{K,q}(n,n)$-modules embeds into the category of K-representations for $GL_n(q)$. The part of the latter category coming from the q-Schur algebra in this way is referred to as the *polynomial representation theory* of $GL_n(q)$ in the non-describing characteristic.

The decomposition triangle connects the respective decomposition matrices of $\mathcal{H}_q$, of $S_{K,q}(n,n)$ and of $KGL_n(q)$. To see precisely how all these ideas link up, it is best to ponder the following example. James [1990b] gave simple rules for computing all p-modular decomposition matrices for the q-Schur algebras $S_q(n,n)$ with $(p,q) = 1$ and $n \leq 10$. But it seems likely that one can also perform calculations with a q-version of the Jantzen Sum Formula for q-Weyl modules: formally it looks similar to that in Jantzen [1978, p. 314]. A version for the q-hyperalgebra appears in Thams [1993], (see also Andersen *et al.* [1991]) .

Example. Take $r = n = 5$, and let $p = \text{char } K$ divide $1 + q$. Hence $e = 2$; see (7.9). The usual Kostka matrix is defined combinatorially (e.g. Macdonald [1979, p. 59]):

$$A = [a_{\lambda\mu}] = \begin{bmatrix} 1 & 0 & 0 & 0 & 0 & 0 & 0 \\ 4 & 1 & 0 & 0 & 0 & 0 & 0 \\ 5 & 2 & 1 & 0 & 0 & 0 & 0 \\ 6 & 3 & 1 & 1 & 0 & 0 & 0 \\ 5 & 3 & 2 & 1 & 1 & 0 & 0 \\ 4 & 3 & 2 & 2 & 1 & 1 & 0 \\ 1 & 1 & 1 & 1 & 1 & 1 & 1 \end{bmatrix}.$$

Here and below we have the row and column labels in increasing lexico-graphic order.

Suppose that somehow we have decomposed the various M_q^λ as direct sums of q-Young modules.

$$\begin{aligned}
M_q^{(5)} &= x_{(5)}\mathcal{H}_q = Y_q^{(5)} \\
M_q^{(4,1)} &= Y_q^{(5)} \oplus Y_q^{(4,1)} \\
M_q^{(3,2)} &= (1-\alpha)Y_q^{(5)} \oplus Y_q^{(4,1)} \oplus Y_q^{(3,2)} \\
M_q^{(3,1^2)} &= 2Y_q^{(4,1)} \oplus Y_q^{(3,1^2)} \\
M_q^{(2^2,1)} &= (1-\alpha)Y_q^{(5)} \oplus 2Y_q^{(4,1)} \oplus Y_q^{(3,2)} \oplus Y_q^{(2^2,1)} \\
M_q^{(2,1^3)} &= 2Y_q^{(4,1)} \oplus Y_q^{(3,1^2)} \oplus Y_q^{(2^2,1)} \oplus Y_q^{(2,1^3)} \\
M_q^{(1^5)} &= (5-\alpha)Y_q^{(2^2,1)} \oplus 4Y_q^{(2,1^3)} \oplus Y_q^{(1^5)}
\end{aligned}$$

where $\alpha = 1$ if $p = 2$, and $\alpha = 0$ otherwise. We record these multiplicity coefficients in the matrix B, so, for example, the first row records the multiplicities of $Y_q^{(1^5)}$:

$$B = [b_{\lambda\mu}] = \begin{bmatrix} 1 & 0 & 0 & 0 & 0 & 0 & 0 \\ 4 & 1 & 0 & 0 & 0 & 0 & 0 \\ 5-\alpha & 2 & 1 & 0 & 0 & 0 & 0 \\ 0 & 1 & 0 & 1 & 0 & 0 & 0 \\ 0 & 0 & 1 & 0 & 1 & 0 & 0 \\ 0 & 2 & 2 & 2 & 1 & 1 & 0 \\ 0 & 0 & 1-\alpha & 0 & 1-\alpha & 1 & 1 \end{bmatrix}.$$

This is called the *matrix of structure constants* for $S_q(n,5)$. Let $\Delta = AB^{-1}$. This produces the following square lower unitriangular matrix

$$\Delta = [\tilde{d}_{\lambda\mu}] = \begin{bmatrix} 1 & 0 & 0 & 0 & 0 & 0 & 0 \\ 0 & 1 & 0 & 0 & 0 & 0 & 0 \\ \alpha & 0 & 1 & 0 & 0 & 0 & 0 \\ \alpha+1 & 0 & 1 & 1 & 0 & 0 & 0 \\ \alpha & 0 & 1 & 1 & 1 & 0 & 0 \\ 0 & 1 & 0 & 0 & 0 & 1 & 0 \\ 1 & 0 & 0 & 1 & \alpha & 0 & 1 \end{bmatrix}.$$

The main theorem states that, in the case char $K|(1+q)$, Δ is the decomposition matrix for the q-Schur algebra and furthermore that knowing Δ is equivalent, via $\Delta = AB^{-1}$, to knowing the q-Young module multiplicities in the M_q^λ. Also Δ is the part of the decomposition matrix for $K\mathrm{GL}_5(q)$ that corresponds to the unipotent representations; moreover, Δ includes the decomposition matrix for $\mathcal{H}_q$, by considering all but the first three columns—those columns indexed by e-restricted partitions. In the special case when char $K = 2$, so that $q = 1$ and $e = p$, Δ is the decomposition matrix for Weyl modules for $GL_5(K)$ over K and its first three columns give the 2-modular decomposition matrix of $K\Sigma_5$.

The following theorem gives the general case.

Theorem 7.6.4 (Dipper, James) *Suppose* $r \leq n$.

(i) *The decomposition matrix Δ for the q-Schur algebra is square and lower unitriangular. Let A be the Kostka matrix and B the matrix of q-Young module multiplicities in the various $x_\lambda \mathcal{H}_q$. Ordering the partitions of r in reverse lexicographic order, A and B are lower unitriangular. Moreover A, B and Δ are related by*

$$\Delta = AB^{-1}.$$

(ii) Δ *is that submatrix of the decomposition matrix for $K\mathrm{GL}_n(q)$ corresponding to the unipotent representations.*

(iii) Δ *includes the decomposition matrix of $\mathcal{H}_q$ as a submatrix. More precisely, one deletes columns indexed by e-singular partitions, where e is the least positive integer i such that $p|(1 + q + q^2 + \cdots + q^{i-1})$.*

If $p|(q - 1)$ (i.e. precisely when $e = p$), then we also have:

(iv) *For $m \geq n$ and K an infinite field of characteristic p then Δ is the decomposition matrix for Weyl modules for $K\Gamma_K$.*

(v) *Hence, by deleting those columns not indexed by p-restricted partitions, we obtain the p-modular decomposition matrix of n.*

Proof For a detailed survey of the proof the reader is advised to consult Dipper's survey article [1991]. The complete argument is extensive; the various parts to be found as follows: for (i) see Dipper and James [1989, 4.10, 4.12, 4.14]; for (ii) see Dipper and James [1989, 4.9]; for a proof of (iii) see the paper of James [1986, 8.1]; it is also a consequence of a result of Martin [1993] who gave a construction of an explicit (dual) q-Specht filtration of the q-Young modules with appropriate multiplicities. (iv) is clear, because in the case $p|(q-1)$, we have that $\mathcal{H}_q \cong KG$ and so $S_q(n,r) \cong S_r(\Gamma)$; (v) is known from our previous work in Chapter 4.

$\square$

7.7 Conclusion

Let me re-issue the warning that there is no published algorithm for computing the decomposition matrices of Σ_n, so the Δ are still unknown. We do however know their block structure (Dipper and James [1989, 6.7]).

From part (ii) of 7.6.4, the category of the q-Schur algebra gives the unipotent part of the matrix for $\mathrm{GL}_n(q)$ in the non-describing characteristic. I must point out that these results can be extended using q^t-Schur algebras to obtain a classification of all irreducible $\mathrm{GL}_n(q)$-modules from 7.1.6, using weights. In fact the full p-modular decomposition matrix for $\mathrm{GL}_n(q)$ with $(p,q) = 1$ can be computed via a combinatorial algorithm provided we know, for all $d, w \in \mathbf{N}$ with $dw \leq n$, the various matrices Δ for the q^d-Schur algebra for the integer w. I have nothing new to say about this; rather I urge you to consult the surveys of James [1990a] and Dipper [1991] for more references.

There are (at least) two ways in which quantum groups play a part in understanding the representations of general linear groups. First, one may replace the residue field by $\mathbf{C}$ and q by a primitive complex eth root of unity. Let the corresponding matrix be $\bar{\Delta}$. It is not true that $\bar{\Delta} = \Delta$ in general, but $\bar{\Delta}$ seems to provide a 'good approximation' to the matrix Δ where q has multiplicative order e mod p. There is a plausible reason why this might be true: the decomposition matrix depends on the characteristic p of K and the minimal polynomial (7.10) of q over the prime subfield, but James [1990b, §4] conjectures that the dependence is actually only on e and p.

A second benign influence of quantum groups comes through the so-called quantum hyperalgebra $\mathcal{U}_q$. The category of polynomial representations of $\mathrm{GL}_n(q)$ is essentially the comodule category of $A_{K,q}(n)$, and as such the objects can be viewed as representations of the q-hyperalgebra. So we might consider computing composition multiplicities of $L_q(\mu)$ in the $V_q(\lambda)$ (thinking of these as $\mathcal{U}_q$-modules). There are conjectures of Lusztig that relate these to Kazhdan-Lusztig polynomials: the old Lusztig Conjecture was for the special case $q = 1$ and char $K = p$; the new one is for $\mathcal{U}_q$ over fields of characteristic zero and q a primitive eth root of unity. Unlike the old one, which was valid for Weyl modules having highest weights far from the alcove walls, the new conjecture is apparently valid for all Weyl modules.

Appendix: a review of algebraic groups

Inevitably in this text, we have needed to use ideas from the theory of semisimple and reductive algebraic groups, though in practice we have been mostly concerned with these ideas as they pertain to GL_n or SL_n. There is an ample literature on algebraic group theory: proofs of all the assertions below may be found either in the standard texts of Humphreys [1987] or Jantzen [1987, II 1, 2]. I urge you to read this Appendix (if at all) only after digesting Chapter 1. We assume throughout that K is algebraically closed.

A.1 Linear algebraic groups: definitions

Let us view the polynomial ring $K[x_1, \ldots, x_n]$ as the polynomial functions on an n-dimensional space, $\mathbf{A}^n(K)$, over K. Regard $\mathbf{A}^n(K)$ as an affine space, as opposed to a vector space, since, given any point $a = (a_1, \ldots, a_n)$ in it we cannot distinguish $K[x_1, \ldots, x_n]$ from $K[x_1 - a_1, \ldots, x_n - a_n]$ (so a becomes the origin with this new viewpoint). By an *affine variety* we mean a collection of points in $\mathbf{A}^n(K)$ given by the simultaneous vanishing of a set of polynomials.

Definition 1 A *linear algebraic group* $\mathbf{G}$ over K is an affine variety V furnished with a compatible group structure. This means that multiplication $V \times V \to V$ and inversion $V \to V$ are morphisms of varieties (which is to say that the coordinates of the map are given by polynomials). Denote the *affine algebra* or *coordinate ring* of V by $K[V]$. One says that $\mathbf{G}$ is *defined over* $k \leq K$ if the equations defining it as a variety and the equations defining multiplication and inversion may be expressed with k-coefficients.

For example, $\mathrm{SL}_n(K)$ can be viewed as the set of zeros of the function
det -1 (so it is a hypersurface in $\mathrm{Mat}_n(K)$). The general linear group
$\Gamma = \mathrm{GL}_n(K)$ can be regarded as a linear algebraic group by taking for its
affine algebra $K[\Gamma]$, the subalgebra of K^Γ generated by $A_K(n) = \langle c_{ij} \rangle$
and $\det^{-1}$. Clearly, Γ is defined over the field $\mathbf{F}_p$ with p elements.

Definition 2 Suppose that $\mathbf{G}$ is defined over the finite field $\mathbf{F}_q$, where
$q = p^e$. The *Frobenius map*, F, with respect to $\mathbf{F}_q$ is the algebraic group
morphism given by raising the coordinates to the qth power.

A.2 Examples of linear algebraic groups

It is easy to construct other examples using the fact that a closed subgroup of Γ is also a linear algebraic group (Humphreys [1987, II.8.6]):

- the non-zero scalars K^*, the subset of $\mathbf{A}^1(K)$ with multiplication $x \times y = xy$ and inversion x^{-1} $(x, y \in K^*)$;
- B^+, the non-singular upper triangular matrices, called the *upper Borel* (subgroup). This is the set of zeros in Γ_K of c_{ij} with $i > j$. Similarly we have B^-, the *lower Borel* (subgroup), containing the non-singular lower triangular matrices. In general, Borel subgroups are defined to be the maximal connected soluble (closed) subgroups of $\mathbf{G}$;
- the non-singular diagonal matrices $T = B^+ \cap B^-$, the set of zeros of c_{ij} where $i \neq j$. Thus $K[T] = K[\varepsilon_i, \varepsilon_i^{-1}]$ (as at the top of p. 18). This group is called the *maximal torus* of Γ. In general, a torus is defined as a group isomorphic to a direct product of copies of K^*.
- $\mathrm{Uni}_n(K)$, the $n \times n$ upper triangular matrices with 1s on the diagonal.

Let $\mathbf{G}$ be a linear algebraic group. By a $\mathbf{G}$-module, we mean a right
$K[\mathbf{G}]$-comodule (V, ω_V). The structure map ω_V is a $\mathbf{G}$-module map
where $\mathbf{G}$ acts on $V \otimes K[\mathbf{G}]$ with trivial action on the left term.

A linear algebraic group is *connected* if its underlying variety cannot
be written as a disjoint union of two proper subvarieties each defined
by the vanishing of some polynomials (in other words, it is connected in
the Zariski topology). The group $\mathbf{G}$ is *unipotent* if it is isomorphic to a
subgroup of $\mathrm{Uni}_n(K)$ for some n. It is *reductive* if it has no non-trivial,
normal (closed) connected unipotent subgroups. (Examples are T and
Γ).

For a reductive group, the maximal subgroups of $\mathbf{G}$ isomorphic to
a torus are those of the Borel subgroups, and are all conjugate; the

common dimension of the maximal tori is called the *Lie rank*. Clearly, the Lie rank of Γ is $n - 1$.

A.3 The weight lattice

The characters of the maximal torus D form a free abelian group $X = X(D) = \mathrm{Hom}(D, K^*)$ under tensor product, of rank equal to $\dim D$. This is called the *weight lattice*. For the case of Γ, the characters of the maximal torus T correspond to tuples $\lambda = (\lambda_1, \ldots, \lambda_n) \in \mathbf{Z}^n$: we shall write λ for both the tuple and the character, the latter of which is given explicitly by

$$\lambda : T \;\to\; K^*$$
$$\mathrm{diag}(t_1, \ldots, t_n) \;\mapsto\; t_1^{\lambda_1} \cdots t_n^{\lambda_n},$$

for $t_i \in K^*$. As we deal mostly with polynomial Γ-modules in this text, we restrict ourselves to *polynomial weights*, i.e. to tuples in $\mathbf{N}_0^n$; as we deal with fixed degrees of homogeneity, we are reduced to a consideration of *polynomial weights of degree* r, i.e. to those in $\Lambda(n, r)$. Notice also that $X(T)$ has canonical basis consisting of the functionals ε_i (the restrictions of the coordinate functions c_{ii} to T).

Now D acts completely reducibly on any finite-dimensional **G**-module M, so we have a decomposition $M = \bigoplus_{\lambda \in X} M^\lambda$ where $M^\lambda = \{v \in M : d \circ v = \lambda(d)v \; \forall d \in D\}$. Call the elements $0 \neq v \in M^\lambda$ *weight vectors of weight* λ, and call M^λ the λ-*weight space*; this allows one to define the *formal character* of M to be the (finite) sum

$$\mathrm{ch}(M) = \sum_{\lambda \in X} (\dim M^\lambda) e^\lambda \in \mathbf{Z}X.$$

Here the integral group ring $\mathbf{Z}X$ has canonical basis consisting of all formal exponentials e^λ ($\lambda \in X$) that multiply according to the rule $e^\lambda e^\mu = e^{\lambda + \mu}$. The case of GL_n is dealt with in 1.5.1, where one identifies the indeterminate X_i with the generator e^{ε_i}.

A.4 Root systems

Now, the classification of connected reductive linear algebraic groups depends on a certain subset of X called a *root system*: this determines the isomorphism type of **G** modulo its largest normal abelian subgroup.

Namely, inside X, the root system Φ of **G** is the set of non-zero characters of D occurring in the restriction to D of the action of the group by conjugation on its Lie algebra (see Humphreys [1987]) For illustration we may take Γ: its Lie algebra is $\mathrm{gl}_n(K)$ (the $n \times n$ matrices). The eigenspaces of T (with non-zero eigenvalues) in its conjugation action on $\mathrm{gl}_n(K)$ are the matrices with a non-zero entry in a single given off-diagonal position, (i, j) say. The character has $\lambda_i = 1$, $\lambda_j = -1$ and $\lambda_k = 0$ $(k \neq i, j)$, i.e. it is the set of $\varepsilon_i - \varepsilon_j$ where $i, j \in \mathbf{n}$ and $i \neq j$. We observe that all roots lie in the hyperplane $\sum \lambda_i = 0$ of $X(T)$.

Root systems thus defined are 'abstract root systems' in the terminology of Humphreys [1987, §27], [1972, Chapter 3]: every root system is a direct sum of indecomposable root systems, and indecomposable root systems are classified by their Dynkin diagrams. The general linear group Γ has a root system that is indecomposable of type A_{n-1}.

A connected group is *semisimple* if the **Z**-span of Φ has finite index in $X(T)$: the example to bear in mind is $\mathrm{SL}_n(K)$. A semisimple group is also reductive.

A.5 Weyl groups

If **G** has maximal torus D, we call $W = N(D)/D$ the *Weyl group* of Φ. It acts naturally on X and hence induces a faithful representation $W \to \mathrm{GL}(X_{\mathbf{R}})$ where $X_{\mathbf{R}} = \mathbf{R} \otimes_{\mathbf{Z}} X$. Let us pick a W-invariant positive definite symmetric bilinear form $(\ ,\ )$ on $X_{\mathbf{R}}$. We identify W as a group of linear transformations of $X_{\mathbf{R}}$, namely the one generated by all reflections

$$s_\alpha : x \mapsto x - \frac{2(x, \alpha)}{(\alpha, \alpha)} \alpha$$

for all $\alpha \in \Phi$ and $x \in X_{\mathbf{R}}$. In the case of Γ, $X = \mathbf{Z}^n$ and we may identify W with Σ_n, for Σ_n acts on X by $w(\varepsilon_i) = \varepsilon_{w(i)}$, $i \in \mathbf{n}$, $w \in \Sigma_n$. The appropriate form on $X_{\mathbf{R}} = \mathbf{R}^n$ is $(\varepsilon_i, \varepsilon_j) = \delta_{ij}$, for $i, j \in \mathbf{n}$, and the reflection $s_{\varepsilon_i - \varepsilon_{i+1}}$ corresponds to the *basic transposition* $(i, i+1)$.

Denote by $\Pi = \{\alpha_1, \ldots, \alpha_\ell\} \subseteq \Phi$ the set of *simple roots*. Thus each root $\alpha \in \Phi$ has a unique expression $\alpha = \sum k_i \alpha_i$ where the k_i are integers all of the same sign. Choose a *positive system* $\Phi^+ \subseteq \Phi$ (those for which $k_i \geq 0$). One can define a partial order on X by decreeing that $\lambda \leq \mu$ if and only if $\mu - \lambda$ is a sum of positive roots, so that λ is positive if (helpfully) $\lambda > 0$. By letting the Borel $B^- \supset D$ determine the set of negative roots, we obtain the set of *dominant weights* of D with respect to Φ^+: namely, the set $X^+ = X(D)^+ = \{\lambda \in X(D) : (\lambda, \check{\alpha}) \geq 0 \ \forall \alpha \in \Phi^+\}$. Here $\check{\alpha} = 2\alpha/(\alpha,\alpha)$, for $\alpha \in \Phi^+$, is the *coroot*.

Now if $\mathbf{G}$ is semisimple, Π is a basis of $X_{\mathbf{Q}}$, so the collection of coroots $\{\check{\alpha}_i : \alpha_i \in \Pi\}$ forms a basis of $X_{\mathbf{Q}}^*$, i.e. there exists $\{\lambda_\alpha : \alpha \in \Pi\}$ such that $(\lambda_\alpha, \check{\beta}) = \delta_{\alpha\beta}$ for all $\alpha, \beta \in \Pi$. These are called the *fundamental dominant weights*. We say that $\mathbf{G}$ is *simply connected* if, in addition, $\lambda_\alpha \in X$ for all $\alpha \in \Pi$ (An example is $\mathrm{SL}_n(K)$). The λ_α for a basis for X in this case.

Example. In particular, for Γ take $\Phi^+ = \{\varepsilon_i - \varepsilon_j : i < j\}$ corresponding to the simple roots $\Pi = \{\alpha_i = \varepsilon_i - \varepsilon_{i+1} : i < n\}$. Then Φ^+ is associated with the upper Borel, B^+, and the natural partial ordering on $X = \mathbf{Z}^n$ defined by Π is precisely the usual dominance ordering. Dominant weights are those elements of $\mathbf{Z}^n$ that are decreasing; dominant polynomial weights are all decreasing tuples in $\mathbf{N}_0^n$ and the dominant polynomial weights of degree r comprise the set $\Lambda^+(n,r)$. The fundamental dominant weights are the $\lambda_i = \varepsilon_1 + \cdots + \varepsilon_i$, $1 \leq i < n$, and their sum is $\sum_{i=1}^n (n-i)\varepsilon_i$.

A.6 The affine Weyl group

The *affine Weyl group*, W_p is the group generated by the affine reflections $s_{\alpha,mp} = s_\alpha + mp\alpha$ for $m \in \mathbf{Z}$ and $\alpha \in \Phi^+$. Let ρ be half the sum of the positive roots, this being an element of $X_{\mathbf{Q}}$ if $\mathbf{G}$ is semisimple. Then W_p (and hence W) acts on $X_{\mathbf{R}}$ via

$$s_{\alpha,mp} \cdot v = s_\alpha(v + \rho) - \rho + mp\alpha$$

for $v \in X_{\mathbf{R}}$. This is known as the *dot action*. In the case of Γ the shift is by the weight $\rho = \frac{1}{2}(n-1, n-3, \ldots, -(n-3), -(n-1))$; but translating ρ by any multiple of (1^n) leaves the dot action unaffected, since (1^n) is invariant under all permutations. In practice one normally shifts by the weight $\delta = (n-1, n-2, \ldots, 0)$.

This action of W_p divides $\mathbf{R}^n$ into a disjoint union of *alcoves*, i.e. the connected components of the complement in $X_{\mathbf{R}}$ of the union of all reflecting hyperplanes,

$$X_{\mathbf{R}} \setminus \bigcup_{\alpha \in \Phi} \bigcup_{m \in \mathbf{Z}} \{v : \langle v + \rho, \check{\alpha}\rangle = mp\}.$$

The union of the closures of these alcoves is $X_{\mathbf{R}}$. One alcove is of particular interest: if $\mathbf{G}$ is semisimple and simply connected let

$$C_0 = \{v \in X_{\mathbf{R}} : 0 < (v + \rho, \check{\alpha}) < p \; \forall \alpha \in \Phi^+\}.$$

This is the *standard alcove* (one checks $C_0 \neq \emptyset$). The closure of C_0 is $\overline{C_0} = \{v \in X_{\mathbf{R}} : 0 \leq (v + \rho, \check{\alpha}) \leq p \; \forall \alpha \in \Phi^+\}$. $\overline{C_0}$ is a *fundamental domain* for the dot action of W_p on $X_{\mathbf{R}}$ since W_p acts simply transitively on alcoves. If $v \in X_{\mathbf{R}}$, let $w_0(v)$ be the element of the W_p-orbit lying in $\overline{C_0}$ (this element is used in Chapter 5).

Special linear groups. Most of what we have said applies to the derived group $\Gamma' =\mathrm{SL}_n(K)$ of Γ. We denote the maximal torus $T_1 = T \cap \Gamma'$, and the Borel $B_1 = B \cap \Gamma'$. Note that $X(T_1) = X(T)/\mathbf{Z}(\varepsilon_1 + \cdots + \varepsilon_n)$, so the canonical map $X(T) \to X(T_1)$ induces an isomorphism of root systems. We may identify $W = \Sigma_n$ with the Weyl group $N_{\mathrm{SL}}(T_1)/T_1$ of SL via the natural isomorphism $N_{\mathrm{SL}}(T_1)/T_1 \to N_{\mathrm{GL}}(T)/T$.

A.7 Simple modules for reductive groups

Let λ denote the 1-dimensional rational B-module on which T acts by weight λ and on which the unipotent radical acts trivially. Inducing this module from B to $\mathbf{G}$ produces a rational $\mathbf{G}$-module

$$H^0(\lambda) = \mathrm{Ind}_B^{\mathbf{G}}(\lambda) = \{f \in K[\mathbf{G}] : f(gb) = \lambda(b)^{-1}f(g) \; \forall g \in \mathbf{G}, b \in B\}$$

where $\mathbf{G}$ acts by left translation as follows: if $g \in \mathbf{G}$ and $f \in H^0(\lambda)$ then $gf : \mathbf{G} \to \lambda$ takes $x \mapsto f(g^{-1}x)$ $(x \in \mathbf{G})$. If $\mathbf{G}$ is semisimple we can interpret this as $H^0(\lambda) = H^0(G/B, \mathcal{L}(\lambda))$, the space of global sections of the homogeneous line bundle $\mathcal{L}(\lambda)$ induced by λ on the quotient variety $\mathbf{G}/B$. Now $H^0(\lambda)$ is finite-dimensional (Donkin [1985, 1.8]), and nonzero provided $\lambda \in X^+$. The next result is due to Chevalley; see Jantzen [1987, II.2.7].

Theorem 3 *If λ is dominant then* $\mathrm{Soc}H^0(\lambda) = L(\lambda)$ *is simple, and* $\{L(\lambda) : \lambda \in X^+\}$ *is a complete set of pairwise non-isomorphic simple rational $\mathbf{G}$-modules.*

Remark. These are the socles of the Schur modules, if $\mathbf{G} = \Gamma_K$ and $\lambda \in \Lambda^+(n, r)$. In 3.2 the $H^0(\lambda)$ are constructed combinatorially for λ a polynomial dominant weight of degree r. I issue a warning that there exist dominant weights which are not polynomial but for which $H^0(\lambda) \neq 0$, and in this case the combinatorial description is not valid.

Now ch $H^0(\lambda)$ is given by Weyl's character formula (Jantzen [1987, II.5.10]), but ch $L(\lambda)$ is still unknown. Taking $\lambda \in X^+$, $H^0(\lambda)$ has composition factor $L(\lambda)$ once, and we have $\mu \leq \lambda$ whenever $[H^0(\lambda) : L(\mu)] \neq 0$ $(\mu \in X^+)$. Knowing the modular character of $L(\lambda)$ is equivalent to computing the decomposition number $[H^0(\lambda) : L(\mu)]$ for $\mu \in X^+$: the Lusztig Conjecture (Lusztig [1980]), for which there is now much evidence, provides a method for solving this. The proposed algorithm involves the computation of inverse Kazhdan-Lusztig polynomials at 1 (if p is large and λ is regular).

Weyl modules enter the picture as follows. Let $\mathcal{G}$ be the complex semisimple Lie algebra of the same type as $\mathbf{G}$ and choose an irreducible finite-dimensional $\mathcal{G}$-module $V_{\mathbf{C}}(\lambda)$ of highest weight $\lambda \in X^+$. Pick a maximal weight vector $v \in V_{\mathbf{C}}(\lambda)$ and define $V_{\mathbf{Z}}(\lambda) = \mathcal{U}_{\mathbf{Z}}v$ using the Kostant $\mathbf{Z}$-form in the universal enveloping algebra of $\mathcal{G}$. The *Weyl module* is now constructed as $V(\lambda) = V_{\mathbf{Z}}(\lambda) \otimes K$, a module for $\mathbf{G}$ (equivalently for $\mathcal{U}_K = \mathcal{U}_{\mathbf{Z}} \otimes K$) of highest weight λ. Now $V(\lambda)$ is the universal highest weight $\mathbf{G}$-module of highest weight $\lambda \in X^+$, in the sense that any other $\mathbf{G}$-module with these properties is a quotient of $V(\lambda)$. The modules $V(\lambda)$ and $H^0(\lambda)$ are dual to each other. For let π_0 be the longest element in the Weyl group and let $\lambda^* = -\pi_0\lambda$. Then $V(\lambda) \cong H^0(\lambda^*)^*$ (this makes sense since $H^0(\lambda)$ is finite-dimensional). Indeed $H^0(\lambda)$ and $V(\lambda)$ are contravariant dual to each other in the sense of 3.4.4. Thus the dual statement to Theorem 3 is the assertion that $V(\lambda)$ has a unique maximal submodule having simple head isomorphic to $L(\lambda)$.

A.8 General linear group schemes

A good account of group schemes and modules over them is to be found in Jantzen [1987, I.2]. We shall be interested in the general linear group scheme: take the free commutative ring $A_{\mathbf{Z}}(n) = \mathbf{Z}[x_{ij}]$ in the n^2 variables x_{ij} $(i, j \in \mathbf{n})$. As in the case of $A_K(n)$, this is a bialgebra under Δ and ε (cf. 1.3.6). The determinant function d is group-like and one localises with respect to $\{(\det x_{ij})^m : m \geq 0\}$ to produce a Hopf algebra $\mathbf{Z}[x_{ij}, \delta]$ where $\delta = d^{-1}$. The antipode γ inverts d and $\gamma(c_{ij}) = \det(c_{ij})^{-1}A_{ji}$ where A_{ji} is the cofactor of c_{ji} in the matrix $[c_{ij}]$.

Definition 4 The *general linear group scheme* over $\mathbf{Z}$, denoted $\Gamma_{\mathbf{Z}}$, is the affine group scheme having coordinate ring $\mathbf{Z}[\Gamma_{\mathbf{Z}}] = \mathbf{Z}[x_{ij}, \delta]$.

If R is a commutative principal ideal domain, we obtain the group scheme $\Gamma_R = \operatorname{Hom}_R(\mathbf{Z}[x_{ij}, \delta], R)$ *by base extension*. This is a group by virtue of the Hopf algebra structure of $\mathbf{Z}[x_{ij}, \delta]$ and we may identify it with $\operatorname{GL}_n(R)$ as follows: to the element φ of $\operatorname{Hom}_R(\mathbf{Z}[x_{ij}, \delta], R)$ we associate the matrix $[\varphi(x_{ij})]_{i,j \in \mathbf{n}}$.

References

Abe, E. (1980). *Hopf Algebras. Cambridge Tracts in Mathematics*, Cambridge University Press, Cambridge.

Akin, K. (1988). On complexes relating the Jacobi-Trudi identity with the BGG resolution. *J. Algebra* **117**, 494–503.

Akin, K. (1989). Resolutions of representations. *Contemp. Math.* **88**, 209–217.

Akin, K. and Buchsbaum, D.A. (1985). Characteristic-free representation theory of the general linear group I. *Adv. Math.* **39**, 149–200.

Akin, K. and Buchsbaum, D.A. (1988). Characteristic-free representation theory of the general linear group II: homological considerations. *Adv. Math.* **72**, 171–210.

Akin, K., Buchsbaum, D.A. and Weyman, J. (1982). Schur functors and Schur complexes. *Adv. Math.* **44**, 207–278.

Andersen, H.H. (1989). The linkage principle and the sum formula for quantum groups. Preprint.

Andersen, H.H., Polo, P. and Kexin, W. (1991). Representations of quantum algebras. *Invent. Math.* **104**, 1–59.

Auslander, M. (1974). Representation theory of artin algebras I. *Comm. Algebra* **1**, 177–268.

Bernstein, I.N., Gelfand, I.M. and Gelfand, S.I. (1976). A category of $\mathcal{G}$-modules. *Funct. Anal. Applic.* **10**, 87–92.

Bessenrodt, C. and Olsson, J.B. (1991/92). On Mullineux symbols. Institut Mittag-Leffler, Report No. 27.

Carter, R.W. and Lusztig, G. (1974). On the modular representations of the general linear and symmetric groups. *Math. Z.* **136**, 193–242.

Chung, J.H. (1951). Modular representations of the symmetric group. *Canad. J. Math.* **3**, 309–327.

Clausen, M. (1979 and 1980). Letter place algebras and a characteristic-free approach to the representation theory of the general linear and symmetric groups I, II. *Adv. Math.* **33**, 161–191 and **38**, 152–177.

Clausen, M. (1991). Multivariate polynomials, standard tableaux and representations of symmetric groups: invariant-theoretic algorithms in geometry. *J. Symbolic Comp.* **11**, 483–522.

Cline, E., Parshall, B.J. and Scott, L.L. (1980). Cohomology, hyperalgebras and representations. *J. Algebra* **63**, 98–123.

Cline, E., Parshall, B.J. and Scott, L.L. (1988a). Algebraic stratification in representation categories. *J. Algebra* **117**, 504–521.

Cline, E., Parshall, B.J. and Scott, L.L. (1988b). Finite-dimensional algebras and highest weight categories. *J. reine angew. Math.* **391**, 85–99.

Cline, E., Parshall, B.J. and Scott, L.L. (1990). Integral and graded quasi-hereditary algebras, I. *J. Algebra* **131**, 126–160.

Crawley-Boevey, W.W. (1990). Lectures on representation theory and invariant theory. Universität Bielefeld Notes 90-004.

Curtis, C.W. (1970). Modular representations of finite groups with split BN-pairs. In *Lecture Notes in Mathematics* **131**, Springer, Berlin, pp. 57–95.

Curtis, C.W. and Reiner, I. (1981 and 1987). *Methods in Representation Theory, Vols. I, II.* John Wiley and Sons.

DeConcini, C., Eisenbud, D. and Procesi, C. (1980). Young diagrams and determinantal varieties. *Invent. Math.* **56**, 129–165.

Deruyts, J. (1892). Essai d'une théorie générale des formes algébriques. *Mém. Soc. Roy. Sci. Liège* **17**, 1–156.

Désarménien, J. (1980). An algorithm for the Rota straightening formula. *Discrete Math.* **30**, 51–68.

Désarménien, J., Kung, J.P.S. and Rota, G.-C. (1978). Invariant theory, Young bitableaux and combinatorics. *Adv. Math.* **27**, 63–92.

Dieudonné, J. (1981). Tableaux de Young et foncteurs de Schur en algèbre et géométrie. *Astérisque* **87–88**, 7–19.

Dipper, R. (1990). On quotients of Hom-functors and representations of finite general linear groups, I. *J. Algebra* **130**, 235–259.

Dipper, R. (1991). Polynomial representations of finite general linear groups in non-describing characteristic. In *Progress in Mathematics* **95**, Birkhäuser Verlag, Basel, pp. 343–370.

Dipper, R. and Donkin, S. (1991). Quantum GL_n. *Proc. L.M.S.* **63**, 165–211.

Dipper, R. and Du, J. (1993). Trivial and alternating source modules of Hecke algebras of type A. *Proc. L.M.S.* **66**, 479–506.

Dipper, R. and James, G.D. (1986). Representations of Hecke algebras and general linear groups. *Proc. L.M.S.* **52**, 20–52.

Dipper, R. and James, G.D. (1987). Blocks and idempotents of Hecke algebras of general linear groups. *Proc. L.M.S.* **54**, 57–82.

Dipper, R. and James, G.D. (1989). The q-Schur algebra. *Proc. L.M.S.* **59**, 23–50.

Dipper, R. and James, G.D. (1991). q-tensor space and q-Weyl modules. *Trans. A.M.S.* **327**, 251–282.

Dipper, R. and James, G.D. (1992). Representations of Hecke algebras of type B_n. *J. Algebra* **146**, 454–481.

Dlab, V. and Ringel, C.M. (1989). Quasi-hereditary algebras. *Illinois J. Math.* **33**, 280–291.

Donkin, S. (1980). The blocks of a semisimple algebraic group. *J. Algebra* **67**, 36–53.

Donkin, S. (1981). A filtration for rational G-modules. *Math. Z.* **177**, 1–8.

Donkin, S. (1983). A note on decomposition numbers for reductive algebraic groups. *J. Algebra* **80**, 226–234.

Donkin, S. (1985). *Rational Representations of Algebraic Groups. Lecture Notes in Mathematics* **1140**, Springer, Berlin.

Donkin, S. (1986). On Schur algebras and related algebras I. *J. Algebra* **104**, 310–328.

Donkin, S. (1987). On Schur algebras and related algebras II. *J. Algebra* **111**, 354–364.

Donkin, S. (1992). On Schur algebras and related algebras IV: the blocks of the Schur algebras. *J. Algebra* (to appear).

Donkin, S. (1993). On tilting modules for algebraic groups. *Math. Z.* **212**, 39–60.

Donkin, S. and Reiten, I. (1993). On Schur algebras and related algebras V: some quasi-hereditary algebras of finite type. *J. Algebra* (to appear).

Doty, S.R. (1991). The symmetric algebra and representations of general linear groups. In *Proc. Hyderabad Conf. on Algebraic groups*, (ed. S. Ramanan), 123–150, Manoj Prakashan, Madras, India.

Doty, S.R. and Walker, G. (1992a). The composition factors of $F_p[x_1, x_2, x_3]$ as a GL(3, p)-module. *J. Algebra* **147**, 411–441.

Doty, S.R. and Walker, G. (1992b). Modular symmetric functions and irreducible modular representations of general linear groups. *J. Pure Appl. Algebra* **82**, 1–26.

Doubilet, P., Rota, G.-C. and Stein, J. (1974). On the foundations of combinatorial theory IX: combinatorial methods in invariant theory. In *Studies in Appl. Math.* **53**, 185–216.

Drinfel'd, V.G. (1986). Quantum Groups. In *Proc. I.C.M., 1986 Vol. 1* (ed. A.M. Gleason), pp. 798–820.

Du, J. (1991). The modular representation theory of q-Schur algebras, II. *Math. Z.* **208**, 503–536.

Du, J. (1992a). The modular representation theory of q-Schur algebras. *Trans. A.M.S.* **329**, 253–271.

Du, J. (1992b). The Green correspondence for the representations of Hecke algebras of type A_{r-1}. *Trans. A.M.S.* **329**, 273–287.

Du, J., Parshall, B.J. and Jian-pan Wang (1991). Two-parameter quantum linear groups and the hyperbolic invariance of q-Schur algebras. *J. L.M.S.*(2) **44**, 420–436.

Erdmann, K. (1990). *Blocks of Tame Representation Type and Related Algebras. Lecture Notes in Mathematics* **1428**, Springer-Verlag, Berlin.

Erdmann, K. (1993a). Schur algebras of finite type. *Quart. J. Math. (Oxford)(2)* **44**, 17–41.

Erdmann, K. (1993b). Symmetric groups and quasi-hereditary algebras. Preprint.

Erdmann, K. and Martin, S. (1993). Quiver and relations for the principal p-block of Σ_{2p}. *Proc. L.M.S.* (to appear).

Erdmann, K., Martin, S. and Scopes, J.C. (1993). Morita equivalence for blocks of the Schur algebras. *J. Algebra* (to appear).

Fettes, S. (1985). A theorem on Ext[1] for the symmetric group. *Comm. Alg.* **13(6)**, 1299–1304.

Fong, P. and Srinivasan, B. (1982). The blocks of the finite general linear and unitary groups. *Invent. Math.* **69**, 105–153.

Frobenius, G. (1896). Über Gruppencharaktere. *Sitz. der Königlich Preussischen Akad. der Wissen. zu Berlin*, 985–1021.

Frobenius, G. (1900). Über die Charaktere der symmetrischen Gruppe. *Sitz. der Königlich Preussischen Akad. der Wissen. zu Berlin*, 516–534.

Grabmeier, J. (1985). Unzerlegbare Moduln mit trivialer Youngquelle und Darstellungstheorie der Schuralgebra. *Bayreuther Math. Sch.* **20**, 9–152.

Green, J.A. (1955). The characters of the finite general linear groups. *Trans. A.M.S.* **80**, 402–447.

Green, J.A. (1976). Locally finite representations. *J. Algebra* **41**, 137–171.

Green, J.A. (1980). *Polynomial Representations of* GL$_n$. *Lecture Notes in Mathematics* **830**, Springer, Berlin.

Green, J.A. (1981). Polynomial representations of GL$_n$. In *Algebra, Carbondale 1980 (Proceedings)* (ed. R.K. Amayo), *Lecture Notes in Mathematics* **848**, Springer, Berlin pp. 124–140.

Green, J.A. (1985/86). Functor categories and group representations. *Portugaliae Math.* **43**, 3–16.

Green, J.A. (1990a). On certain subalgebras of the Schur algebra. *J. Algebra* **131**, 265–280.

Green, J.A. (1990b). Schur algebras. Unpublished lectures given at Bielefeld.

Green, J.A. (1991a). Schur algebras and general linear groups. In *Groups, St. Andrews 1989 Vol. I* (eds. C.M. Campbell and E.F. Robertson), *L.M.S. Lecture Note Series* **159**, Cambridge University Press, Cambridge, pp. 155–210.

Green, J.A. (1991b). Classical invariants and the general linear groups. In *Progress in Mathematics* **95**, Birkhäuser Verlag, Basel, pp. 247–272.

Green, J.A. (1992). Combinatorics and the Schur algebra. Preprint.

Harris, J.C. and Kuhn, N.J. (1988). Stable decompositions of classifying spaces of finite abelian p-groups. *Math. Proc. Camb. Phil. Soc.* **103**, 427–449.

Higman, G. (1967). Representations of general linear groups and varieties of p-groups. In *Proc. Internat. Conf. Thy of Groups, A.N.U. Canberra, 1965*, 167-173, Gordon and Breach, New York.

Hoefsmit, P.N. (1974). Representations of Hecke algebras of finite groups with *BN*-pairs of classical type. Ph.D. thesis, University of British Columbia.

Humphreys, J.E. (1972). *Introduction to Lie Algebras and Representation Theory*. *Graduate Texts in Mathematics* **9**, Springer-Verlag, Berlin/New York.

Humphreys, J.E. (1987). *Linear Algebraic Groups. Graduate Texts in Mathematics* **21**, Springer-Verlag, Berlin/New York/ Heidelberg.

Humphreys, J.E. (1992). *Reflection Groups and Coxeter Groups*. Cambridge Studies in Advanced Mathematics **29**, Cambridge University Press, Cambridge.

Humphreys, J.E. and Jantzen, J.C. (1978). Blocks and indecomposable modules for semisimple algebraic groups. *J. Algebra* **54**, 494–503.

Iwahori, N. (1964). On the structure of a Hecke ring of a Chevalley group over a finite field. *J. Fac. Sci. Univ. Tokyo Sect. I* **10**, 215–236.

James, G.D. (1978a). *The Representation Theory of the Symmetric Group. Lecture Notes in Mathematics* **682**, Springer, Berlin.

James, G.D. (1978b). On a conjecture of Carter concerning irreducible Specht modules. *Math. Proc. Camb. Phil. Soc.* **83**, 11–17.

James, G.D. (1980). The decomposition of tensors over fields of prime characteristic. *Math. Z.* **172**, 161–178.

James, G.D. (1981). On the decomposition matrices of the symmetric groups III. *J. Algebra* **71**, 115–122.

James, G.D. (1983). Trivial source modules for symmetric groups. *Arch. Math. (Basel)* **41**, 294–300.

James, G.D. (1984). *Representations of General Linear Groups. L.M.S. Lecture Note Series* **94**, Cambridge University Press, Cambridge.

James, G.D. (1986). The irreducible representations of the finite general linear group. *Proc. L.M.S.* **52**, 236–268.

James, G.D. (1990a). Representations of S_n and GL$_n$ and the q-Schur algebra. In *Topics in Algebra*, Banach Centre Publications **26**, pp. 303–316.

James, G.D. (1990b). The decomposition matrices of $GL_n(q)$ for $n \leq 10$. *Proc. L.M.S.* (3) **60**, 225–265.

James, G.D. and Kerber, A. (1981). *The Representation Theory of the Symmetric Groups.* Encyclopedia of Mathematics **16**, Cambridge University Press, Cambridge.

James, G.D. and Murphy, G.E. (1979). The determinant of the Gram matrix for a Specht module. *J. Algebra* **59**, 222–235.

Jantzen, J.C. (1973). Darstellungen halbeinfacher algebraischer Gruppen und zugeordnete kontravariante Formen. *Bonner Math. Schrift.* **67**, Bonn.

Jantzen, J.C. (1980). Weyl modules for algebraic groups. In *Finite Simple Groups II, Durham 1978,* (ed. M.J. Collins) Academic Press, London/New York.

Jantzen, J.C. (1987). *Representations of Algebraic Groups.* Academic Press, Orlando, Florida.

Jantzen, J.C. and Seitz, G.M. (1992). On the representation theory of the symmetric groups. *Proc. L.M.S.* **65**, 475–504.

Jimbo, M. (1986). A q-analogue of $U(gl(n+1))$, Hecke algebras and the Yang-Baxter equation. *Lett. Math. Phys.* **11**, 247–252.

Klyachko, A.A. (1983/84). Direct summands of permutation modules. *Sel. Math. Sov.* **3(1)**, 45–55.

Krop, L. (1986 and 1988). On the representations of the full matrix semigroup on homogeneous polynomials I, II. *J. Algebra* **99**, 370–421 and *J. Algebra* **102**, 284–300.

Lusztig, G. (1980). Some problems in the representation theory of finite Chevalley groups. In *The Santa Cruz Conf. on Finite Groups. Proc. Sympos. Pure Math.* **37**, (eds. B. Cooperstein and G. Mason) pp. 313–317.

Macdonald, I.G. (1979). *Symmetric Functions and Hall Polynomials.* Oxford University Press, Oxford.

Manin, Y.I. (1988). *Quantum Groups and Non-commutative Geometry.* Publ. Centre de Recherches Mathématiques, Université de Montréal.

Martin, S. (1989). On the ordinary quiver of the principal block of certain symmetric groups. *Quart. J. Math. (Oxford)(2),* **40**, 209–223.

Martin, S. (1991). Ext^1 for general linear and symmetric groups. *Proc. Roy. Soc. Edinburgh* **119A**, 301–310.

Martin, S. (1993). Filtrations for q-Young modules. Preprint.

Martins, M.T.F.O. (1982). A theorem on decomposition numbers. *Comm. Algebra* **10**, 383–392.

Mead, D.G. (1972). Determinantal ideals, identities and the Wronskian. *Pacific J. Math.* **42**, 165–175.

Meier, N. and Tappe, J. (1976). Ein neuer Beweis der Nakayama-Vermutung über die Blockstruktur Symmetrischen Gruppen. *Bull. L.M.S.* **8**, 34–37.

Mullineux, G. (1979). Bijections of p-regular partitions and p-modular irreducibles of the symmetric group. *J. L.M.S.* (2) **20**, 60–66.

Parshall, B.J. (1987). Simulating algebraic geometry with algebra II: stratifying algebraic representation categories. In *The Arcata Conf. on Repns. of Finite Groups. Proc. Sympos. Pure Math.* **47**, (ed. P. Fong), pp. 263–281.

Parshall, B.J. (1989). Finite-dimensional algebras and algebraic groups. *Contemp. Math.* **82**, 97–114.

Parshall, B.J. and Scott, L.L. (1988). Derived categories, quasi-hereditary algebras and algebraic groups. In *Proc. Ottawa-Moosonee Workshop in Algebra 1987, Mathematics Lecture Note Series* **3**, Carleton Univ. and Univ. d'Ottawa, pp. 1–105.

Parshall, B.J. and Jian-pan Wang (1991). *Quantum Linear Groups. Memoirs of the A.M.S.* **89**.

Ringel, C.M. (1991). The category of good modules over a quasi-hereditary algebra has almost split sequences. *Math. Z.* **208**, 209–225.

Robinson, G. de B. (ed.) (1977). The collected papers of Alfred Young (1873–1940), *Math. Exp.* **21**, University of Toronto Press, Toronto.

Rota, G.-C. (ed.) (1976/77). Appendix to *Théorie combinatoire des invariants classiques. Série de mathématiques pures et appliquées* **1/S-01**, I.R.M.A., Strasbourg.

Rotman, J.J. (1979). *An Introduction to Homological Algebra*. Academic Press, London.

Santana, A.P. (1993). The Schur algebra $S(B^+)$ and projective resolutions of Weyl modules. *J. Algebra* (to appear).

Schaper, K.-D. (1981). Charakterformeln für Weyl-Moduln und Specht-Moduln in Primcharakteristik. Diplomarbeit, Universität Bonn.

Schur, I. (1901). Über eine Klasse von Matrizen, die sich einer gegebenen Matrix zuordnen lassen. In *I. Schur: Gesammelte Abhandlungen Vol. I* (eds. A. Brauer and H. Rohrbach), Springer-Verlag, Berlin (1973), pp. 1–71.

Schur, I. (1927). Über die rationalen Darstellungen der allgemeinen linearen Gruppe. In *I. Schur: Gesammelte Abhandlungen Vol. III* (eds. A. Brauer and H. Rohrbach), Springer-Verlag, Berlin (1973), pp. 68–85.

Scopes, J.C. (1991). Cartan matrices and Morita equivalence for blocks of the symmetric groups. *J. Algebra* **142**, 441–455.

Scott, L.L. (1973). Modular permutation representations. *Trans. A.M.S.* **175**, 101–121.

Scott, L.L. (1987). Simulating algebraic geometry with algebra I: the algebraic theory of derived categories. In *The Arcata Conf. on Repns. of Finite Groups. Proc. Sympos. Pure Math.* **47** (ed. P. Fong), pp. 271–281.

Soergel, W. (1990). Construction of projectives and reciprocity in an abstract setting. Preprint.

Stanley, R.P. (1971). Theory and application of plane partitions: Part I. *Stud. in Appl. Math.* **50**, 167–188.

Steinberg, R. (1951). A geometric approach to the representations of the full linear group over a Galois field. *Trans. A.M.S.* **71**, 274–282.

Sullivan, J.B. (1977). Simply connected groups, the hyperalgebra, and Verma's conjecture. *Amer. J. Math.* **100**, 1015–1019.

Sweedler, M.E. (1969). *Hopf Algebras*. Benjamin, New York.

Thams, L. (1993). Two classical results in the quantum mixed case. *J. reine angew. Math.* **436**, 129–153.

Thrall, R.M. (1942). On the decomposition of modular tensors, I. *Ann. of Math.* **43**, 671–684.

Towber, J. (1979). Young symmetry, the flag manifold and representations of GL(n). *J. Algebra* **61**, 414–462.

Weyl, H. (1973). *The Classical Groups: their Invariants and Representations*. Princeton University Press, Princeton NJ.

Woodcock, D.J. (1992). A vanishing theorem for Schur modules. Preprint.

Woodcock, D.J. (1993). Straightening codeterminants. Preprint.

Xi, C. (1992). The structure of Schur algebras $S_k(n,p)$ for $n \geq p$. *Canad. J. Math.* **44**, 665–672.

Xi, C. (1991). On representation types of q-Schur algebras. Preprint.

Wong, W.J. (1971). Representations of Chevalley groups in characteristic p. *Nagoya Math. J.* **45**, 39–78.

Wong, W.J. (1972). Irreducible modular representations of finite Chevalley groups. *J. Algebra* **20**, 355–367.

Zelevinskii, A.V. (1987). Resolvents, dual pairs and character formulas. *Funct. Anal. Applic.* **21**, 152–154.

Index of Notation

227

Index of Terms

229